Foreword

To say that this is an unusual book would be an understatement on several counts. It is an extraordinary *tour de force* of scientific writing, covering as it does an enormous panorama of facts and theories about climate change and nuclear war. It contains summaries and critical analyses of a wide range of relevant research in both the Soviet Union and the western world, a scholarly achievement that is unmatched, to my knowledge. It will be certain to fascinate scientists who have been working in this broad area by presenting a number of stimulating new ideas about the factors that influence climate, past and future — ideas that for the most part have not been much publicized outside the Soviet Union.

I will explain some of the foregoing assertions. First, the author, Prof. Kirill Kondratyev, is well known to atmospheric scientists throughout the world for his scientific leadership and his amazingly prolific pen. Several of his books and articles in Russian have been translated into English so that they could reach a wider audience, and he has also published extensively in the English-speaking scientific literature. The subjects covered by him at one time or another, to mention a few, have been meteorological rockets and satellites, climate change, the effects of dust particles (aerosols) on the radiation balance of the lower atmosphere, the spread of deserts, the chemistry and radiation balance of the stratosphere, the atmospheres of the other planets, and the atmospheric effects of nuclear weapons. It is therefore evident that he is one of the few scientists who could write a book such as this one with its broad sweep, covering the work of others as well as his own.

v

This treatment appears at a time in the history of the geosciences when we are earnestly seeking to study the global system that determines our climate as one interactive whole. Isolated pigeonholes of knowledge about the earth are no longer in fashion. Thus, the international community of environmental scientists is organizing itself to attack multidisciplinary problems of global change on a scale that befits the task, and in the Western world we hear about TOGA (the Tropical Ocean Global Atmosphere Program) and WOCE (the World Ocean Circulation Experiment). We learn from Kondratyev about the Soviet counterparts to these ambitious international field programs, notably the Sections Program (organized by Soviet Academy President G. I. Marchuk), which is designed to study the energetically active zones of the ocean (called EAZO) and their short-term effects on climate. It is clear that the main thrusts of all these programs are in the same direction.

The introductory chapter of this book is also a kind of summary of many of the important points that Kondratyev wants to make. In particular, it indicates where he thinks emphasis should be placed in our research on global change and in the use of satellites to monitor the state of the globe. (We note that he does not include biological or ecological research in a significant way.)

A reader untrained in atmospheric science may find the treatment a bit technical in parts, but (thanks to careful editing and clarification of difficult parts by Michael Glantz) the important messages can be gleaned by a layman. I do believe, however, that Kirill Kondratyev had his scientific colleagues in mind when he wrote this book. He does not shy away from mathematical explanations when they are called for, and the detail into which he goes in places will probably appeal mostly to specialists. Indeed, for those of us in this field of reasearch his reviews of several of the scientific questions and debates that have been raging for the past few years will be most illuminating. In a very real sense, this book is a kind of milestone, marking the turning points and forks in the road that were being faced in the late 1980s.

As Kondratyev points out, the subject of climate change is so complex that we must draw from a wide variety of sources of information. Thus the book discusses the effects of nuclear weapons, the effects of volcanic eruptions, balloon, rocket, and satellite observations of solar radiation, the behavior of the atmospheres of Mars and Venus, the Tunguska event of 1908 (see below), and much more—plus, of course, the theoretical numerical models that have been devised to relate the physical and chemical factors involved in climate change. At first this menu seems too large to digest, but it all fits together in the end to tell an important story.

I mentioned that there were some stimulating new ideas here, and indeed

CLIMATE SHOCKS:
NATURAL AND ANTHROPOGENIC

WILEY SERIES IN CLIMATE AND THE BIOSPHERE

Edited by Michael H. Glantz and Robert E. Dickinson

THE GEOPHYSIOLOGY OF AMAZONIA: VEGETATION AND CLIMATE INTERACTIONS
Robert E. Dickinson Editor

CLIMATE SHOCKS: NATURAL AND ANTHROPOGENIC
K. Ya. Kondratyev

CLIMATE SHOCKS: NATURAL AND ANTHROPOGENIC

K. Ya. Kondratyev

Laboratory for Remote Sensing
Institute for Lake Research
Leningrad, USSR

Translated from the Russian by
A. P. Kostrova

WILEY

A WILEY-INTERSCIENCE PUBLICATION

JOHN WILEY & SONS

NEW YORK   CHICHESTER   BRISBANE   TORONTO   SINGAPORE

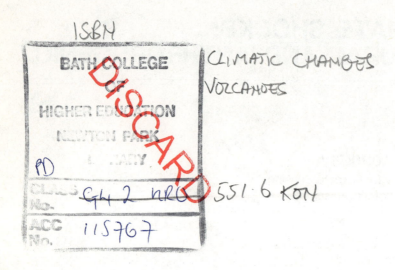
Copyright © 1988 by John Wiley & Sons, Inc.

All rights reserved. Published simultaneously in Canada.

Reproduction or translation of any part of this work
beyond that permitted by Section 107 or 108 of the
1976 United States Copyright Act without the permission
of the copyright owner is unlawful. Requests for
permission or further information should be addressed to
the Permissions Department, John Wiley & Sons, Inc.

Library of Congress Cataloging in Publication Data:
Kondrat'ev, K. ĪA. (Kirill ĪAkovlevich)
   Climate shocks.

   (Wiley series in climate and the biosphere)
   "A Wiley-Interscience publication."
   1. Climatic changes.   2. Greenhouse effect,
Atmospheric.   3. Volcanoes.   4. Nuclear explosions.
I. Title.   II. Series.

QC981.8.C5K67   1987   551.6   87-21551
ISBN 0-471-83019-4

Printed in the United States of America

10 9 8 7 6 5 4 3 2 1

there are. Of course, no idea is ever completely new, since science progresses step by step in piecing observations and theory together, each new insight based on the insights of the past. Nevertheless, here are some propositions or hypotheses that have so far not had the scrutiny they deserve—not in the Western literature, at least. I will just mention three that struck me, and others will no doubt discover some more.

For example, those of us who have been struggling with the proper explanation of the observed global temperature trend over the past 100 years or more have been frustrated by a number of features of the temperature record. One of these features is the apparent cooling trend that prevailed, especially in the Northern Hemisphere, between about 1945 and 1965. The overall long-term trend has, of course, been a warming of 0.5 to 0.7 K from 1880 to the present, and it is tempting to ascribe this to the greenhouse effect of carbon dioxide and the other infrared absorbing gases that have been steadily increasing in the same time frame. *But* how do we explain that 20-year cooling trend—volcanic activity? a decrease in solar radiation? something else?

That "something else," according to Kondratyev and his colleague, G. A. Nikolsky, may have been the nitrogen dioxide (NO_2) injected into the stratosphere by the succession of large yield nuclear tests that began in the late 1940s and ended, for the most part, in the early 1960s. Nitrogen dioxide absorbs solar radiation, and its enhanced presence in the stratosphere for a period of two decades could have reduced the sunlight reaching the surface by a few percent. The authors present data which suggests that this is just what happened, and point to the fact that the temporary cooling trend was only observed in the northern hemisphere, which is where nearly all of the atomic tests were conducted. I will not enter into the debate that is bound to ensue about this idea, but the data and the theoretical arguments mustered by Kondratyev and Nikolsky throw important new light on this puzzle.

A closely related subject is the question of what will happen to stratospheric ozone following a nuclear exchange. Kondratyev discusses the "conventional wisdom" that calls for a profound decrease in stratospheric ozone due to destruction by the oxides of nitrogen (often referred to collectively as NO_x), a decrease that could result in greatly enhanced lethal solar ultraviolet radiation reaching the earth's surface. However, a search for such a decrease of stratospheric ozone after the nuclear tests was inconclusive, he says, and the lack of any substantial decrease may have been due to other by-products of the explosions such as water vapor, which could reduce the ozone depletion effect due to NO_x alone. Again, I will not engage in this debate, but call attention to it as another example of the way in which Kondratyev illuminates a fascinating subject.

A perusal of the book's table of contents reveals that almost an entire chapter is devoted to the Tunguska event of 1908, an explosion in the atmosphere that laid to waste a large area of Siberia and created an atmospheric pressure wave recorded around the world. Was it a comet or a very large meteor that entered our atmosphere and disintegrated? Kondratyev and his colleagues have obviously been studying this event in great detail and reconstructing the phenomena that occurred in 1908 in the light of our current knowledge of the physics and chemistry of the upper atmosphere. They demonstrate that, indeed, one can learn a great deal about the effects of a great high-altitude explosion, be it cosmic or nuclear, from this experience of 80 years ago.

The third general area that I will mention will be of particular interest to the growing cadre of global climate modelers who are concerned with the "greenhouse effect" of increasing carbon dioxide and other infrared-absorbing trace gases. Kondratyev, an expert on radiative transfer in the atmosphere, understands very well what theoreticians at home and abroad are saying about the global warming in store for the earth. However, he prefers to remain a bit skeptical about all those forecasts of a drastic climate change until we know more about the system with which we are dealing. He warns that there may be some "antigreenhouse effects" that have not been adequately recognized. One of his caveats, for example, is the possible role of biological activity in the oceans that could limit the increase of carbon dioxide in the atmosphere. Said Kondratyev to his editor: There has been "too much orthodoxy in the CO_2 and climate problem and irresponsible predictions of climate change. I am glad the situation has greatly changed recently in this respect" (letter from Kondratyev to Glantz of 6 June 1986). Let us heed his words!

Most of the material in this book is not "perishable," in the sense that the ideas might be destined to become obsolete with time. Nevertheless, we note that the first draft of the manuscript was prepared at the end of 1985, and only part of it has been revised to reflect publications in 1986. With the inevitable delays due to editing and communicating back and forth between Leningrad and Boulder, Colorado, such a lag is entirely understandable. Those who are actively engaged in the climate change and nuclear winter controversies may note that the most recent references are missing, but they will probably agree also that this is not a serious flaw. Kondratyev himself is entirely *au courant*, and has recently plugged several of the gaps that arose due to the advancing front of science.

Finally, a very big word of praise is due the author, the editor, and the publisher for having the energy and the patience to make this book a reality. I am sure that it was an enormous task for Kirill Kondratyev to write this large and elaborate book in English, even granting that much of the material was at hand. And I happen to know that Michael Glantz undertook another enormous task in helping to edit the text into a publishable form. Finally, John Wiley & Sons had to have faith that all this labor would result in a remarkable treatise by a leading Soviet scientist on a subject of inestimable importance to the world. Their faith was rewarded.

WILLIAM W. KELLOGG
Senior Scientist (Retired)
National Center for
Atmospheric Research

Preface to the Series

The Climate and the Biosphere series focuses on the interaction of atmospheric processes and the biosphere. In recent years there has been a growing awareness of the extent to which atmospheric processes affect ecological systems and society. Human activities, such as deforestation and the burning of fossil fuels, have likewise directly or indirectly modified the atmosphere. Changes in the atmospheric processes, in turn, affect ecological processes and societal activities.

Several scientific issues involving the atmosphere have been identified in the past as being crucial to society. Each of these has become a major concern to a set of scientists and policymakers not only in the United States but in other developed as well as developing countries. They include:

- Nuclear winter
- El Niño-Southern Oscillation phenomenon
- Carbon dioxide increases, trace gas production and their effects on global atmospheric temperatures and ultimately on rainfall regimes
- Stratospheric ozone depletion
- The effects of different land-use practices (overgrazing, irrigation, deforestation, groundwater depletion) on atmospheric processes (especially on climatic elements such as temperature and precipitation)
- Climate modification schemes in theory and in practice
- The causes of persistent and prolonged drought in Africa and other drought-prone regions

Scientific research continually discovers and confirms, modifies, or rejects theories about the interactive mechanisms between the climate and the biosphere. Climate and the Biosphere is an important area of concern not only to the scientific community but to policymakers as well, because only with a proper understanding of these interactions can human activities be modified to mitigate or avert adverse interactions while selectively reinforcing the positive ones.

The series is intended to encompass different methodologies and disciplines. It will focus on the interactions between the atmosphere and various other elements of the broadly defined biosphere, including societies.

MICHAEL H. GLANTZ
ROBERT E. DICKINSON

Preface

The principal aim of this volume is to offer an overview of present-day understanding of the possible impact on climate of multiple atmospheric nuclear explosions. The previous basic publication on this subject, the two-volume *Environmental Consequences of Nuclear War,* published by the Scientific Committee on Problems of the Environment (SCOPE 28), was devoted entirely to the consideration of one concept, nuclear winter, although the authors of SCOPE 28 have not accepted such a label.

Climate Shocks attempts to discuss the problem on a broader basis, with emphasis on the analysis of original observations in the USSR following several nuclear tests in the late 1950s and early 1960s. These observations showed that a temperature drop of about 0.3°C during the 1960s resulted from an NO_2 increase in the stratosphere, which in turn was due to nuclear explosions. These explosions may be considered as a real model of nuclear winter, but on a small scale.

The first two chapters of my book consider two natural "analogs" of the effects of nuclear disturbances: the impact on climate of the increase of atmospheric CO_2 and other greenhouse gases and the impact on climate of volcanic eruptions. Of course, both analogs are far from realistic, but they can be, nevertheless, of some value. This is especially true for the volcanic eruptions analog for the following reasons: (1) the influence of stratospheric aerosols on climate is analyzed (as it is in the case of nuclear winter); (2) the monitoring of the volcanic stratospheric aerosol propagation has shown that the global distribution of aerosols is very inhomogeneous and the speed of aerosol propagation outside the latitude belt of a volcano is rather slow. The second circumstance is of great significance for an evaluation of the validity of the nuclear winter concept. It is quite clear that what may happen in reality must look more like nuclear climatic chaos than nuclear winter.

This book was completed about two years ago. The author wishes to acknowledge that a number of new publications have appeared since then that are not discussed in this book, although their findings do not contradict the discussion herein. Some of these more recent publications are listed as additional references at the end of this volume.

K. YA. KONDRATYEV

Leningrad, USSR

Acknowledgments

Many people helped with the preparation of this manuscript for publication. Clearly, Ms. Kostrova deserves my deepest thanks for translating the manuscript into English as it was being prepared. I also would like to thank Michael Glantz, the series editor and head of the Environmental and Societal Impact Group at the National Center for Atmospheric Research, Boulder, Colorado, for editing the manuscript and seeing it through to publication, and William Kellogg and V. Ramaswamy for their valuable scientific reviews of the manuscript. In addition, I would like to thank Ursula Rosner, Barbara McDonald, and Jan Stewart for typing several working drafts of the manuscript and Maria Krenz for her editorial and administrative management of the book. I would also like to give my special thanks to the editors at Wiley, who supported this publication from the outset and showed their strong interest in this research.

Contents

CLIMATE SHOCKS:
NATURAL AND ANTHROPOGENIC

Introduction

Because of their vital significance for human life, the problems of climate variability and climate change are of central concern to the scientific community today. For example, substantial changes in the environment caused by industrial activity (1-4) and their possible impacts on climate (5-12) have become a universal concern, because variations in climate can seriously affect agricultural and industrial productivity (5,10). As another example, serious attention has recently focused on the possible effects on the stratosphere of supersonic aviation, space vehicles, and chlorofluoromethanes (3,7). The relatively high degree of stability of aerosol pollution in the stratosphere underscores the problem of the possible adverse impact of such a stratospheric aerosol layer and the need to avoid such anthropogenically induced undesirable climate changes (13). The influence on weather and climate of nuclear tests has been discussed widely (14). Also of primary importance is climate forecasting with due consideration of anthropogenic effects (5,6,9,10,15,16). The possible impact of nuclear war on the atmosphere and on climate also deserves special consideration. Such issues as these make more urgent the study of the physical factors of climate and its changes, further developments of climate theory, and in particular, determination of the relative contributions of natural and anthropogenic atmospheric pollution to present-day climate changes.

Fig. 1 illustrates some factors responsible for the formation of climate. Figs. 2 and 3 characterize peculiar features of the formation of the heat and water balances of the Earth's surface.

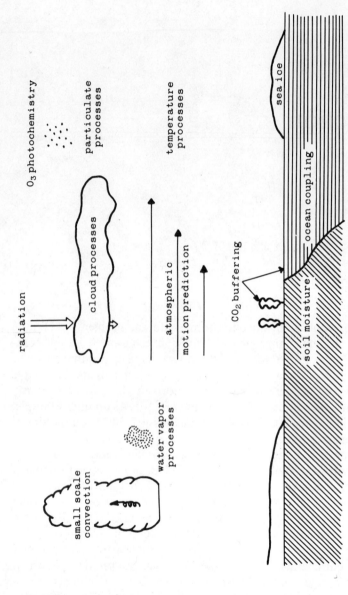

FIGURE 1. Factors responsible for climate formation. (After Ref. 12.)

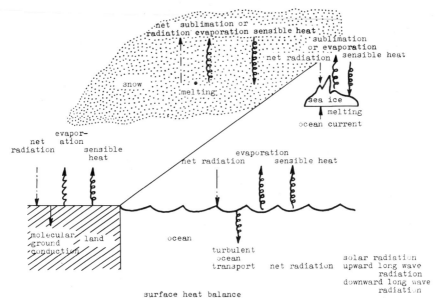

FIGURE 2. Physical processes responsible for the formation of the heat balance for four types of surfaces: bare land, snow cover, open ocean, and sea ice. (After Ref. 22.)

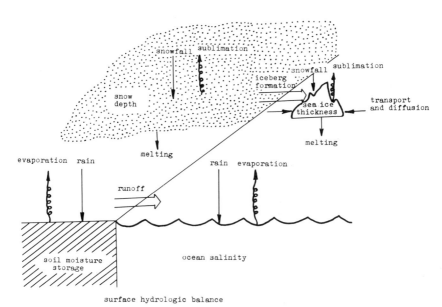

FIGURE 3. Physical processes responsible for the formation of the water balance for four types of surfaces: bare land, snow cover, open ocean, and sea ice. (After Ref. 22.)

Numerous elements of climatic change and their complicated interactions determine the degree of difficulty in formulating an adequate theory of climate and in identifying the most important factors responsible for climatic variations (4,5,9,10,12,15). In this regard special attention is given in this book to a discussion of the problem of climate predictability (15). It is still not at all clear to what extent it is possible to forecast the climate.

A study of the Earth's paleoclimates (5,6,10,15), the climatic features of other planets (17-19), and the specific character of the climate of large industrial cities (1) caused by atmospheric pollution and transformed surface characteristics is of great importance for understanding present-day climate changes.

Various data enable us to form certain hypotheses about the evolution of the Earth's climate during its geological history. Meteorological observations during the last two centuries have provided observational evidence for climate changes in the recent past. There is evidence, for example, that during the period between 1880 and 1940 the mean annual global temperature, as measured by the land station network, increased by 0.6°C and then decreased (5,6,10).

According to the 1976 WMO statement on climate change (20), there exists a "most probable" trend of continued cooling, although the total effect of anthropogenic factors can be expected to produce a warming. This statement requires confirmation, however, since, as the WMO statement noted, our "contemporary understanding of climatic fluctuations is rudimentary" (20). To know the causes of climate change and, in particular, its relation to anthropogenic atmospheric pollution is vitally important.

An assumption of the exponential increase in the industrial output of heat or of the effect of increasing the atmospheric content of carbon dioxide leads to the conclusion of a possible climate warming. The warming trend observed in some parts of the globe, beginning at the end of the 1960s, prompted M. I. Budyko, among others, to forecast a significant global warming by the beginning of the next century.

Putting aside the long-period internal fluctuations within the atmosphere-ocean-land-cryosphere-biosphere system, one can distinguish the following three factors among the many that determine present-day climate variations: (i) variations in the solar constant; (ii) transformation of the Earth's surface properties, and (iii) variations in the gaseous and aerosol composition of the atmosphere. As far as the solar constant is concerned (the flux of solar energy outside the atmosphere), its short-term variations do not exceed a few tenths of one percent. No records exist directly demonstrating long-period variations in the solar constant. There are, however, many proposed correlations between solar activity and climate, but as yet there are no convincing explanations of the physical mechanisms by which the solar activity can affect climate.

Changes in the Earth's surface properties may be of considerable importance, mainly in terms of microclimates or in terms of short-lived anomalies in the climates of isolated regions. This refers, for example, to the effect of oil slicks in the Arctic seas or of the increase in the Earth's surface reflectivity (albedo) in semiarid regions.

Thus, interpretation of climatic changes centers around the effect of variations of atmospheric composition and anthropogenic activity. The physics of this problem consists of the influence of variations of the atmospheric composition on the greenhouse effect (the heating up of the Earth's atmosphere) and on the atmosphere's radiative regime. Related problems of top priority include an improved understanding of the effects on the climate of cloudiness, aerosols, and optically active minor gaseous components.

Table 1 lists data on the possible factors of climatic change and their characteristic time scales. The problem is that there is as yet no clear idea about which of these factors are most essential, although it goes without question, for example, that man's impact on the environment is becoming more and more pronounced: the level of atmospheric pollution is increasing steadily.

Recent paleoclimatic studies have suggested the following conclusions: (i) changes of the climate are "stepwise" and followed by rather rapid transitions; (ii) the climate of the past 2000 years resembles that of the late glacial period; (iii) many catastrophic climate changes have taken place in the past and are possible in the future. It is important to note that climate changes can be location-specific (i.e., regional or local) in different parts of the world, even in adjacent areas. For example, during the last 100 years a temperature increase has been observed in the eastern part of the United States, while in the same period a decrease was observed in the western part of the United States.

To understand the causes of present-day climate changes we must include studies of the data on the state of soils and vegetation, as well as geographical investigations. Analysis of the effects of increasing CO_2 concentrations in the atmosphere and the atmospheric dustloading that results from industrial air pollution is of great importance. Also important is the continued development of the theory and numerical modeling of the biogeochemical cycles of certain environmental components (e.g., carbon, nitrogen, sulfur) that determine the physicochemical transformations of various atmospheric components. These components can affect the climate and are largely transformed by the anthropogenic factors; therefore, the problems of environmental chemistry are becoming critically important to an improved understanding of variations in climate.

TABLE 1 Possible Factors of Climate Change by the Year 2000

Original Source	Factor	Probability of Significant Variation	Probability of Substantial Climatic Impact of Factor Variations	Major Climatic Effect	Characteristic Scales of Climate Variations
Sun	Variations in the solar constant	Low	High	Cooling–warming	Months and longer
	Variations in the UV emission	High	Low–moderate	Not clear	Days and longer
Sun-moon	Tidal perturbations	High	Moderate	Variation in cloud and precipitation (1–10%)	Days, weeks, and longer
Volcanoes	Injection of aerosol to the atmosphere	High	Moderate–high	Cooling (0.1–1°C)	Years and longer
Anthropogenic	Increase in atmospheric CO_2	High	Moderate–high	Warming (1°C)	Secular trend
	Increase in atmospheric dust load	Moderate	Low–moderate	Warming–cooling	Days and longer
	Increase of CFCs in the atmosphere	Moderate[a]	Moderate	Warming	Decades
	Decrease in ozone due to the oxides of nitrogen, CFCs, etc.	Moderate[a]	Moderate	Increase in UV radiation (10%)	Decades
	Thermal pollution	High	High (local effects)	Warming, local clouds and storms	Decades
	Changes in the surface due to agriculture	Moderate	Moderate (regional effects)	Variations in temperature and precipitation	Months and longer
Oceans	Variations in the ocean surface temperature	High	Moderate–high (regional effects)	Variations in temperature and precipitation	Months and longer
Cryosphere	Variation in the area of ice and snow cover	High	Moderate–high (regional effects)	Variations in temperature and precipitation	Months and longer
Biosphere	Polar ice caps variations	Low	High	Sea-level rise, possible glaciation	Years and longer
	Changes in vegetation	Moderate	Moderate (regional effects)	Variations in temperature and precipitation	Years and longer

[a]Assuming limited or constant CFC production beginning in 1980.

Ocean and Climate. To study the large-scale atmospheric processes responsible for weather and climate changes, we must consider the phenomena occurring in the global atmosphere not only in its interactions with the landmass, but especially in its interactions with the ocean, a gigantic heat reservoir. Attention to the atmosphere-ocean interaction (manifesting itself in heat, water, and momentum exchange) is of key importance for the solution of the problems of long-range forecasts of weather and climate changes.

The system of adjoint equations of thermohydrodynamics and the specially developed theory of perturbations, used by Marchuk et al. (8,9) to study the processes of long-term weather anomalies and climate changes manifested as air temperature variations over large territories, has shown that short- and long-range forecasts of temperature anomalies are determined mainly by processes taking place in different regions of the World Ocean, called the energetically active zones of the ocean (EAZO). Such a theoretical analysis, verified by global maps of the World Ocean's heat budget based on observational data, and the previously determined empirical relationships between the state of the ocean and resultant weather, has confirmed the view that EAZO is the most important factor determining long-term weather anomalies and climate changes. This concept is particularly important because the available observational means are still inadequate to routinely obtain necessary meteorological and oceanographic information on a global scale. The Global Weather Experiment has made it possible for the first time to obtain global information for a one-year period. This information has been used to test theoretical models of the atmospheric general circulation and of climate.

Discovering the EAZO has drastically changed the situation: the concentration of available observational means (mainly ships) for the purpose of monitoring processes taking place in the energetically active zones (e.g., EAZO migration and variability) and the use of satellite data permit the development of deeper insights into the processes governing the long-term weather anomalies and climate changes and provide long-range weather forecasters with important information. This has led to the development of the "Sections" program (23), the major objective of which is to develop the long-term (monthly to seasonal) forecasts of mean monthly anomalies of some meteorological parameters (first of all, air temperature) averaged over large regions. Simultaneously, the development of methods of weather forecasting for periods from two seasons to two years (considered a short-term climate forecast) was undertaken. Based on the EAZO concept, the "Sections" program (23) includes both field measurements in different key climatic regions and numerical modeling experiments using observational data.

More graphically, the purpose of the "Sections" program is to study the effect of long-term variations in the thermodynamic state of the ocean on atmospheric processes (as well as the feedback relationships between the atmosphere and processes in the ocean), which results from the ocean-atmosphere interaction concentrated in the EAZO of the ocean. For this purpose, consideration of the variability of the heat transport by the major systems of oceanic currents (an important factor in the formation of large-scale anomalies in the atmosphere) is of great importance.

An analysis of data on the heat budget of the World Ocean has shown, for example, that in January in the Northern Hemisphere maximum heat fluxes from the ocean to the atmosphere are observed in the western Sargasso Sea, east of Newfoundland, in the Norwegian Sea, and east of Japan. Although the maximum heat transfer area of the Newfoundland and Norwegian zones constitutes only 25% of the area of the North Atlantic, their contribution to the ocean surface-to-atmosphere total heat flux north of 40°N reaches 47%.

In the Southern Hemisphere in July the zones of maximum heat release are located in the region of the Drake Passage, south of Africa, in the region of the Tasman Sea, and south of 60°S, to the Antarctic. The energetically active zones are characterized by the highest variability of ocean surface temperatures. Many of these zones are connected with warm currents flowing from low to high latitudes, including the Gulf Stream and its continuation in the North Atlantic, as well as the Kuroshio Current and its continuation in the North Pacific Ocean.

The existence of the EAZO in the Drake Passage is explained by interaction between the Antarctic circumpolar current (encircling the Antarctic) and the Peru and Brasilia Currents. The EAZO south of Africa is connected with the Agulhas Current, and in the Tasman Sea with the East Australian Current. Of particular interest is the tropical belt of the World Ocean, where gigantic amounts of solar radiation are absorbed, with subsequent transport of heat to high latitudes. From observational data in the 10–30°N latitudinal belt, the amount of heat released in the upper layer of the ocean, as a result of solar radiation absorption, reaches 350 W/m^2.

The EAZO include the zones of migration of the polar ice edge and the regions of powerful monsoon formation in the Indian Ocean and in other regions of the World Ocean as well. The role of the different EAZO (from the point of view of their contribution to the formation of long-term weather anomalies) varies substantially both within a given year and from year to year.

Scientific justification for the "Sections" program is based on numerical modeling of the atmospheric general circulation. For the first time in planning such a large-scale experiment, the observational systems and the requirements for data are substantiated based on numerical experiment results, on an account of the specific character of the problem to be solved, and on the need for information to check the central hypothesis which serves as the basis for a theoretical model of long-term forecasts of the weather and of climate changes. Such a multifaceted approach toward achieving the objectives of the "Sections" program is very promising.

Programs like Tropical Ocean-Global Atmosphere (TOGA) and World Ocean Circulation Experiment (WOCE) provide the international coordination required to solve the question of how the World Ocean contributes to the formation of climate. For example, two major objectives of TOGA are (i) studies of the nature of the interannual variability of the tropical oceans and the global atmosphere; and (ii) an analysis of the mechanisms responsible for the interannual variability, as well as predicting this variability (24). These principal aspects of TOGA also cover major large-scale processes studied within the World Meteorological Organization's World Climate Programme (WCP).

It is very important to note that not only are specialized observation programs needed, but so too are long-term regular observations for periods longer than a decade. Various ground-based and ocean-based observational means must be used, and remote sensing will undoubtedly play a key role. Because of the specific character of TOGA connected with studies of interannual variability, errors of measurements averaged in time (month or season) and space (to $10° \times 10°$ latitude) will be of great importance.

The Effect of Mountains. Large-scale orographic effects influence the atmospheric general circulation through closely interacting dynamic and thermal processes (with mountains playing a major role from the point of view of dynamics), and the thermal regime is affected by mountains acting as an elevated heat source. Gates (25) undertook a review of theoretical studies and simplified models of such effects.

A major manifestation of the effect of orography on atmospheric dynamics is the formation of planetary-scale leeward waves, whereas the basic thermal effect is connected with the intensification of the sources of latent and sensible heat over an elevated area. Use of atmospheric General Circulation Models (GCMs) enables one to analyze the combined effect of orography on the dynamics and on the thermal regime of the atmosphere. In this connection, results obtained with the use of the atmospheric GCMs are considered (25).

A major effect of mountains in winter, when the impact of orography is most efficient due to more intensive westerly transport, is manifested through the intensification and the shift (toward the prevailing flow) of troughs, and of quasistationary long waves over the largest continents. In summer the effect is through increasing southern flow and precipitation to the southeast of large mountains. In most cases an interaction takes place between forced planetary-scale waves and transient free waves, with cyclogenesis in leeward zones most active in the regions near the Rocky Mountains, the Alps, and the southern Andes. Probably, the resonance between forced and free waves is a reason for the formation of blocking situations. The orographically driven planetary waves are apparently an important factor in the propagation of energy to the stratosphere and therefore are to some extent responsible for sudden warmings of the stratosphere. From the phenomenological point of view, the Southern Hemisphere mountains localize the zones of cyclogenesis in their leeward part and focus the trajectories of cyclones in the northeast direction.

Gates (25) discussed results of new experiments carried out with and without data from orography in winter and summer, using a two-level GCM. Results of these experiments confirmed previous conclusions and demonstrated the role of mountains in maintaining atmospheric circulation and respective rainfall distribution in the mid- and tropical latitudes.

A special analysis of orographic effects in East Asia showed that the Tibetan Plateau substantially affects the formation of a regional summer climate because of its influence on the southeast Asian monsoon. Furthermore, the Tibetan Plateau favors the formation of the Siberian anticyclone in winter, but orography only weakly affects the total kinetic energy of the atmosphere.

Greenhouse Effect of the Atmosphere. It is well known that the radiative regime of the atmosphere is determined largely by such optically active components as water vapor, carbon dioxide, ozone, and aerosols, and that the greenhouse effect is one of the major mechanisms by which radiative factors affect climate. Although relatively transparent to solar radiation, the atmosphere to a considerable extent prevents heat loss via longwave thermal emission into space from the Earth's surface. Only in the so-called atmospheric transparency windows (within the range of 8 to 14 μm) can emission through a cloudless atmosphere to the surface become much less than the thermal emission of the Earth's surface. For this reason it is important to study the optical properties of the atmosphere in the transparent "windows" and to investigate the physical nature of the greenhouse effect and its variability in relation to climate. An analogous situation in general, but one that differs considerably in detail, takes place on other planets (e.g., Venus and Mars).

The traditional approach to the greenhouse effect, and particularly to its variations, is based on an estimation of the contribution made by carbon dioxide. One of the most popular hypotheses of climate change, both historically and at present, relates temperature variations to the variations in the carbon dioxide greenhouse effect. However, it is well known that the atmospheric absorption spectrum in the transparency window is determined not only by the effect of carbon dioxide, but also by such gaseous components as water vapor, ozone, and many other minor gaseous components, as well as aerosols such as atmospheric dust. From the viewpoint of the theory of climate changes, the leading role belongs to optically-active components with long-term trends. In particular, the chlorofluorocarbons are atmospheric components of entirely anthropogenic origin and are steadily increasing in quantity (21,26).

Typically, cloudiness is a stabilizing factor in the formation of the equilibrium climate on Earth, limiting the internal feedback between the tropospheric temperature increase due to the greenhouse effect of the optically active gaseous components.

Calculations of a possible warming of the Earth's atmosphere as a result of increasing atmospheric CO_2 concentrations have shown that this effect is most strongly pronounced in cold polar regions, where a doubled CO_2 concentration can lead to a surface warming that exceeds the mean global value two- to threefold (11).

Trace Gases. When considering possible climatic changes due to anthropogenically induced variations in the chemical composition of the atmosphere, it is necessary to account for the interaction (synergism) of various climate-forming factors. For example, an increase in the chlorofluorocarbon content of the atmosphere can substantially change the content and distribution of atmospheric ozone. Calculations show that the temperature in the stratosphere is very sensitive to variations in the vertical structure of ozone, and this in turn affects the distribution of other atmospheric components as well as ozone itself. Therefore, it is important that variations in the ozone concentration be measured systematically, along with stratospheric temperature.

If a 25% decrease in the ozone concentration leads to a 0.45 K drop in the surface temperature for the mean planetary model of the atmosphere, then a nonuniform decrease in the ozone concentration causes an antigreenhouse effect of $\Delta T_s = -0.25$ K. As Bojkov (27) noted, it is important to take ozone variations into account not only in the stratosphere but also in the troposphere, since variations in troposheric ozone often lead to inverse effects compared to perturbations in the concentration of stratospheric ozone.

The complicated nature of the greenhouse effect requires regular monitoring of various minor optically active gaseous components of the atmosphere. To accomplish such monitoring, various techniques have been developed. For example, the content of these minor gaseous components in the atmosphere can be measured from satellites by measuring the outgoing atmospheric thermal emission at different wavelengths (28).

There is also a more efficient way to use satellite-borne instruments: measuring the extinction of solar radiation on its way from the sun to the satellite through the atmosphere at sunrise and sunset (the so-called occultation measurements). Such an experiment on the retrieval of the atmospheric minor gaseous components was realized for the first time by the manned orbiting station Salyut-4, equipped with a complex of sun spectrometers that registered the solar radiation at various wavelengths at sunrise and sunset with respect to the orbiting station, when the sun's rays penetrated the atmosphere. Measuring the absorption of solar radiation by the atmosphere, one could obtain, for example, information about the water vapor and ozone content in the atmosphere (28). Abundant information about stratospheric aerosols has been obtained from SAGE and Nimbus-7 occultation measurement data (29). Space-based lidar sounding (30) opens up wide possibilities for monitoring the structural parameters of the atmosphere, minor gaseous components, and atmospheric dust.

The Carbon Cycle. Of great importance for the understanding and eventual forecasting of climate is information about soils and vegetation. Studies on the global biogeochemical cycles of various environmental components, especially carbon, are becoming increasingly urgent (2).

If climate changes can be caused by increasing amounts of CO_2, as a result of human activities, an explanation is needed about how the observed CO_2 concentration is formed in the atmosphere in order to understand its changes in the future.

Studies of the global carbon cycle that attempt to estimate the components of this cycle and to forecast changes in future atmospheric CO_2 concentrations have led, thus far, to controversial conclusions. As for the present global carbon cycle, the role of the biosphere (e.g., deforestation) and of the World Ocean is still not very clear. According to Gorshov's model (2) of the carbon cycle, the ocean is so powerful a sink for carbon dioxide that, even after burning all existing fossil fuels, the CO_2 concentration in the atmosphere cannot increase by more than 35% to 40%. Forecasts about CO_2 dynamics in the future are also hindered by uncertain estimates about energy use by humankind for the next several decades. All this contributes to the uncertain character of the forecasts of future climate, and underscores the urgent need to monitor environmental changes.

The controversy arises because we cannot explain how the global CO_2 budget is formed and how it will change in the future. To explain this, it will be necessary to have considerably more adequate data than currently exist about vegetation cover and its dynamics on a global scale.

CO_2 and the Study of Other Planets. For an improved understanding of the processes taking place on Earth, much useful information can be obtained from studying other planets, such as Venus, Jupiter, and Mars (18,19,31). The need for such studies is based on the fact that planets of the solar system may be considered as models of climate formation under conditions other than those on Earth. Differences between planets are determined, first of all, by specific composition of planetary atmospheres. For example, Earth's atmosphere consists mainly of nitrogen and oxygen, yet various minor components, notably water vapor, determine its optical properties; whereas on Mars and Venus a major component is optically active carbon dioxide. While on Earth a nitrogen-oxygen atmosphere has formed, the thick atmosphere of Venus has become predominately one of CO_2 with negligible water vapor.

The problem of climate is so broad that it makes the studies of other planets very important. Such studies could also be very significant for the investigation of the possible impacts of nuclear war on climate.

Nuclear Winter. Recently, widespread interest in the possible impact of a nuclear war on climate has resulted in the rather rapid development of an extensive literature (see, e.g., Refs. 32–64). Estimating the possible climatic impacts of major postnuclear disturbances of the gaseous and aerosol composition of the atmosphere is exceptionally difficult. A thorough analysis of the contributions by various interactive factors is necessary as well as an analysis of the role of numerous feedbacks, using not only numerical modeling techniques but also possible natural analogues and data from observations that followed nuclear tests during the late 1950s and early 1960s. Despite the limited value of natural analogues, they do contain certain reference points that can help to reduce the uncertainties of numerical modeling that result from various key assumptions.

Thompson (62) noted that conclusions drawn from numerical modeling (indicating that a nuclear exchange would result in a catastrophic decrease of surface air temperature by about $30°C$, i.e., nuclear winter, caused by atmospheric pollution due to urban and forest fires) were questioned by some experts. The reason they were questioned was that such conclusions were based on a very approximate scheme of reality. When solving such a complicated and important problem, modeling results must be critically analyzed by experts from various fields of research.

Concern has developed for reliable quantitative estimates of the possible climatic implications of a nuclear war. This concern has led to the application of three-dimensional climate models that were not specifically designed to simulate the impact on climate of an extreme smoke loading of the atmosphere. Of primary importance in this effort is an adequate consideration of the nonlinear interaction of radiation processes and the propagation, transformation, and fallout of aerosols, the specific features of which are difficult to predict.

One of the most serious uncertainties relates to the amount of smoke aerosols that can be formed in the atmosphere one day after a nuclear exchange. Available estimates of smoke generation are based on estimates of factors difficult to forecast, such as the total area of regions set on fire, the amount and properties of burning material and resulting smoke; diffusion processes and fallout of aerosol; and so forth. These factors are particularly difficult to simulate during the first hours and days of fires, when the smoke clouds would be very dense and the processes of coagulation would be very active. Analysis of conditions related to the formation of smoke during large forest fires in the past can play an important role here.

It is also difficult to estimate the direct radiative effects of such an aerosol because of a limited data base on the optical characteristics of smoke. Recent calculations have highlighted the importance of considering the diurnal aspects of insolation. Also significant is the effect of stratospheric aerosols (not previously considered), which maintain (and intensify) cooling near the surface and reduce tropospheric heating by half.

These and other considerations are a good illustration of the serious difficulties of developing an adequate model of the climatic consequences of a nuclear war. Although precise simulation experiments are impossible, approximate estimates of an ecological catastrophe are extremely important and necessary.

REFERENCES

1. M. E. Berlyand and K. Ya. Kondratyev, 1972. *Cities and Climate of the Planet.* Leningrad: Gidrometeorizdat.
2. V. G. Gorshkov, 1982. The possible global budget of carbon dioxide, *Nuovo Cimento, Ser. 1* **5C**(2), 209–222.
3. K. Ya. Kondratyev, 1981. *Stratosphere and Climate,* Uspekhi nauki i tekhniki, Meteorologia i klimatologia, Vol. 6. Moscow: VINITI.
4. K. Ya. Kondratyev, 1982. *World Climate Research Programme: State, Perspectives, and the Role of Observations from Space,* Uspekhi nauki i tekhniki, Meteorologia i klimatologia, Vol. 8. Moscow: VINITI.
5. E. P. Borisenkov, 1976. *Climate and Its Changes.* Moscow: Znanie.
6. K. Ya. Kondratyev, 1980. *Radiative Factors of the Present Day Global Climate Changes.* Leningrad: Gidrometeorizdat.
7. K. Ya. Kondratyev and D. V. Pozdnyakov, 1976. Stratosphere and freons, *Izv. Akad. Nauk SSSR Fiz. Atomos. Okeana* **12**(7), 683–695.

8. G. I. Marchuk, K. Ya. Kondratyev, and V. P. Dymnikov, 1983. *Some Problems in Climate Theory,* Uspekhi nauki i tekhniki, Meteorologia i klimatologia, Vol. 7. Moscow: VINITI.

9. G. I. Marchuk et al., 1984. *Mathematical Modeling of the General Circulation of the Atmosphere and Ocean,* Leningrad: Gidrometeorizdat.

10. M. I. Budyko, 1984. *Evolution of the Biosphere.* Leningrad: Gidrometeorizdat.

11. National Academy of Sciences, 1982. *Carbon Dioxide and Climate: A Second Assessment.* Washington, D.C.: National Academy Press.

12. J. Smagorinsky, 1981. Climate modeling, *Proc. Tech. Conf. on Climate in Asia and Western Pacific, Dec. 15–20 1980,* WMO Publication 578. Geneva: WMO, pp. 139–156.

13. M. I. Budyko, 1977. *Present Day Climate Change,* Leningrad: Gidrometeorizdat.

14. K. Ya. Kondratyev, 1977. *Present Day Climate Changes and Their Determining Factors,* Uspekhi nauki i tekhniki, Meteorologia i klimatologia, Vol. 4. Moscow: VINITI.

15. GARP Joint Organizing Committee, 1975. *The Physical Basis of Climate and Climate Modeling,* GARP Publications Series 16, Geneva: WMO.

16. National Academy of Sciences, 1978. Geophysical predictions, *Studies in Geophysics,* Washington, D.C.: National Academy Press.

17. K. Ya. Kondratyev and N. I. Moskalenko, 1977. *Thermal Emission of Planets.* Leningrad: Gidrometeorizdat.

18. K. Ya. Kondratyev, 1977. *Meteorology of Planets.* Leningrad: LGU.

19. K. Ya. Kondratyev and G. E. Hunt, 1982. *Weather and Climate on Planets.* Oxford: Pergamon Press.

20. World Meteorological Organization, 1976. WMO Statement on the impact of man's activity on the ozone layer, and on some possible geophysical consequences, *WMO Bull.* **25**(1), 74–79.

21. V. Ramanathan, R. J. Cicerone, H. B. Singh, and J. T. Kiehl, 1985. Trace gas trends and their potential role in climatic change, *J. Geophys. Res.,* **90**(D3), 5547–5562.

22. J. G. Charney, 1975. Dynamics of deserts and drought in the Sahel (Symons Memorial Lecture), *Quart. J. R. Meteorol. Soc.* **101**(428), 193–202.

23. G. I. Marchuk (Ed.), 1984. *The Programme for Investigation of the Ocean-Atmosphere Interaction to Study Short Term Climate Variations, Programme "Sections,"* Uspekhi nauki i tekhniki, Atmosfera, okean, kosmos, Vol. 1. Moscow: VINITI.

24. World Meteorological Organization, 1983. *Report of the JSC/CCCO Study Group on Interannual Variability of the Tropical Oceans and the Global Atmosphere (TOGA), 13–16, Oct. 1982, Princeton, World Climate Paper 49.* Geneva: WMO.

25. W. L. Gates, 1984, *The Effects of Large-Scale Mountains on the Atmospheric General Circulation and Climate with Special Reference to Eastern Asia,* Climatic Research Institute Report 50. Corvallis, Oreg.: Oregon State University.

26. W.-C. Wang, D. J. Wuebbles, W. M. Washington, R. G. Isaacs, and G. Molnar, 1986. Trace gases and other potential perturbations to global climate, *Rev. Geophys.* **24**(1), 110–140.

27. R. D. Bojkov, 1984. Tropospheric *Ozone, Its Change and Possible Radiative Effects, WMO Special Environmental Report 16*. Geneva: WMO.

28. K. Ya. Kondratyev (Ed.), 19 *Environmental Studies from Manned Orbital Stations*. Leningrad: Gidrometeorizdat.

29. M. P. McCormick, 1983. Aerosol measurements from Earth orbiting spacecraft, *Adv. Space Res.* **2**(5), 73–865.

30. V. E. Zuev, 1983. Laser sounding of aerosols using airborne and space facilities, *Adv. Space Res.* **2**(5), 39–47.

31. J. S. Lewis and R. G. Prinn, 1984. *Planets and Their Atmospheres, Origin and Evolution*. London: Academic Press.

32. National Academy of Sciences, 1975. *Long-Term Worldwide Effects of Multiple Nuclear-Weapon Detonations*. Washington, D.C.: National Academy Press.

33. L. Dotto and H. Schiff, 1978. *The Ozone War*. New York: Doubleday.

34. Anon., 1982. Nuclear War: the Aftermath, *Ambio* **11**(2–3), 75–1765.

35. J. Falk, 1982. *Global Fission: The Battle over Nuclear Power*. Oxford: Oxford University Press.

36. S. A. W. Gerstl and A. Zardecki, 1982. *Reduction of Photosynthetically Active Radiation under Extreme Stratospheric Aerosol Loads*, Geological Society of America Special Paper, 190, pp. 201–210.

37. A. Goudie, 1982. *The Human Impact: Man's Role on Environmental Change*. Cambridge, Mass.: MIT Press.

38. A. M. Katz, 1982. *Life after Nuclear War*. Cambridge, Mass.: Ballinger.

39. B. Martin, 1982. Critique of nuclear extinction, *J. Peace Res.* **19**(4), 289–300.

40. J. Peterson and D. Hinrichsen (Eds.), 1982. *Nuclear War: The Aftermath*. Elmsford, N.Y.: Pergamon Press.

41. A. B. Pittock, 1982. Nuclear explosions and the atmosphere, *Aust. Phys.* **19**(11), 189–192.

42. V. V. Aleksandrov and G. L. Stenchikov, 1983. On the modeling of the climatic consequences of nuclear war, *Proc. on Applied Mathematics, The Computing Centre of the USSR Academy of Sciences, Moscow*.

43. M. Denborough (Ed.), 1983. *Australia and Nuclear War*. Sydney: Groom Helm Australia.

44. P. R. Ehrlich et al., 1983. Long-term biological consequences of nuclear war, *Science* **222**(4630), 1293–1299.

45. G. S. Golitsyn and A. S. Ginzburg, 1983. *Climatic Consequences of a Possible Nuclear Conflict and Some Natural Analogs: A Scientific Study*. Moscow: Committee of the Soviet Scientists on Peace Defense against Nuclear Threat.

46. Yu. A. Izrael, 1983. Ecological consequences of a possible nuclear war, *Meteorol. Gidrol.* **10**, 5–10.

47. C. Paine, 1983. The aftermath of nuclear war, *Science* **220**(4599), 812–813.

48. Anon., 1983. The atmospheric consequences of nuclear warfare, *Bull. Am. Meteorol. Soc.* **64**(11), 1302–1311.

49. R. P. Turco, O. B. Toon, T. Ackerman, J. B. Pollack, and C. Sagan, 1983. Nuclear War: global consequences of multiple nuclear explosions, *Science* **222** (4630), 1283–1292.

50. V. V. Aleksandrov and G. L. Stenchikov, 1984. On a computing experiment that models the climatic consequences of a nuclear war, *J. Comput. Math. Math. Phys.* **24**(1), 140–144.

51. V. V. Aleksandrov and N. N. Moiseev, 1984. A nuclear conflict as viewed by climatologists and mathematicians, *Vestn. Akad. Nauk SSSR* **11,** 65–76.

52. I. J. Barton and G. W. Paltridge, 1984. Twilight at noon overestimated, *Ambio* **13**(1), 49–51.

53. C. Covey, S. H. Schneider, and S. L. Thompson, 1984. Global atmospheric effects of massive smoke injections from a nuclear war: results from general circulation model simulations, *Nature* **308,** 21–25.

54. P. J. Crutzen, 1984. Darkness after a nuclear war, *Ambio* **143**(1), 52–54.

55. P. J. Crutzen and I. E. Galbally, 1984. Atmospheric effects from postnuclear fires, *Clim. Change* **6,** 323–364.

56. D. M. Elsom, 1984. Climatic change induced by a large scale nuclear exchange, *Weather* **39**(9), 268–271.

57. Yu. A. Izrael, 1984. *Ecology and Environmental Control.* Moscow: Gidrometeorizdat.

58. E. S. Knorre, 1984. A conference on disarmament, *Sci. USSR,* **1,** 2–10, 104–109.

59. K. Ya. Kondratyev and G. A. Nikolsky, 1984. *Possible Impacts of a Nuclear Conflict on the Atmosphere and Climate,* Moscow: International Projects Center, GKNT.

60. J. Raloff, 1984. Nuclear winter research heats up. *Sci. News,* **126**(12), 182.

61. C. Sagan, 1984. Nuclear war and climatic change—guest editorial, *Clim. Change* **6**(1), 1–4.

62. S. L. Thompson, 1984. An evolving "nuclear winter"—guest editorial, *Clim. Change* **6**(2), 105–108.

63. G. F. White and J. London, 1984. Nuclear winter scenario, *Science* **224**(4645), 110–113.

64. K. Ya. Kondratyev, S. N. Baibakov, and G. A. Nikolsky, 1985. Nuclear war, atmosphere, and climate, *Sci. USSR,* **2.**

Chapter 1

The Greenhouse Effect of the Atmosphere and Climate

1.1 CARBON DIOXIDE AND CLIMATE

The increasing scale of society's industrial activities draws attention to possible anthropogenic factors that affect climate (11–88). Of fundamental interest in this regard is that aspect of the problem concerned with the impact of the increasing atmospheric concentration of CO_2 and the greenhouse effect. A discussion of this problem began at the turn of this century and most recently reappeared in the 1950s. It is once again at the center of attention, sparking scientific controversy as well as opposing points of view. The controversy prompted the Global Climate Group of WMO's Commission for Atmospheric Sciences to state (71):

> A concern is expressed for the fact that during recent years various national and international organizations convened quite a number of conferences on the CO_2 problem, and sometimes scientifically unsubstantiated alarmistic statements were made. Such a non-coordinated approach can have negative consequences for the reputation of science, and therefore the convocation of WMO/UNEP/ICSU joint conferences like that of the experts on the assessment of the CO_2 impact on climate changes and their consequences (Villach, Austria, November 1980) is more expedient.

Despite numerous efforts in numerical modeling of CO_2's climatic impact, the level of understanding of the problem still remains insufficient. Until now, all studies have demonstrated a well-known scenario: an increase in carbon dioxide concentrations intensifies the greenhouse effect, which, in turn, leads to a global climate warming. It has been known for some time, however, that this simple scenario is a crude approach to estimating the greenhouse effect due only to CO_2 and is far from realistic. Many unsolved problems connected with different aspects of the physics of climate change remain, particularly as they concern considerations of the effects of oceans and clouds. Of particular importance is water vapor, which contributes much to the formation of the greenhouse effect. Therefore, an account of the feedbacks that permit an adequate simulation of water vapor dynamics in the course of climate change is essential.

Recent studies shed light on difficulties associated with identifying a more realistic assessment the impact of CO_2 on climate. Such difficulties are related to (i) studies of the multifaceted nature of the greenhouse effect (e.g., the growing importance of various minor optically active atmospheric constituents, the effect of hot bands, the temperature dependence, etc.) (7,24,37,39,40,65,73,82,86,87,89 – 99) and (ii) the discussion of poorly studied aspects of the global carbon cycle, as well as the numerical modeling of climate changes that result from constituent changes of the atmosphere. Therefore, of great interest are considerations by Smagorinsky (80,81), who discussed two key climate-forming factors: (i) the effect of increased CO_2 concentrations and (ii) sea surface temperature (SST) anomalies. Although in the past numerical modeling has led to the conclusion that a doubling of CO_2 concentration would lead to a global climate warming between 1.5 and 4.5°C, more recent climate model experiments suggest that the correct value may be closer to the lower level.

Uncertainties in available estimates are determined by a great number of assumptions on which numerical modeling is based. The greatest drawback is a rough account of the role of the ocean in the formation of the global carbon cycle and in the impact of CO_2 on climate (100). Problems related to the varying optical properties of cirrus clouds and snow cover are also unsolved (101). The contribution of numerous optically active minor gaseous components and aerosols to the greenhouse effect has to be more reliably estimated. In addition, the identification of CO_2-induced climate changes based on observational data is masked by natural short-term climate variations.

The solution to the problem of climatic implications of an increase in CO_2 concentration in the atmosphere requires, first of all, an assessment of the dynamics of the global carbon cycle. What is presently known is an approximate relationship between the total anthropogenic release of CO_2 to the atmosphere and an increase in the CO_2 content in the atmosphere. At least 50% (maybe 70%) of the amount released to the atmosphere go to other natural CO_2 reservoirs, including forests, ocean surface layers (down to 1 km), and deep layers of the ocean.

At present the fact of the increases in the anthropogenically induced CO_2

concentrations in the atmosphere is clearly established. A successful development of climate theory has created the basis for comparatively realistic estimates of the impact of this increase on climate. A group of experts convened by the U.S. National Academy of Sciences published a report in 1979 where results from studies on the CO_2 impact on climate were summarized. Some years later a second, similar report on the status of the problem was also completed. The Second Report of the U.S. National Academy of Sciences (9) provides an accurate picture of the development of contemporary studies of CO_2 and its impacts on climate. In this report, the following aspects were considered in detail: (i) the role of the ocean; (ii) the reliability of the conclusions about the impact of CO_2 on climate, drawn from analysis of local observational data on the land surface heat balance; (iii) the contribution by minor optically active components and by aerosols; (iv) the filtering out of the "signal" of the CO_2 impact on climate from observational data.

An important role has been emphasized (9) for simple climate models (energy-balance, radiative-convective) in an identification of the most important fields of application and interpretation of results obtained with the use of sophisticated 3-D models. Therefore, of great importance is a coincidence of estimates of the sensitivity of mean global temperatures to increasing CO_2 concentrations obtained from simple and sophisticated models. An important factor that determines the reliability of the estimates is a consideration of the effect of CO_2 on the whole surface-atmosphere system (not only radiative heating of the surface), and of all the most substantial climate-forming processes and respective feedback mechanisms. (Neglecting feedback mechanisms explains the unreliable estimates of climate sensitivity based on data of local observations of the heat balance.)

Of central importance is the atmosphere-ocean interaction, which slows down the effect of CO_2 on climate. Approximate assessments have shown that with a doubled CO_2 concentration it takes 3 years to raise by 2°C the equilibrium temperature of a 50-m mixed layer, 30 years for a 500-m layer, and 300 years for the entire global ocean to a depth of 5 km. (This phase shift is much greater, because in a climate warming the ocean-atmosphere sensible and latent heat fluxes grow, which delays attaining the equilibrium temperature.) The assessments mentioned above should, however, be considered tentative, because the basis for the assessments, according to which additional heat coming to the ocean behaves like a passive substance not affecting the processes of exchange responsible for the downward transport of this heat, is generally incorrect but may be accepted for the upper layer of the ocean, where vertical mixing is determined mainly by winds and seasonal gyres and where horizontal circulation is caused by wind stress and not by the surface temperature gradient.

Numerical modeling with an approximate interactive ocean-atmosphere model based on an assumption of a doubled CO_2 concentration and a respective 4-W/m^2 increase in energy coming to the ocean surface has led to the conclusion that a 2°C rise in the tropical temperature and a 4°C rise in subpolar temperature correspond to an equilibrium state approached in 150 years. About 50% of the warming in the tropics occurs in the first 5 years. The increase in temperature then slows down (80% of warming in 100 years). In the high latitudes a 50% warming is reached in 25 years.

The effect of polar ice occurs when its melting in the summer prevents climate warming, which is clearly manifested only in winter. The understanding of the relationship between the formation of sea ice and deep waters remains rudimentary. So, too, is an understanding of climatic factors that govern the extent of ice cover in the Northern and Southern Hemispheres. Particular attention must be given to the role of radiation processes and mixing in the ocean in the formation of different annual changes in the extent of sea ice in the Northern and Southern Hemispheres. Until this becomes clear, even a qualitative assessment of the effect of sea ice on the reaction of climate to increases in the atmospheric concentration of CO_2 and the formation of deep water remains difficult.

Two important aspects of the cloud-radiation feedback remain uncertain: (i) the effect of the CO_2-induced climate warming on variations in the amount, top height, optical properties, and spatial structure of clouds; and (ii) the feedback effect of these changes on climate. The supposition that variations in the cloud cover negligibly affect the sensitivity of climate to the external perturbations are premature and unsubstantiated. No doubt, a climate model cannot be considered reliable if it is unable to reproduce accurately the observed annual change of cloudiness and Earth's Radiation Budget (ERB) components.

With regard to the problem of model validation, it is important to compare the results obtained not only with averaged fields of various parameters but also with statistical characteristics (moments) of higher order. It is also important to undertake the testing, based on field experiments, of the adequacy of techniques for parameterization of various physical processes (e.g., cloudiness-radiation interaction, the formation of snow and ice cover, heat and moisture exchange, and so forth). Because the "calibration" of existing climate models may be insufficient, the use of paleoclimate data and studies of climate on other planets may prove to be very important.

In simple climate models an account of the ice albedo feedback may be decisive for the prediction of climate warming in the high latitudes. With the appearance of open water, however, evaporation increases, and this can lead to a more intensive formation of stratus clouds, to an increase in the Earth's albedo, and hence to a compensation of the CO_2-induced warming. Numerical experiments using 3-D models have verified the growth of cloud cover by several percent in regions where increased CO_2 concentrations lead to the melting of sea ice; a change in the system's albedo in the regions mentioned above is four times lower than in the surface albedo. As far as results of numerical modeling of the CO_2-induced climate changes are concerned, mean global and mean zonal temperature values can be considered reliable. However, results of calculations of the geographical distribution of wind, soil moisture, cloud cover, and solar radiation cannot be considered realistic and should be interpreted only as conditional scenarios.

Comparing the numerous estimates of the mean global warming that might result from a doubling of CO_2, the value $3 \pm 1.5°C$ appears most probable. Gates et al. (18,102) and Mitchell (53) gave a much lower level of warming (0.2 to 0.3°C), but this is probably explained by a prescribed sea surface temperature which strongly limits variations in surface air temperature. Stratospheric cooling in the layer from 30 to 45 km reaches 7 to 11°C.

A doubled CO_2 concentration causes a marked intensification of the hydrologic cycle. Calculations of mean zonal temperature values give a high-latitude warming that is two to three times higher than in the tropics, where such temperature values are limited by an effect of moist convection, "spreading" the warming throughout the whole atmosphere. Other effects on the mean zonal fields of climate parameters that can only be judged qualitatively (quantitative estimates are not reliable) are as follows: (i) a marked increase of mean annual runoff takes place in high latitudes as a result of the intensification of precipitation resulting from the intrusion of moist warm air masses north of 60°N; (ii) earlier snowmelt and later snowfall due to the high-latitude warming; (iii) reduced soil moisture in the Northern Hemisphere summer in middle and high latitudes (north of 35°N), explained by an earlier beginning of the period of relatively strong evaporation (following snowmelt) and by the weakening of summertime precipitation; and (iv) a reduced area of polar ice cover.

Available observational data do not permit one to assess the reliability of the conclusions mentioned above and to identify the "signal" of the anthropogenically induced climate warming (103–106). The lack of convincing evidence for the warming to take place sometime in the future can be explained by the following three circumstances: (i) the ocean-caused delay in the manifestation of warming; (ii) difficulties in filtering out the CO_2 signal against other factors (a change in the extra-atmospheric insolation, in the aerosol content, etc.); and (iii) limited observational data. In this connection an extensive and long-range program of climate monitoring, as well as revealing the parameters most sensitive to the CO_2 signal, is of great importance. Apparently, the latter are: mean zonal surface air temperature in summer (except for high latitudes, where the signal/noise ratio gets higher); mean zonal summer temperature in the stratosphere and mesosphere, where strong anthropogenic cooling should be manifested at a low natural variability; the temperature of surface and deep layers of the ocean; and the extent of the polar ice cover.

1.2 THREE-DIMENSIONAL NUMERICAL MODELING OF THE CO_2 EFFECT ON CLIMATE

Early achievements in the application of 3-D atmospheric general circulation models (AGCM) to evaluate the impact of CO_2 on climate are associated with studies by Manabe et al. (47–50). Manabe (51) has summarized the results of 15 years of numerical climate modeling activities at Princeton University's Geophysical Fluid Dynamics Laboratory to evaluate the effect of an increase in CO_2 concentration in the atmosphere on climate. Basic results from calculations based, as a rule, on the assumption of the fourfold increase of the CO_2 concentration (to separate the CO_2 signal more reliably from the noise of the natural variability of climate) are as follows: a climate change consists of a tropospheric warming and stratospheric cooling. The former is caused by the increased emissivity of the atmosphere, by a respective upward shift of the "emission level" to space, and by the need for higher temperatures to maintain a zero radiation budget of the Earth. The latter is caused by intensified radiative cooling of the stratosphere.

As noted earlier, the mean annual climate warming near the Earth's surface in the high latitudes is two to three times greater than in low latitudes. The warming over the Arctic ocean and its environment is characterized by a strong annual change: maximum in winter, minimum in summer. In low latitudes there is a much smaller annual change.

With a CO_2-induced warming an intensification of the hydrological cycle occurs. Mean global evaporation is raised by 7% and is the result of changes in the surface radiation budget: an increase in the atmospheric thermal emission should be balanced by heat spent on the latent and sensible heat exchange, with the latter dominating the evaporation in high latitudes (in low latitudes the situation would be just the opposite).

Other consequences of an increased CO_2 concentration include a decrease in the areal extent and thickness of sea ice and an earlier melting of snow cover; as a result, in summer the soil moisture would dry up in the mid- and high latitudes of the Northern Hemisphere. In addition, the mean latitudinal band of precipitation would shift northward as a result of the transport of moist and warm air masses to the higher latitudes. The mean annual runoff would also intensify in the high latitudes.

In order to validate the calculated climate changes induced by an increased CO_2 loading of the atmosphere, it is important to determine the level of the signal/noise ratio (S/N). If $S/N \gg 1$, the CO_2-induced changes exceed the level of the internal noise of the model (internally caused variability) and consequently, the CO_2 signal can be reliably recognized. The S/N ratio was calculated for mean monthly and zonally averaged values of the surface air temperature and soil moisture with the use of numerical modeling results for a period of 20 years ($1 \times CO_2$) and 16 years ($4 \times CO_2$) and subsequent averaging over 10 and 3 years, respectively.

The mean monthly meridional S/N profiles for temperatures over the continents are characterized by a maximum in the midlatitudes of the Northern Hemisphere summer (about 10 to 20), when the internal variability of temperature is at a minimum (comparisons with observational data enables one to draw a conclusion about the similarity between the internal noise of the model and the natural variability of the climate). Naturally, over the oceans (where internal temperature variability is weaker) the S/N ratio is higher than over the continents, and, on average, for Northern Hemisphere conditions the meridional S/N profile is close to that corresponding to the continents.

In polar regions the S/N ratio is small, which is related to the existence there of maximum internal variability. On the other hand, high S/N values in the case of soil moisture [about 2 to 4 ($S/N = 1.45$ is equivalent to the statistical significance at a 95% confidence level)] reflect the summertime drying-up of soils in the mid- and high latitudes of the Northern Hemisphere, a decrease in humidity near 25°N in winter, and an increase in humidity in the 60°N region during the fall, winter, and spring seasons. In sum, the problem of detection of anthropogenically induced changes in soil moisture is much more complicated than recognition of the CO_2 signal in the field of the surface air temperature.

Estimates of the effect of increased CO_2 concentrations on climate have been obtained mainly by comparing equilibrium climates that correspond to the present-day and projected future levels of increased CO_2 concentrations. Calculations have shown that the latitudinal profiles of climate change under conditions of a stepwise and a gradual increase of the CO_2 concentration must be qualitatively similar. However, for more reliable conclusions, climate models must be further improved, especially with regard to the interaction between the atmosphere and the ocean.

Manabe (51) noted that existing models have a substantial drawback. They incorrectly simulate the thermocline, making it too deep in low latitudes. This can lead to an erroneous estimation of the effect of the gradual increase in the CO_2 concentration on climate. Numerous drawbacks are also characteristic of the atmospheric component of climate models, which are manifested, for example, by the unsatisfactory simulation of the hydrologic cycle, particularly with regard to orography. Although a successful simulation of the annual change of mean hemispherical values of the ERB components with the use of fixed cloudiness climate models can be interpreted as a manifestation of the lack of sensitivity of climate to variations in the global cloud cover, the problem of an interactively changing cloudiness remains urgent.

One of the principal drawbacks of existing climate models is their neglect of the cloudiness-radiation interaction as one of the most important factors governing the dynamics of the climatic system. For this reason Thompson and Schneider (83, 84) believe that at present the conclusions about the impact of cloudiness on the sensitivity of climate to an increased CO_2 concentration must be treated as preliminary. Even the sign of the possible impact of CO_2-induced cloud variations on surface air temperature is unknown. If one assumes, for example, that there would be even a small increase in the amount of cloud cover, then, with an increase in CO_2 concentration, it would lead to an increase in the Earth's albedo and to a climate cooling, since an increase in the cloud cover causes a decrease in the Earth's radiation budget (96,97). The feedback determined by changes in atmospheric water content, during the process of climate change, is also of great importance. Estimates have shown, for example, that depending on the response of the hydrologic cycle and atmospheric water content to a CO_2 doubling, the climate warming can vary between 0.5 and 10°C.

Of key importance is an adequate accounting of the ocean-atmosphere interaction. During the past few years preliminary steps have been taken to solve this problem. The inadequacy of observational data and the insufficient understanding of the processes of mixing in the ocean (the phenomenon of intermittent turbulence illustrates the complexity of these processes) are serious constraints on the construction of realistic interactive ocean-atmosphere models and are basic sources of uncertainties in climate modeling. The present ideas about feedback mechanisms determined by formation of sea ice and deep-water circulation are still quite rudimentary.

In this regard Schlesinger (76,77) emphasized the importance of the choice of subgrid parameterization techniques for the following processes: sensible heat exchange, moisture and momentum exchange between the surface and the atmosphere; turbulent transport of heat, moisture, and momentum in the atmosphere due to dry and wet (cumulus clouds) convection; phase transformations of water in the atmosphere; transfer of short- and longwave radiation; formation of clouds and their radiative properties; formation and melting of snow cover; and heat and moisture exchange in soils. To obtain reliable estimates of the CO_2 climatic implications, it is necessary to consider the interactive system of the ocean (including sea ice) and the atmosphere, taking into account the ocean's mixed layer of varying thickness and sea currents. When estimating the increased CO_2 impact on climate, one must consider the noise from the models themselves to which a signal must be related, which represents the difference between the disturbed and control cases. The need to increase the S/N ratio has prompted scientists to estimate the effects of a quadrupled CO_2 concentration.

Available models greatly differ with regard to (i) geography and orography, which vary from a conditional "sector" distribution of land and oceans to a more realistic reproduction of geography (and orography); (ii) oceans (from the "swamp" ocean to a mixed-layer model); (iii) insolation (from prescribed mean annual insolation to consideration of the annual course); and (iv) vertical resolution.

All of the GCMs lead to the conclusion about a tropospheric warming and stratospheric cooling with an increased CO_2, but important details of these effects significantly differ. For example, the Geophysical Fluid Dynamics Laboratory (GFDL) model shows that warming increases with height, reaching a maximum at an altitude of 10 km, whereas the NCAR model shows the same pattern but the maximum is weaker. A similar situation occurs with regard to an intensification of an atmospheric warming which increases with latitude.

According to the NCAR and GFDL models, there is an interaction with clouds. The major patterns of warming are quite similar in experiments with specified clouds, as well as experiments with interactive models. Calculations with the Oregon State University (OSU) model gave an intensification of climate warming from the tropics to subtropics, then an attenuation toward the midlatitudes, and again, an intensification of warming in the high latitudes, where the temperature rise is only twice as high as in the tropics. The GFDL sector models, however, revealed in their 1975 and 1980 studies a four- to fivefold temperature increase in the atmospheric warming.

There is strong longitudinal variability in the effect of a doubled CO_2 in the temperature field. In some cases, however, experiments have not been calculated for long enough to reach an equilibrium, and the period of averaging (the last 100 days) is, perhaps, too short to obtain highly statistically significant results. A comparison of data from the OSU model obtained with specified distributions of SST and sea ice, taking into account the ocean-atmosphere interactive system, has shown that to the first approximation the effect of warming turns out to be underestimated by an order of magnitude compared to studies with interactive oceans.

Data from the GFDL model show that the basic result of considering the seasonal change of insolation consists of decreasing the mean annual values of warming at all latitudes from $0.5\,°C$ in the tropics to $4\,°C$ at latitudes higher than $70\,°$. Effects of a CO_2 doubling and quadrupling are qualitatively similar, but as a rule, with a quadrupled CO_2 the warming is less than twice as large.

Table 1.1 characterizes the results obtained by various authors with regard to the mean global surface air temperature in the control case ($1 \times CO_2$); the difference between calculated and observed temperatures ($1 \times CO_2 - OBS$), equal to $14.2°C$; and temperature increases with a doubled ($2 \times CO_2$) and quadrupled ($4 \times CO_2 - 1 \times CO_2$) CO_2 concentration. As shown, minimum values of a warming were obtained with the OSU model (Mitchell, Gates, and others) at fixed distributions of SST and sea ice, but the same OSU model with an interactive account of these factors gave an increase in warming by about an order of magnitude.

It follows from Table 1.1 that most of the models overestimate the mean global temperature. The higher the overestimation, the more substantial the CO_2-induced climate warming (Wetherald and Manabe obtained an opposite result when estimating the climate sensitivity to variations in the solar constant). According to the GFDL model, in the case of $2 \times CO_2$ precipitation increases occur almost everywhere in the latitudinal belts higher than $45°$, but they do not exceed 1 mm/day. The geographical distribution of the impact on precipitation is nonuniform. It can be illustrated by the fluctuating character of the mean zonal profiles of precipitation variations. In the case of $2 \times CO_2$ the relative mean global variation ranges for different models from -2.5% (Mitchell) to 7.8% (Manabe and Wetherald, 1975). The higher the model-overestimated temperature, the higher the relative increase in precipitation. The mean global profile of variations in the soil moisture content also fluctuates.

OSU model estimates have led to the conclusion that the time period for reaching an equilibrium mean global temperature is about 300 days. The same period is needed to reach an equilibrium geographical temperature distribution. These estimates, however, refer only to the model with a "swamp" ocean.

In analyses of simulation experiment results, of paramount importance is an evaluation of their statistical significance. Table 1.2 illustrates the estimates of statistical significance of variations in the mean global values of temperature, precipitation, and soil moisture content, using the OSU model as an example. The level of statistical significance P is determined as the probability that the signal/noise ratio (Z) can be greater than any randomly calculated ratio (Z_{SIM}), based on the Gaussian distribution. Here

$$Z = Z_{SIM} - \frac{\overline{X}(2 \times CO_2) - X(1 \times CO_2)}{\{1/N[V^2(2 \times CO_2) + V^2(1 \times CO_2)]\}^{1/2}}$$

where V is the variability of respective parameters and N is the sample size. If, for example, $|Z_{SIM}| = 1$ and 3, then a random inequality $|Z| > |Z_{SIM}|$ can hap-

TABLE 1.1 Global Mean Air Temperature Values (°C) and Their Variations Caused by a Doubling and Quadrupling of the CO_2 Concentration

Model	$1 \times CO_2$	$1 \times CO_2$–OBS	$2 \times CO_2 - 1 \times CO_2$	$4 \times CO_2 - 1 \times CO_2$	
Manabe and Wetherald (1975)	20.9	6.7	2.9	—	GFDL
Manabe and Wetherald (1980)	21.3	6.9	3.0	5.9	GFDL
Manabe and Stouffer (1980)	14.8	0.6	—	4.1	GFDL
Wetherald and Manabe (1980)[a]	16.0	1.8	—	6.0	GFDL
	16.8	2.6	—	4.8	
Washington and Meehl (1983)[b]	11.7	-2.5	1.1	3.4	NCAR
	11.5	-2.7	1.4	—	
Mitchell (1981)[c]	12.3	-0.9	0.2	—	
Gates, Cook, and Schlesinger (1981)[c]	14.8	0.6	0.2	0.4	OSU
Schlesinger (1982)	17.9	3.7	2.0	—	OSU

[a]The first (second) line without (with) account of the annual course of insolation.
[b]The first (second) line with (without) account of the interactive cloudiness.
[c]Sea surface temperature held constant.

TABLE 1.2   Statistical Significance of Variations in the Mean Global Values of Climate Parameters (the OSU Model) with CO_2 Doubling

Parameter	Air Temperature (°C)	Precipitation (mm/h) Natural Logarithm	Soil Moisture Content (cm)
Experiment	19.87	1.05	3.41
Control	17.88	1.00	3.43
Difference	2.00	0.05	−0.02
Signal/noise	32.1	9.6	−0.4
Statistical significance (%)	0.0	0.0	71.9
95% Confidence interval	0.12	No estimate	0.09

pen with the probability 31.74% and 0.26%, respectively. In case of a 2°C global warming, a 95% confidence interval constituting 0.12 means that the probability of warming by less than 1.88°C or greater than 2.12°C is 5%.

Thus the estimates of the global warming are statistically reliable, whereas the estimates of changes in soil moisture related to a CO_2 warming are unreliable. The statistical significance of the results is decreased in the transition to calculations of mean zonal values and geographical distribution.

With regard to the temperature field, the results are statistically significant at a level below 1% over the largest part of the ocean, but over land this level rises above 10%. The latter also applies to the global distribution of precipitation, the variations of which turn out to be statistically unreliable. This applies even more strongly to soil moisture.

Schlesinger (77) recommended the following to provide for the comparability and reliability of calculated climate changes: (i) a sufficiently long integration that ensures reaching an equilibrium; (ii) estimates of the statistical significance of modeling results; (iii) consideration of such climate indicators as characteristics of the cryosphere (snow and ice cover); (iv) a broad application of the energy balance and radiative-convective models (using GCMs mainly to analyze the reasons for the difference in the estimates); and (v) simulation modeling that takes into account the atmosphere-ocean interaction within the interactive ocean-atmosphere system.

A major difficulty for numerical climate modeling is the complicated interaction of numerous feedbacks that operate in the climate system. The most important problem in studies of climate formation is the analysis of contributions of the various factors and feedback mechanisms that determine climate. Thus far, no convincing way has been found to assess the effect of individual parameters on the formation of the temperature field, one of the most important climatic indicators. But such an analysis can be applied to the Earth's radiation budget and its components.

In this connection Mitchell et al. (52) undertook numerical modeling of climate at recent normal CO_2 levels (e.g., 320 ppm) and at doubled CO_2 levels using a zonal dynamic statistical mean annual climate model developed by the Lawrence Livermore National Laboratory. Results of this modeling form the basis for the estimation of the sensitivity of ERB (R) and its components—absorbed solar radiation (S) and outgoing longwave emission (F)—to variations in surface air temperature (T), total water content of the atmosphere (W), total cloud amount (C), surface albedo (A), and carbon dioxide concentration (CO_2). Table 1.3 illustrates the results of calculations for the Northern Hemisphere (NH) and the Southern Hemisphere (SH), as well as for the entire globe (G). Variations (δ) in ERB and its components are calculated with the use of the radiation routine of the model as differences corresponding to disturbed and undisturbed conditions (only with respect to a subsequent parameter). Naturally, a climate warming will lead to an increase in F but will not change S, and an increased atmospheric water content will cause an increase in S and a decrease in F (the latter determines the growth of the ERB).

There is a hemispheric difference in the effect of water content on the absorbed radiation but there is symmetry with respect to outgoing emissions. This is explained by the fact that the solar radiation absorption takes place mainly in the lower troposphere, whose water content is much greater in the Southern Hemisphere than in the Northern Hemisphere.

With a CO_2 doubling, the NH cloud cover somewhat decreases and in the SH it increases markedly, which is manifested in respective changes of the absorbed radiation (either an increase or a decrease). With regard to the outgoing emission, the latitudinal redistribution of clouds and changes in cloud top heights are also clearly manifested. Variations in surface albedo are

TABLE 1.3 Variation in ERB and Its Components under the Influence of Various Factors (W/m²)[a]

ERB		T	W	C	A	CO_2
				Area		
δS	NH	0.000	0.569	1.906	0.501	0.000
	SH	0.000	0.120	−3.048	0.285	0.000
	G	0.000	0.344	−0.571	0.393	0.000
δF	NH	4.359	−1.823	1.570	0.000	−2.454
	SH	2.700	−1.657	−0.041	0.000	−2.640
	G	3.530	−1.740	0.764	0.000	−2.547
δR	NH	−4.359	2.392	0.335	0.501	2.454
	SH	−2.700	1.777	−3.007	0.285	2.640
	G	−3.530	2.085	−1.336	0.393	2.547

[a]See text for definition of symbols.

caused only by variations in the surface soil moisture. In this case (a CO_2 doubling) the extent of sea ice cover did not change and the albedo feedback mechanism did not appear.

Table 1.4 presents estimates of the difference between variations caused by the combined effect ot two factors and by each factor separately, which characterize either synergism or negative synergism (changes either in one direction or in opposite directions) of the influence of the factors under consideration.

It follows from Table 1.4 that, for example, with respect to the influence of T and W on the global ERB a negative synergism is observed: a combined effect (-1.388 W/m^2) turns out to be less than the sum of individual effects (-1.455 W/m^2). In this case an increase in T causes an increase in F, and an increase in W leads to a decrease in F but an increase in S. Note that for all eight coupled combinations the basic contribution to the formation of mutually dependent effects (synergism) in the estimation of mean global parameters comes from two latitudes: 60°N and 40°S. Mitchell et al. (52) emphasized that their results were based on numerical modeling of climate changes caused by a CO_2 doubling. With other climate disturbances, different relationships between feedback mechanisms would be observed.

The continued efforts to apply complex 3-D climate models to assess the effect of increased CO_2 concentrations on climate are of great importance. Thus, as one example, based on the use of an improved GISS (Goddard Institute of Space Studies) climate model, Rind and Lebedeff (107) estimated the variations caused by a doubled CO_2 concentration in parameters of the hydrologic regime over the United States and Canada.

The GISS nine-level model has a horizontal resolution of 8° latitude × 10° longitude. A new feature of the model is an account of varying SST estimates from the equation of the sea surface heat balance (with precalculated radiation budget, latent and sensible heat fluxes, and prescribed heat transport by currents) and varying ice cover extent (ice forms at a water temperature below $-1.6°$C and melts at a positive temperature).

Parameterization of some physical processes (especially of heat transport in the ocean and of cloud formation) as well as a rough spatial resolution determine the preliminary character of numerical modeling results and the inexpediency of treating their output as a forecast (a reliable climate forecast is still not possible). Serious limitations are set by consideration of an equilibrium climate caused by a steplike increase in the CO_2 concentration. Furthermore, a serious limitation of past climate model experiments is that they only consider an "equilibrium climate" rather than a gradual increase of the CO_2 concentration.

A comparison of the global distribution of mean global precipitation (calculated for a 5-year period with fixed SST) and its interannual variability

TABLE 1.4 Impact of Two Mutually Dependent Factors on ERB (W/m²)

ERB	Area	Factors							
		T, W	T, C	T, CO_2	W, C	W, CO_2	W, A	C, A	C, CO_2
$\underline{\delta S}$	NH	0.000	0.000	0.000	-0.195	0.000	-0.008	-0.012	0.000
	SH	0.000	0.000	0.000	-0.016	0.000	0.000	-0.028	0.000
	G	0.000	0.000	0.000	-0.106	0.000	-0.004	-0.020	0.000
$\underline{\delta F}$	NH	-0.106	-0.213	-0.420	-0.005	0.033	0.000	0.000	-0.002
	SH	-0.009	0.068	-0.335	0.057	0.023	0.000	0.000	-0.006
	G	-0.057	-0.073	-0.377	0.026	0.028	0.000	0.000	-0.004
$\underline{\delta R}$	NH	0.106	0.213	0.420	-0.190	-0.033	-0.008	0.012	0.002
	SH	0.009	-0.068	0.366	-0.073	-0.023	0.000	-0.028	0.006
	G	0.057	0.073	0.377	-0.131	-0.028	-0.004	-0.020	0.004

[a]Symbols are defined in the text.

for the control case of the GISS model with observational data revealed much in common: little precipitation in desert regions (e.g., the Sahara, Gobi, and Australian deserts) and along the west coast of the continents (South and North America, southern Africa); intensive rainfall in the tropical forest zones (central Africa, the Amazon), and in the central tropical Pacific. Rainfall amounts calculated for the Bay of Bengal and New Guinea, as well as for some regions of northern Africa, were overestimated. Observed interannual rainfall variability agrees with that derived by the model, except that it was underestimated for the tropical Pacific and Atlantic Oceans, caused, apparently, by a prescribed fixed SST in these regions rather than the real ENSO-induced interannual SST variability. In an analysis of variations in the North American hydrologic regime, calculations were made for the period of 35 years but with an averaging over just the last 5 years.

Comparisons with observational data showed that on the whole, the GISS model gives quite a realistic pattern of the observed rainfall distribution. Exceptions are as follows: a systematic overestimation of rainfall intensity (by about 50 to 100%) in a latitudinal band centered at 100°W and 11°W, stretched from Canada southward to Mexico; an underestimation by nearly half of rainfall in the state of Tennessee. These discrepancies are probably caused by a prescribed estimation of soil moisture. In this regard it must be remembered that comparisons of numerical modeling results for individual spatial grid cells with local observations are preliminary. Both the calculated and observed distributions of evaporation are qualitatively similar (evaporation in a region west of the Mississippi River was overestimated by 30% and underestimated in Tennessee). The calculated runoff has been overestimated by 200% in the same cells where rainfall has been overestimated. Calculations for a double CO_2 concentration were also made for 35 years, but averaged only for the last 10 years.

The global mean climate warming calculated by GISS constituted 4.16°C, while in the continental United States it varied from 4.2°C (in the eastern United States) to 4.9°C (in the central and western United States), with the annual warming manifested through an intensification by 40% of a warming in winter (as compared to summer). A 15 to 20% intensification of rainfall takes place in northern and northwestern North America, whereas other regions are characterized by variations of different signs. On the whole, a double CO_2 concentration gives an 11% intensification of the mean global rainfall due to increased evaporation from the ocean surface as a result of higher temperatures. The distribution of evaporation, which increases by 15 to 20% and more in the northern and western regions of North America, exhibits similar features. Runoff is similar to the distribution of the rainfall-minus-evaporation difference. Soil moisture content changes little.

Calculations of the variability of the drought index, determined from the contrast of the rainfall-minus-evaporation difference for the control case and for a double CO_2, led to the conclusion that there is a mean annual increase in the duration of the drought (with a doubled CO_2) in the south and east, but that in the northwest, wetness increases. Vegetation water stress revealed a decrease with increasing latitude. Variations in the length of the vegetation's growing season are rather substantial. The growing season increases everywhere, especially (up to 50%) in the northern regions. The CO_2-induced variations in the amount of cloud cover are small.

Rind and Lebedeff (107) performed a detailed analysis of the features of the annual change of the foregoing parameters in various cells of the GISS spatial grid, locally very specific but of little statistical significance. Variations in the temperature and humidity observed during the last 100 years enable one to conclude that there has been a substantial intensification of rainfall in the eastern United States during the climate warming.

The most important objective of the subsequent stages of numerical modeling will be a consideration of the real evolution of the climate system with a gradually increasing CO_2 concentration. To obtain statistically reliable results, repeated experiments are needed; as are estimates of the contribution by various forcing mechanisms (atmosphere and ocean, cloudiness and radiation, hydrosphere and biosphere), the ice cover dynamics, and other factors. Rind and Lebedeff (107) have characterized an approach taken by scientists at GISS to assess past and future global climate changes that resulted from variations in the concentration of the optically active trace gases of anthropogenic origin (carbon dioxide, chlorofluorocarbons, methane, nitrogen monoxide) and of volcanic aerosols.

The study of climate sensitivity to various external factors is a problem of primary importance. To characterize the sensitivity of the global climate determined by the mean global surface air temperature T, Hansen et al. (108) proposed a total feedback coefficient f expressed as

$$\Delta T_{eq} = f \Delta T_0,$$

where ΔT_{eq} is the equilibrium variation in T and ΔT_0 the variation needed to reestablish the broken radiative equilibrium, without feedback (FB) taken into account. With

$$\Delta T_{eq} = \Delta T_0 + \Delta T_{FB}$$

and the gain coefficient of the system

$$g = \Delta T_{FB} / \Delta T_{eq},$$

we have

$$f = 1/(1 - g).$$

The parameters f and g are determined by a set of various physical processes. With the feedback mechanisms independent, the total value will be

$$T_{FB} = \sum_i \Delta T_i,$$

with the coefficient g being additive:

$$g = \sum_i g_i$$

and

$$g = g_1 + g_2,$$

whereas

$$f = f_1 f_2 / (f_1 + f_2 - f_1 f_2);$$

that is, f cannot be considered multiplicative. It follows that with a strong positive feedback, an addition of one moderate positive feedback can cause a considerable increase in the total (f), that is, climate sensitivity.

Using a modified GISS climate model, Hansen et al. (108) estimated the sensitivity of climate and the role of various feedbacks for the following prescribed boundary condition changes: (i) the solar constant increased by 2% (2% SC) and a doubled CO_2 concentration ($2 \times CO_2$); (ii) lower boundary conditions corresponding to the Wisconsin glaciation (18,000 years BP). Calculations were also made using a 1-D radiative-convective model. Empirical estimates of climate sensitivity were also obtained taking into account the mean global temperature change between 1850 and 1980, together with data on variations in the atmospheric content of various "greenhouse" gases.

In the case of $2 \times CO_2$ (or of 2% SC), GISS model calculations gave a global climate warming of about 4°C, which corresponds to $f = 2$ to 4 for feedback processes with characteristic time scales of 10 to 100 years. The combined contribution by most substantial feedbacks due to water vapor ($f_{H_2O} \sim 1.6$) and snow-ice cover ($f_{si} \sim 1.1$) gives $f \sim 2$. The cloud feedback gives $f \sim 1.3$, which determines an increase in the total of f of up to 3 to 4 (as a result of the nonlinear interactions of the feedback mechanisms).

Although estimates about the effects of cloudiness cannot be reliable because of a rough schematic parameterization of the processes of interactive cloud cover formation in the GISS model, some aspects that determine the possible sign of the cloud feedback are realistic, including an increase of the cirrus cloud amount in the tropics and an elevation of the average cloud top height with a climate warming and intensified penetrative moist convection. Apparently, the cloud feedback is, at the least, weakly positive. (Verification of this conclusion using data based on global cloud cover observations is critically important.)

A recent study by Somerville and Remer (109) showed that not only "apparent" but also "latent" manifestations of the cloud feedback are possible. Although cloud feedback is a major source of errors in the present-day climate models, many models ignore these feedbacks. In the radiative-convective models that are widely used to estimate increased CO_2 effects on climate, the cloud amount, the bottom and top cloud heights, as well as optical properties of clouds are fixed. The 3-D atmospheric GCMs, which give the most reliable estimates of the CO_2 effect on climate, enable one to calculate the cloudiness changes as a simultaneous product of the parameterization of convection and of the field of relative humidity.

The results obtained provide a quantitatively reliable estimate of the global cloud cover and its annual change [e.g., the minima of cloudiness in arid subtropical regions, the maxima in the intertropical convergence zone (ITCZ), the midlatitudinal cloudiness of persistent cyclogenesis]. There are, however, substantial differences because of the less than sufficient reliability of the parameterization of processes in the atmospheric boundary layer.

Even the satisfactory modeling of cloud cover characteristics typical of present-day climate does not attest to the possibility of a realistic prediction of cloud cover variations and an adequate account of cloud feedbacks that might occur as a result of the anthropogenic impacts on climate. This assures a level of uncertainty in the forecasts of CO_2-induced climate changes. Here the effect of cloudiness can be substantial, even in the case of a stable distribution of cloudiness and a constant cloud amount and height.

According to Somerville and Remer (109), a feedback mechanism may be manifested through the temperature dependence of the cloud water content (WC). The importance of this mechanism was assessed using an advanced global radiative-convective model whose improvement incorporates a moist-adiabatic convective adjustment (instead of a fixed vertical temperature lapse rate), and an empirical dependence of WC on temperature, which determines the optical thickness of clouds τ. The parameter

$$\tau = \tau_c [1 + K (T - T_c)],$$

where T is the cloud-cover temperature, K is the empirical coefficient, and the subscript c refers to conditions of the control experiment when cloud feedback is not considered. Calculations for control experiment conditions yielded a CO_2 doubling-induced mean global surface air temperature increase of 1.74 K. Such an underestimated climate warming (as compared to the results of 3-D climate models) is typical of a radiative-convective model that does not consider the albedo and other feedbacks.

Consideration of cloud feedback using the most realistic values of the coefficient K (0.04 or 0.05) has led to a cutting in half of a CO_2-induced warming because of an increase in cloud water content (hence the optical thickness and cloud albedo) at increased temperatures, and because of the albedo effect on the radiative balance which would prevail over the greenhouse warming (the latter applies to all types of clouds except cirrus). If these results are confirmed by more accurate subsequent calculations, the need to consider the given cloud feedback in 3-D climate models will be clear. This will be a difficult problem, however, because the temperature dependence of cloud water content is specific to different conditions.

Hansen et al. (108) obtained estimates of temperature changes based on the assumption of boundary conditions corresponding to the ice age climate, with values of the feedback coefficient (f) for the continental ice $f_{ci} \sim 1.2$ to 1.3, sea ice $f_{si} \sim 1.2$, and vegetation $f_v \sim 1.05$ to 1.1. These estimates were obtained from data on the variability of the Earth's radiation budget. Empirical estimates of the total f determined by "rapid" (e.g., time scales of 10 to 100 years) feedback mechanisms are between 2 and 4. These estimates were found by comparing the cooling due to either slow or specific changes (e.g., continental ice, CO_2, vegetation cover) with the total cooling of 18,000 years ago.

High values of f indicate that the transition to the glacial or interglacial climate could have taken place with relatively small climate disturbances. A mean global warming by about 0.5°C for the last 130 years corresponds to the 2.5 to 5°C climate response to a doubled CO_2 concentration, provided that (i) the temperature increase has been caused by an increased atmospheric content of "greenhouse" gases; (ii) the CO_2 concentration in 1850 was about 270 ± 10 ppm; and (iii) the heat input to the upper layers of the ocean is mixed downward like passive tracers with a vertical mixing coefficient of about 1 cm^2/s.

Results obtained by Hansen et al. (108) show that in all cases (on all time scales) f exceeds unity, but its estimates are very uncertain due to an unreliable consideration of the feedback of clouds. The authors show a strong dependence on f of the oceanic thermal relaxation. An e-fold change in the time constant of the reaction of an isolated oceanic mixed layer is approximately 15 years. This period is long enough for a substantial heat exchange to take place between the mixed and deep layers of the ocean. At an f of 3 to 4 the characteristic time for the SST reaction to the thermal disturbance reaches 100 years, provided that the disturbance is so small in scale that it does not bring about any changes in the rate of heat exchange with deep ocean layers.

The reliability of climate forecasts for time scales of 10 to 100 years is critically dependent on our understanding of the dynamics of the ocean and, particularly, of the processes that determine the reaction of the mixed layer to climate changes near the ocean's surface. It is still unclear whether there exists some threshold for the effect of climatic changes on the ocean's dynamics (from the viewpoint of an intensified interaction between the mixed and deeper layers of the ocean) that promotes a more rapid formation of an equilibrium climate.

In connection with the problem of considering the ocean-atmosphere interaction, Newell and Hsiung (110) discussed factors that determine air temperature variations during the current period and from the ice age through the Holocene and to the present climate. Major factors that control temperature variations in a low-latitude free atmosphere are sea surface temperature anomalies determined by atmospheric circulation in the midlatitudes. Atmospheric circulation and temperature on the same time scales in the midlatitudes varied because of (i) the effect of the Milankovich mechanism, which controlled the high-latitude incoming radiation and the extent of ice cover (as Milankovich suggested); and (ii) the effect of insolation in the summer midlatitudes on the accumulation of heat by the ocean, with its transfer to the atmosphere during the next winter.

The above is based on consideration of a new mechanism of the effect on climate of the variability of insolation due to orbital changes: the interactive effects of changes in insolation in the high and midlatitudes can explain midlatitude temperature variations, which control, in turn, variations in the low latitudes due to wind-driven oceanic upwelling. Such simultaneous combined effects could have occurred in the period of the last glaciation and the Holocene. For example, an increase in insolation over the polar region and an increase in the ocean's heat content in summer could have determined an attenuation of the temperature gradient in the midlatitudes during the Holocene. An accompanying increase in the west-to-east transport must have been followed by a reduced transport in the opposite direction (under conditions of the preservation of angular momentum) and hence reduced upwellings in tropical waters, raising tropical temperatures. During the Holocene the low-latitude temperature variations had been restricted to a maximum of 30°C by the heat required for evaporation, but the latitudinal extent of the zone of maximum values had exceeded the normal one, as happened during the 1982–1983 El Niño event.

The CO_2 content in the atmosphere is dependent on the climate state through modulation of the upwelling in low latitudes. An intensified upwelling is followed by an ejection of biogenic components, which raises the rate of formation of calcium carbonate and reduces the ocean-atmosphere CO_2 transport in low latitudes. Therefore, the CO_2 concentration decreased in the glaciation periods, when the upwellings were stronger, and increased during the Holocene, when the upwellings were weaker. Since the absence (or insufficiency) of upwellings isolates a CO_2 source in deep waters, there appears to be an effect of saturation when the CO_2 concentration increases. The fact is that the atmospheric CO_2 content may be controlled by the climate.

Since the heat expenditure on evaporation confines an SST increase to a maximum of 30°C, a slight increase of insolation in the tropics or a doubling of the atmospheric CO_2 concentration cannot affect the maximum tropical temperature.

1.3 APPLICATION OF SIMPLE CLIMATE MODELS

Uncertainties inherent in climate models do not permit one to consider modeling results to be other than conditional scenarios which only reflect the laws of the model climate.

Idso (28), for example, emphasized that the successful explanation of the secular trend of the mean global temperature based on simple (e.g., energy balance) climate models is only a result of adjusting model parameters. This conclusion has also been suggested by other controversial explanations proposed by various authors, particularly by those considering the effects on climate of volcanic eruptions and solar activity.

Rather convincing are considerations that illustrate insufficient reliability of observational data and the conditional character of comparisons of the calculated with the observed secular trend of mean global surface air temperatures (28). The main point is that now there is no global data set which would enable one to obtain correct mean global values. As for the Northern Hemisphere, an averaged linear trend of cooling slightly exceeding 0.1°C per decade was observed here during 1975–1980. Calculations by Hansen et al. (21) identified a warming for this period of about 0.25°C. In fact, instead of a "dramatic" warming, the climate cooled (e.g., a contradiction between observations and calculations). According to satellite data, an increase of the snow cover extent by about 3×10^6 km² was observed in the Northern Hemisphere from 1966 to 1980 and has subsequently decreased again. Conclusions based on comparison with the observed variability of the World Ocean level have also been contradictory. For example, while some observers contend that the 12-cm rise of the level observed during this century *should* be ascribed to the effect of a climate warming, others estimate that a climate warming should be followed by a fall in the level of the World Ocean.

With regard to the comparison of the observed trends of the CO_2 concentration and climate (e.g., surface air temperature), two periods of CO_2 increase are observed: approximately before 1945 (when the increase was slow) and after 1945 (a period of accelerated increase). Yet temperature data identify a clearly manifested climate warming before 1945 and the cessation of the warming (or the onset of a cooling) after 1945. Thus from this comparison it follows that the CO_2 increase was followed by climate cooling.

Combined accounts of increased CO_2 concentrations, volcanic eruptions, and solar activity undertaken by Hansen et al. (21) and Gilliland (119) have led to substantially different conclusions. For example, according to Hansen et al., the CO_2 effect began in 1880 whereas according to Gilliland, it became manifested after 1925. According to Hansen et al., the eruptions of such volcanoes as Krakatoa (1883), Soufrier-Santa Maria (1902–1904), and Agung (1963) had considerable influence on the climate, whereas Gilliland referred to the eruptions of Askya (1875), unidentified eruptions (1885–1886), and those of Katmai (1912) and Surtsey (1963–1965) as having had a major influence on climate. Estimates of the effect of solar activity also differ drastically.

An analysis of the variability of such climatic indicators as stratospheric temperature, the extent of the sea ice cover and of the continental snow cover, and the sea level of the World Ocean was unconvincing. No doubt, a climate warming did take place during the last century. This, however, does not contradict the possibility of a CO_2-induced global warming.

With regard to the study by Hansen et al. (21), which explains the secular trend of the mean global surface air temperature as a result of the combined effect of increased CO_2 concentrations, volcanic eruptions, and solar activity considered in a 1-D climate model, an overview of this problem has been reproduced in the form of letters to the editor (45).

MacCracken drew attention to some contradictions connected with the explanation of the secular trend of climate given by Hansen et al. The interpretation of the climate cooling that occurred from the late 1930s to the early 1960s as a result of the effect of solar activity is unconvincing (e.g., it is not clear why this cooling took place only to the north of 23.6°N). A serious drawback of the Hansen et al. study is the absence of an analysis of the effect of volcanoes on climate on hemispheric and regional scales. A better explanation about the fact that the effect of the 1963 Agung eruption was stronger in the Northern Hemisphere than in the Southern Hemisphere, the usual reference of the thermal inertia of the Southern Hemisphere oceans, is not, by itself, convincing.

It also remains unclear how real the intensification of the CO_2 effect in polar regions that has been predicted by numerical modeling would be, since an analysis of observational data has led to contradictory conclusions. Serious uncertainties remain concerning the estimation of a CO_2-induced climatic impact: (i) reliable CO_2 concentration trend data are available only from 1957 onward; (ii) mutually compensating simplifications in the model suggest the value of 2.8 K for a climate warming, resulting from a doubled CO_2 concentration, to be reliable only within a factor of 2; (iii) a conditionally assumed ratio of the areas of land and oceans requires testing; and (iv) using 1880 as a reference point means setting the initial conditions as being anomalously cool climate because of the effect of major volcanic eruptions that occurred at that time.

Idso has noted that the global warming of the 1960s, considered by Hansen et al., has been detected mainly from data of only a few stations of the Southern Hemisphere. An analysis of data, however, for the low latitudes and the Northern Hemisphere revealed the opposite trend. An averaging of the data used by Hansen et al. for low latitudes for the last 55 years gives an almost unchanged temperature. For high latitudes of the Northern Hemisphere, beginning with 1975, a negative trend on the order of 0.1 °C per year is observed. This is of particular importance for a realistic assessment of the growth of the calculated warming in high latitudes as a result of increased CO_2 concentrations.

The probable global climate warming modeled for the period 1975 – 1980 was about 0.25 °C, and in high latitudes of the Northern Hemisphere 0.5 to 1.25 °C. Since, in reality, the mean temperature had lowered by 0.5 °C, the difference of 1.0 to 1.75 °C from calculated data should be considered as a result of the climate model imperfection and some other factors. Often a close coincidence of different theoretical estimates of the global warming, due to a doubled CO_2 concentration, is interpreted as an indication of their reliability, but in fact this interpretation is misleading. According to Hansen et al., for example, the warming by 2.2 °C is caused in part by CO_2 (1.2 °C) and in part by a contribution by H_2O to the intensification of the greenhouse effect (1.0 °C). Ramanathan, however, gives 0.5 °C and 1.7 °C, respectively.

Disagreeing with MacCracken and Idso, Hansen et al. believe that their work testifies convincingly to a realistic impact of the greenhouse effect on the global temperature trend and to an inevitable, almost unprecedented, climate change to be expected during the next century. Observations clearly verify an intensification of both high-latitude temperature increases and decreases. Differences in the estimates of the global warming components result not from the nature of the problem but from a determination of the parameters considered. As for the choice of the reference point and of the ratio between land and ocean areas, it does not in fact show up in the results. Thus objections by MacCracken and Idso are not at all convincing.

Robock (120) noted that the interannual variability of the surface air temperature may be caused by various processes classified according to whether they were connected with external effects, internal regular oscillations, or random processes resulting primarily from dynamic instability of the atmosphere. The most probable external factors are large-scale volcanic eruptions and increased CO_2 concentrations in the atmosphere (to date, there is no convincing evidence for the effect of solar activity on climate). Regular changes in the atmospheric general circulation (AGC) in the North American region have been related to ENSO. A considerable contribution to atmospheric variability is made by its internal instability. With the varying intensity and location of cyclones and jet streams a random variability of heat fluxes occurs in the atmosphere, leading to a random variability in air temperature.

To understand and forecast long-term climate variability, it is important to analyze the role of each of the AGC mechanisms and to detect their "signals" in surface air temperature. With this in mind, Robock (120) undertook an analysis of air temperature monthly means for 1891 – 1981 in the Northern Hemisphere on a $5° \times 10°$ grid. The parameter "volcanically weighted temperature anomaly" was taken as a criterion of the effect of volcanoes on climate:

$$TW = \Sigma T \cdot DVI(t - L)/\Sigma DVI(t - 1) - 1/91 \, \Sigma T, \qquad (1.1)$$

where TW is the volcanically weighted anomaly, T is the temperature anomaly (from data for the whole period), DVI the dust veil index characterizing the power of eruption (after Mitchell), t the time (in years), and L the time shift (from 2 to 4 years). The last term characterizes the mean anomaly of temperature for 1881 – 1975 and is very small for all the months. An analysis of calculations of the annual change of zonally averaged values of TW for negative L did not uncover any regularities, but at $L = 0$ negative anomalies prevail, being most substantial in polar regions during winter.

Maximum response of the atmosphere to a volcanically induced disturbance is observed about 2 to 3 years after an eruption and is manifested by cooling everywhere except in the Arctic in summer. These results agree well with energy-balance calculations and can be explained as being a result of the effect of sea ice thermal inertia (ice thermal inertia feedback) and, to a lesser extent, of the snow-albedo feedback. The reaction of the temperature field to ENSO oscillations (pressure anomalies in Darwin, Australia, are its indicator) is manifested either synchronously or with a moderate time shift of about 6 months. Although in this case the zonally averaged reaction is close to zero, it results from mutual compensation of strong changes of the opposite sign at different longitudes.

Thus, if the signal of external effects is most clearly detected in mean zonal temperature values, the spatial structure of the temperature field can serve as an indicator of internal variability. If, by analogy to (1.1), we calculate the mean weighted values of temperature using its instantaneous mean hemispherical values as weighting coefficients, the trends are revealed that can be considered as caused by increased CO_2 concentrations, which are, as a rule, positive, except for northeastern and northwestern North America and Asia, where the trends are of the opposite sign. This interpretation must, however, be specified with an account of the effect of long-term ocean-atmosphere interaction.

Speaking about the estimates of the CO_2 climatic implications given in (54, 111 – 113), Idso emphasizes that the estimates are based on observational data. Newell and Dopplick (54,55) based their conclusions on the view that the ocean surface temperature in the tropics cannot exceed 30°C, because the heat spent on evaporation at higher temperatures exceeds the surface radiation budget. An account of this and of some energy-balance considerations had led to an estimate of 0.3°C/(W/m²) for the surface air temperature sensitivity to the radiation budget over the ocean surface, and an estimate of 0.1°C/(W/m²) for the entire globe. Using this estimate in calculations of a mean global warming due to radiation changes resulting from a CO_2 doubling, we get 0.25°C. A close value, but in another way, was obtained by Idso (111 – 113), who noted that even over a dry land surface the surface air temperature cannot exceed some specific level, which excludes the possibility of a runaway greenhouse effect as had happened on Venus. Of course, empirical estimates of the CO_2 climatic impact obtained by Newell and Dopplick (54) and Idso (111 – 113) are open to question. However, given the inadequacies of present-day numerical modeling of climate changes, there are no grounds to neglect empirical estimates and to base one's views on conclusions from simulation modeling.

As for the impact of CO_2 increases on the biosphere, Idso (28) provides data which testify to a positive effect of CO_2 increases. These data show an increased vegetation productivity (especially crops), an increase in the efficiency of moisture absorption by plants, and increased resistance of plants to disease, among other positive consequences. A CO_2 doubling (or tripling) would probably lead to a 33% (or 67%) increase in the productivity of certain crops.

Cess and Potter (114) undertook a new analysis of the basis of the conclusion about the occurrence of much weaker (10 times) sensitivity of surface temperature (T_s) to an increased CO_2 concentration than that derived from numerical climate modeling results.

The following equation is for the surface heat balance:

$$F + LH + SH = Q,$$

where F is the net longwave radiation; LH and SH represent the heat expenditure on evaporation and turbulent heat exchange (latent and sensible heat fluxes), respectively; and Q is the absorbed shortwave radiation. With a perturbed energy balance (for the CO_2 climatic impact it represents an increase of the atmospheric thermal emission, F), we obtain

$$\Delta T_s = \lambda \cdot G, \tag{1.2}$$

where

$$1/\lambda = dF/dT_s + d(LH)/dT_s + d(SH)/dT_s - dQ/dT_s. \tag{1.3}$$

Cess and Potter (114) note that previous estimates of the sensitivity coefficient (λ) have been incorrect because of an inadequate account of the heat balance components as well as the substitution of partial derivatives for total derivates (on the assumption that all parameters, except for that in question, remain constant). So, for example, Möller had

$$1/\lambda(M) = dF/dT_s - dQ/dT_s;$$

the estimates by Newell and Dopplick were made with the relationship

$$1/\lambda(ND) = dF/dT_s + \partial(LH)/\partial T_s + \partial(SH)/\partial T_s.$$

Idso used the formula

$$1/\lambda(I) = G(IR)/\Delta T_s + dF\downarrow/dT_s,$$

where $F\downarrow$ is the atmospheric thermal emission. Table 1.5 lists the estimates with an assumed doubled CO_2 concentration of various parameters obtained by numerical modeling based on a mean annual version of the LLNL two-level 2-D dynamic statistical climate model with a swamp ocean. With data of Table 1.5 and Eq. (1.3),

$$\lambda = 1.10°C/(W\ m^{-2})$$

and according to Eq. (1.2),

TABLE 1.5   Reaction of a Climate Model to a Doubled CO_2 Concentration

Parameter	ΔT_s	G	$dF\downarrow/dT_s$	dF/dT_s	dQ/dT_s	$d(LH)/dT_s$	$d(SH)/dT_s$	$\partial(LH)/\delta T_s$	$\delta(SH)/\delta T_s$
Value	1.87°C	1.69 W/m²	5.87	−0.24	−0.39	1.11	−0.35	51.64	18.26

$$\Delta T_s = 1.87°C,$$

which corresponds to data in Table 1.5

However, neglecting various heat-balance components leads to the following estimates:

$$\Delta T_s(M) = 11.52; \quad \Delta T_s(ND) = 0.02; \quad \Delta T_s(I) = 0.25°C.$$

Also, it is important to note that λ, determined by Eq. (1.2), depends on the type of climatic disturbance. For example, calculations for a 2% increase of the solar constant gave $\lambda = 0.70°C/(W/m^2)$. Therefore, it is more correct to estimate λ for the surface-atmosphere system. Cess and Potter (114) emphasized the preliminary character of the present climate theory and the probable existence of the feedback mechanisms not before considered, mechanisms that determine the possibility of further sensitivities of climate to various types of disturbances.

Adem and Garduño (1,115) undertook a numerical modeling of climate changes (e.g., surface temperature) caused by a CO_2 doubling, with the use of a hemispheric thermodynamic climate model (TCM) which can be considered (from the viewpoint of spatial resolution) as intermediate between an energy-balance model and more complete 3-D model. The TCM involves a parameterization of the dynamics and the seasonal cycle.

In this case the climatic system components are: a 10-km atmospheric layer, a 50- to 100-m oceanic surface layer, and an infinitesimal continental layer. Account is also taken of the possibility of a single-layer cloudiness, and snow and ice cover, whose extent is calculated. Other calculated variables are mean temperature of the atmosphere, surface temperature, heat flux divergence, and wind. The coefficient of horizontal turbulent mixing, 3×10^{10} cm²/s, is prescribed. The threshold air temperature that causes the appearance of the snow or ice cover (and an appropriate surface albedo change) is 0°C.

Surface temperature calculations start in August (in a time step of 1 month) and continue until the temperature difference for this month of the current year and previous years turns out to be ≤0.01°C. A parameterization of radiation processes is based on an account of the CO_2 effect on the thermal emission transfer only in the wavelength intervals 12 to 14 μm and 16 to 18 μm. Global radiation is calculated using the Savino-Ångström formula.

A comparison of calculated surface temperature fields for normal conditions and for a doubling of CO_2 revealed a certain intensification of the climate warming effect in the high latitudes. If the mean annual hemispheric warming is 0.7°C, in the 75 to 80° latitudinal belt it reaches 1.5°C, with the annual changes varying from 0.8°C in winter to 2.6°C in summer (the annual change of warming is characterized by a maximum in summer).

An assessment of the role of the interaction of the atmosphere and the ocean in the formation of a climate warming shows that an increase in the thickness of the mixed layer from 25 to 100 m causes a decrease in the hemispheric warming in summer, but has practically no influence on the mean global warming (Table 1.6).

However, with the annual change of SST prescribed (the mixed layer is infinitely thick), a sharp decrease in warming takes place and the mean annual warming becomes equal to 0.26°C. This value closely corresponds to that previously found by Gates (102) at a fixed annual change of SST (0.31°C).

Values of a climate warming obtained by Adem and Garduño (1) are much below 3.0 ± 1.5°C, which is considered by most experts as the most probably CO_2-induced temperature increase. An analysis of the reasons for these differences has underscored the importance of a correct account of the surface heat balance. So, for example, neglecting the nonradiative components of the balance leads to a considerable intensification of the warming effect: the mean annual hemispheric value increases to 1.4°C. More accurate calculations made later by Adem and Garduño (115) confirmed the initial conclusions. The NH warming due to a CO_2 doubling constituted 1.0°C.

The high sensitivity of climate to variations in the solar constant (and to radiative disturbances, in general) is well known. It is revealed by energy-balance climate models and determined by a positive feedback between temper-

TABLE 1.6   Increase in SST with a Doubled CO_2 (ΔT_s) at Different Depths of the Mixed Layer

Depth of the Mixed Layer (m)	Winter	Spring	Summer	Fall	Year
00	0.22	0.33	0.29	0.22	0.26
25	0.59	0.85	0.80	0.60	0.71
100	0.59	0.76	0.76	0.61	0.68

ature and albedo variations, resulting from modifications in the extent of snow and ice cover. Also, the energy-balance models predict that a manifestation of the albedo feedback is intensified with a decrease in the solar constant (SC) to such a degree that the effect of this feedback becomes dominant (as compared to other factors in climate formation) and causes an unstable increase in the extent of snow and ice cover, which leads to the onset of the "white Earth" regime.

Interest in such considerations is determined by the fact that modern climate is close to altering an equilibrium. Energy-balance estimates of the solar constant decrease, leading to a transition to global glaciation, produced values from <2% to more than 7%. The growing importance of albedo feedback for the climate sensitivity at its cooling is revealed by data of Manabe and Wetherald (49) obtained with the use of a 3-D global climate model, although these authors have not undertaken a detailed study of polar caps (here a rapid increase of the albedo feedback effect starts at a SC decrease down to 1340 W/m², and in the case of the 1310-W/m² solar constant a slow and continuous process of glaciation begins which goes on until the end of the integration period). An impression is thus produced that 3-D model conclusions agree with energy-balance considerations.

To analyze the situation more thoroughly, Held et al. (116) studied the role of albedo feedback by means of numerical modeling of the climate sensitivity to variations in the solar constant using a two-level global spectral model. The basic feature of the calculations is a prescription of a zonally averaged surface albedo (A_s) depending on surface temperature (T_s):

$$A_s = 0.1 (T_s > 273 \text{ K})$$
$$A_s = 0.1 + 0.6(273 - T_s)/20 \ (253 < T_s < 273 \text{ K})$$

and

$$A_s = 0.7 \ (T_s < 253 \text{ K}).$$

These surface albedo values correspond to the following albedo estimates for the Earth-atmosphere system (in parentheses): 0.1 (0.34) and 0.7 (0.68). An assumption of a continuous transfer from the ice-free (0.1) to the "white" surface (0.7) is determined by a desire to avoid the instability of a small polar cap appearing in a stepwise albedo profile model (smoothing does not, however, influence the stability of an extended polar cap).

To simplify the model (with surface air temperature and humidity its prognostic variables), a 3° grid confined to ±84° latitudes is assumed, a "swamp-ocean" boundary condition is prescribed (a flat surface with zero heat capacity at 100% saturation), and the spectrum of zonal waves is cut a wavenumbers 0, 3, and 6. Results have led to conclusions that do not agree with energy-balance considerations. Calculations have shown that the latitude of the southern boundary of the polar cap, which is relatively insensitive to the solar constant at a solar constant change $\Delta Q < +1\%$, is very sensitive in the range $+1\% < \Delta Q < -2\%$, and then, again, its sensitivity decreases at $-2\% < \Delta Q < -5.5\%$. When $\Delta Q = 6\%$, global glaciation sets in.

If the solar constant decreases by several percent, thereby shifting the latitudinal albedo gradient at the edge of the polar cap to the equator from $60°$ latitude, the climate becomes exclusively sensitive to the solar constant. This sensitivity, however, does not indicate the approaching instability of the extended polar cap and the onset of the "white Earth" regime. Instead, with a further decrease in the solar constant, the sensitivity of climate to the solar constant weakens and the polar cap remains relatively stable until the solar constant decreases by 5%. Only with the further decrease of the solar constant does the "white Earth" regime set in.

Approximate models are a suitable means for simulation of the climatic effects of cloud feedback. Hunt (27) summarized assessments of the contribution by the variability of cloud cover characteristics (which may occur with a climate warming) to the formation of the thermal regime, compared to that which might accompany a CO_2-induced temperature increase. A similar but more limited study was carried out by Reck (68), who confined himself to the use of a modified 1-D radiative-convective model (all calculations were made for $35°$N latitude, which can be considered as representative from the viewpoint of the characteristic of a mean global situation). Hunt obtained quantitative estimates which convincingly illustrate the extreme complexity and multifaceted nature of this problem.

A key problem linked to the estimation of a CO_2-induced climatic impact is in revealing the role of variations in the amount, type, and top height of clouds that result from climate warming and is manifested through respective feedback mechanisms. Table 1.7 presents results of calculations made by Hunt (27) for land surface temperature sensitivity (T) to a 10% increase of the cloud amount at lower (L), middle (M), and upper (U) levels in the cases $1 \times CO_2$ and $2 \times CO_2$. Cloud parameters are given in Table 1.8.

Table 1.7 suggests that an increase in the amount of lower- and middle-level clouds causes a climate cooling, whereas an increase in the amount of upper-level clouds causes a warming. The effect of lower-level clouds is so substantial that practically complete compensation for climate warming at a 10% increase of the cloud amount takes place (from 0.302 to 0.332), as a result of a CO_2 doubling.

Since the processes of formation and development of clouds are only crudely parameterized even in the most complex GCMs (of course, a 10%

TABLE 1.7 Sensitivity of Land Surface Temperature to Cloud Amount (ΔT_s, Temperature Change)

		Temperature (K)		
	Control Case	+10%L	+10%M	+10%U
$1 \times CO_2$				
T_s } ΔT_s }	293.41	291.73 −1.68	293.22 −0.19	293.77 +0.36
$2 \times CO_2$				
T_s } ΔT_s }	295.23	293.42 −0.81	295.00 −0.23	295.60 +0.37

Source: Ref. 27.

increase of the cloud amount is not necessarily statistically significant), the tentative character of theoretical estimates of the CO_2-induced climatic impact becomes quite apparent. Calculations suggest that a 10% increase in the amount of upper clouds affects the stratospheric temperature in particular. This may bring about substantial changes in the troposphere-stratosphere interaction. Table 1.9 illustrates surface temperature sensitivity to an increase in cloud albedo [from 0.69 to 0.87 (L), 0.48 to 0.73 (M), and 0.21 to 0.44 (U)].

Naturally, an increased cloud albedo causes a strong climate cooling, especially in the case of lower clouds (to compensate for the warming caused by a doubled CO_2, it is sufficient to raise the lower cloud albedo from 0.69 to 0.74, i.e., by 5%). It is important to note that like variations in the cloud amount, a change in the cloud albedo significantly affects the stratospheric temperature (hence the troposphere-stratosphere interaction here can also be substantially affected). Calculations for the interactive atmosphere-ocean system (including an account of the oceanic 30-level mixed layer) gave nearly the same results as for the land, with varying amount and albedo of clouds. Of interest is a strong sensitivity of the mixed-layer thickness to the albedo of clouds. This shows the importance of taking into account the ocean-atmosphere interaction when estimating the climatic effect of changes of various factors.

TABLE 1.8 Basic Parameters of Clouds

Parameter	L	M	U
Height	2.7–1.7	4.1	10.0
Amount	0.302	0.079	0.181
Albedo	0.69	0.48	0.21
Absorptivity	0.035	0.02	0.005

TABLE 1.9. Sensitivity of Land Surface Temperature to Cloud Albedo

	Temperature (K)			
	Control Case	L	M	U
$1 \times CO_2$				
T_s	293.41	287.41	291.59	288.67
ΔT_s		−6.19	−1.81	−4.74
$2 \times CO_2$				
T_s	295.34	289.34	293.33	290.69
ΔT_s		−5.89	−1.90	−4.54

Table 1.10 shows estimates of the effect of an increase in the lower-level cloud-top height by 0.4 km and of the middle- and upper-level cloud-top heights by 0.5 km.

As expected, an increase of the cloud height causes a climate warming due to the resulting decrease of outgoing emission. Thus, in contrast to the effects of the CO_2-induced growth in the amount and albedo of clouds, the influence of an elevation in the cloud-top height is manifested as a positive feedback mechanism. The same is true of an attenuation of the energy exchange between the surface and the atmosphere, described as the abatement of winds near the surface (such an attenuation can result from a decrease in the meridional temperature gradient). The effect of a warming due to a CO_2 doubling turns out to be equivalent to the effect of the reduction by half of wind speed (from 7.5 m/s to 3.75 m/s). Important consequences of wind abatement include a substantial decrease in the mixed-layer depth, a decrease in the meridional heat transport in the ocean (the latter is a critically important climatic factor), and an attenuation of the CO_2 exchange between the atmosphere and ocean.

Assessments of the role of some feedback mechanisms in the formation of climate obtained by Hunt (27) should, of course, be considered as condi-

TABLE 1.10 Sensitivity of Land Surface Temperature to Cloud-Top Height

	Temperature (K)			
	Control Case	L	M	U
$1 \times CO_2$				
T_s	293.41	294.08	293.53	293.39
ΔT_s		0.67	0.12	−0.02
$2 \times CO_2$				
T_s	295.23	295.77	295.34	295.27
ΔT_s		0.54	0.11	0.04

tional in view of the schematic character of the climate model applied. However, these assessments convincingly illustrate the complexity of the climatic system and underscore the need for accurate consideration of various feedbacks, which is still an intractable problem. This shows the conditional and unreliable nature of estimates of the CO_2-induced impact on climate based on presently available climate models. Numerical modeling results reveal the importance of the atmosphere-ocean interaction, which underscores the need for the development of interactive climate models.

1.4 THE MULTIFACETED NATURE OF THE ATMOSPHERIC GREENHOUSE EFFECT

The complexity of anthropogenically induced changes in the gas composition of the atmosphere determines the multicomponent nature of the atmospheric greenhouse effect, studied in detail by Kondratyev and Moskalenko (39–41). Their calculations have shown that the mean global atmosphere of the Earth provides the greenhouse effect of $T = 33.2$ K (the difference between "planetary temperature" and "surface temperature"), with the following contributions by optically active gaseous components at current concentrations: H_2O, 20.6 K; CO_2, 7.2 K; N_2O, 1.4 K; CH_4, 0.8 K; O_3, 2.4 K; and NH_3, CFCs, NO_2, CCl_4, O_2, CF_4, and N_2, 0.8 K. The data obtained verify that the radiative regime of the atmosphere is governed by the amount of water vapor, carbon dioxide, and ozone present in it, with the prevailing influence of water vapor on the greenhouse effect. Having strong, broad rotation-vibration bands, water vapor absorbs most of the thermal emission from the Earth's surface, lowering the outgoing emission. Furthermore, part of the emission from the surface in the 4.8-μm, 8 to 13-μm, and 18-μm transparency windows is lost. The greenhouse effect of water vapor is intensified by the absorption bands of CO_2 and ozone.

Consider the effect of variations in the optical characteristics of various atmospheric components on the greenhouse effect. Many processes taking place in the atmosphere exhibit a close correlation (synergism). A rise in the tropospheric temperature, for example, leads to an increase in its water vapor content. The spectral transmission functions of the main absorbing components (CO_2 and water vapor) strongly depend on temperature, especially in the wings of the bands. The absorption properties of these components increase considerably with temperature. Thus an increase in surface temperature is usually accompanied by an intensified greenhouse effect.

Calculations confirm that increasing CO_2 concentrations are the most important factor with respect to the anthropogenic impacts on climate. However, increased pollution of the atmosphere as a result of human activities leads to increasing concentrations of other atmospheric components, such as SO_2, CO, halocarbons, HNO_3, and hydrocarbon compounds. Of great importance is an accounting of aerosols either directly ejected into the atmosphere or resulting from gas-to-particle transformations. All the enumerated components have absorption bands in the IR spectral region with their contributions to the greenhouse effect.

To study the effect of the gaseous and aerosol atmospheric components, Kondratyev and Moskalenko (39–41) undertook calculations of radiation fluxes for model atmospheres with different chemical compositions. The calculation scheme involved all the components that affect the radiation absorption in the atmosphere, including water vapor (with account of continuum absorption), carbon dioxide, nitrogen and oxygen (pressure-induced absorption), methane, nitric oxide, sulfur dioxide, nitric acid vapors, ethylene, acetylene, ethane, formaldehyde, chlorofluorocarbons, CCl_4, CF_4, ammonia, and aerosols of different chemical composition and size distribution.

Basic optical components—water vapor and carbon dioxide—govern the radiative regime of the troposphere. Minor components—ozone, methane, nitrous oxide, nitric oxide, carbon monoxide, nitric acid vapors, sulfur dioxide, halocarbons, nitrogen dioxide—affect the radiation transfer markedly only in the atmospheric transparency windows. The spectral structure of the absorption bands of these gases is now being thoroughly studied. The role of minor components increases in the formation of the radiative regime of the stratosphere. When calculating the greenhouse effect, it is most important to consider the radiation absorption by the 6.4-μm rotational-vibrational band and the rotational bands of water vapor; by CO_2 4.3- and 15-μm bands; O_3 4.74- and 9.6-μm bands; N_2O 4.5-, 7.8-, and 8.6-μm bands; CH_4 3.3- and 7.6-μm bands; and pressure-induced absorption in N_2 4.3-μm and O_2 6.5-μm bands at collisions of molecules N_2-O_2, O_2-N_2, N_2-N_2, and O_2-O_2. The role of atmospheric rotational-vibrational absorption bands 4.67 μm (CO); 5.3 μm (NO); 6.3 and 10.6 μm (NH_3); 4.75, 8.8, and 20 μm (SO_2); and 4.9 μm (CO_2) in the troposphere is not very great. In the stratosphere it is important to take into account the radiation transfer in the absorption bands 5.6, 11.2, 20 μm (HNO_3); 5.6, 7.35, and 8.69 μm (N_2O); 9.22 and 11.82 μm ($CClF_3$); 8.68, 9.13, and 10.93 μm (CCl_2F_2); 7.14, 0.85, and 13.66 μm ($CFCl_3$); 12.99 μm (CCl_4); and 10 μm (CF_4).

Many of the gaseous components that absorb in the thermal emission spectral region are products of human industrial activities, and therefore their concentration can rapidly increase with time. An example based on halocarbons can be considered representative. If the mixing ratio of CFCs F-11 and F-12 is now about $(1-2) \times 10^{-10}$, by the end of the century, at present level of atmospheric pollution, their concentrations may constitute 2×10^{-9}. The process of chlorofluorocarbon accumulation in the stratosphere is favored by their long residence time in the atmosphere (121).

A doubled concentration of nitrous oxide leads to a 0.7 K rise of mean temperature, and with double ammonia and methane the surface air temperature is raised by 0.1 and 0.3 K, respectively. A 20-fold increase in the chlorofluorocarbon content may lead to a warming of 0.6 to 1 K, and the total greenhouse effect resulting from doubled N_2O, CF_4, SO_2, and HNO_3 reaches 1.2 K.

When considering possible climatic changes due to the anthropogenically induced changes in the chemical composition of the atmosphere, it is necessary to take into account the relationships between various climatological factors. For example, an increase in the chlorofluorocarbon content can substantially change the atmospheric ozone concentration field.

Calculations have shown (39–41) that the greenhouse effect caused by ozone is very sensitive to variations in its vertical profile and is manifested not only through the direct impact of ozone but also, indirectly, through other atmospheric components. The warming can also result from changes in the stratosphere's vertical temperature profile. If a uniform decrease of 25% in the ozone concentration leads to a surface temperature decrease of 0.45 K for a mean planetary atmospheric model, a nonuniform decrease in ozone concentrations causes an "antigreenhouse" effect of $\Delta T_s = 0.25$ K. It is important to consider ozone variations not only in the stratosphere but in the troposphere as well, since tropospheric ozone variations often lead to the opposite effects from results determined by stratospheric disturbances of the ozone concentration field.

In calculating the greenhouse effect it is important to consider the atmospheric constituents CF_4 and CCl_4, with strong bands in the region of 10 and 13 μm. Increased future industrial releases of CF_4 and CCl_4 can lead to an intensification of the greenhouse effect by $\Delta T_s = 0.8$ to 1 K. Based on our calculations, an increase in the NO_2 concentration leads to a weak antigreenhouse effect that depends on the peculiarities of atmospheric vertical temperature profiles and NO_2 concentrations.

The effect of aerosols on the atmospheric radiative regime is manifested through the mechanisms of radiation scattering and absorption by aerosols. Depending on the optical density of aerosols, and their size distribution and chemical composition, conditions are created in which aerosols either raise or lower the albedo of the planet. The ambiguity of conclusions can be explained by the variety of aerosol properties that result from the large spatial and temporal variability of the concentration field and size distribution of the aerosols, as well as by the dependence on conditions of their generation of the chemical composition of aerosols.

For small-sized fractions (submicron) of aerosols the absorption coefficient σ_{aa} always exceeds the scattering coefficient $\sigma_{\dagger aa}$, and in the wavelength region $\lambda > 1$ μm the attenuation coefficient is determined mainly by absorption of radiation by aerosols. Therefore, submicron aerosols heat the atmosphere at the expense of solar radiation and emit in the longwave spectral region, absorbing the thermal emission from the surface. Thus the submicron background aerosol (depending on the type of aerosol), acts like an absorbing gaseous component and intensifies the atmospheric greenhouse effect, raising the surface temperature by 3 K for mean global conditions. An *increase* in the tropospheric aerosol concentration by a factor of 1.5 as a result of human impact will lead to a surface warming of 1.7 K.

The foregoing estimates of the greenhouse effect resulting from the atmospheric aerosol submicron fraction can in part be compensated for by an antigreenhouse effect of the large-sized aerosol fraction (e.g., sea salt, dust, and ice particles). The antigreenhouse effect is also created by the stratospheric sulfate aerosol layer. However, the author assumes that on average, the planetary greenhouse effect due to the strong absorption of the submicron aerosol fraction probably prevails over the antigreenhouse effect of large aerosol particles. Atmospheric pollution due to human industrial activity causes an increase in the concentration of strongly absorbing aerosol particles, which does not lead to substantial variations of the planetary albedo but intensifies the screening of outgoing thermal emissions from the surface.

Clouds are a most important factor determining the radiative regime of the Earth. An increase in cloudiness leads to a higher albedo and decreases the solar radiation flux that reaches the Earth's surface. Yet cloudiness lowers the radiative temperature corresponding to the outgoing emission. Under nighttime conditions the greenhouse-effect mechanism of cloudiness is most effectively manifested, and in daytime (e.g., at positive surface temperatures) the cooling due to increased reflected solar radiation dominates (as a rule) the greenhouse effect of cloudiness. However, in winter (with snow-covered surfaces) the presence of cloudiness does not raise the albedo substantially. In these conditions the greenhouse effect prevails.

Calculations by Kondratyev and Moskalenko (40) have shown that an increase in the upper-level cloud cover for a mean planetary model of the atmosphere leads to the antigreenhouse effect caused by increased planetary albedo, with decreasing shortwave radiation flux coming to the troposphere. The resulting cooling of the surface and troposphere is not compensated for completely by warming due to the varying thermal emission of the planet. Variations in the surface temperature caused by increased upper-level cloud cover (the upper troposphere, lower stratosphere) depends strongly on the choice of the cloud cover model.

In a continuation of previous estimates of the contribution by various components to the greenhouse effect, Lacis et al. (42), using a radiative-convective equilibrium model, calculated the mean global surface air temperature increase due to the greenhouse effects caused by the anthropogenic ejections into the atmosphere of carbon dioxide, CFCs, ammonia, and nitric oxide during the decade of the 1970s as well as with prescribed doubled concentrations of these gaseous components.

Results by Lacis et al. (42) (Table 1.11) show that the total contribution by methane, nitric oxide, and CFCs to the climate warming during the period 1970–1980 constituted (taking into account approximate estimates) 50 to 100% of the CO_2 contribution (see also Ref. 121).

The following approximation is possible with the gas concentration expressed in ppm (except for CFCs, whose concentrations are given in ppb) and at the concentrations: CH_4, <5 ppm; N_2O, <1 ppm; CF_2Cl_2 and CCl_3F, <2 ppb, and ΔCO_2, <300 ppm (ΔCO_2 is an excess concentration of CO_2 with respect to its values, equal to 300 ppm):

$$\Delta T(^\circ C) = 0.57(CH_4)^{0.5} + 1.8(N_2O)^{0.6}$$
$$- 0.057 CH_4 \cdot N_2O + 0.015 CCl_3F + 0.18 CCl_2F_2$$
$$+ 2.5 \ln [1 + 0.005\Delta CO_2 + 10^{-5}(\Delta CO_2)^2].$$

The third term describes the effect of the overlapping of the bands of methane and nitrous oxide. These estimates were obtained without consideration of the atmosphere-ocean interaction, which slows down the manifestation of the intensified greenhouse effect through a climate warming. The consideration of atmosphere-ocean interaction, based on the assumption that it is enough to take an oceanic mixed layer 100 m thick, led to the conclusion that in this case the increase in surface air temperature by the year 1980 should have been half as much, that is, about 0.1 °C. Taking into account pre-1970 atmospheric pollution by CO_2 and other greenhouse gases, the climate warming for 1970–1980 can be estimated at 0.1 to 0.2 °C. Since this level of temperature variation is close to the observed mean standard deviations of

TABLE 1.11 Climate Warming Due to the Greenhouse Effect of Selected Gaseous Components of the Atmosphere[a]

Component	Changes during 1970–1980			Arbitrary Change		
	a_0 (ppb)	Δa (ppb)	ΔT (°C)	a_0 (ppb)	Δa (ppb)	ΔT (°C)
CO_2	325,000	12,000	0.14	300,000	300,000	2.9
CH_4	1,500	150	0.032	1,600	1,600	0.26
N_2O	295	6	0.016	280	280	0.65
CCl_3F	0.045	0.135	0.020	0	2	0.29
CCl_2F_2	0.125	0.190	0.034	0	2	0.36

[a] a_0 and Δa are the initial concentration of the given component and the concentration increase, respectively; ΔT is the temperature increase.

Source: Ref. 42.

10-year temperature values, it is still impossible to detect an anthropogeni-cally induced climate warming. However, if we proceed from the premise that the total warming during the past 20 years (1970s and 1980s) reached 0.2 to 0.3°C, and identification of warming can be realized from observational data by the end of the 1980s. These results testify to the urgency and necessity of the global monitoring of the optically active minor gaseous components of the atmosphere, which contribute markedly to the atmosphere greenhouse effect.

Human activities contribute a variety of optically active minor gaseous components to the atmosphere which contribute to the formation of a greenhouse effect and can affect the climate to a certain extent. A WMO report (73) noted that the related compounds involve gases included in the following groups:

1. *Carbon:* CO_2 and CH_4
2. *Oxygen:* O_3
3. *Nitrogen:* N_2O, NO_2, N_2O_5, HNO_3 vapor, NH_3
4. *Sulfur:* SO_2, COS, CS_2
5. *Halogens:* Freon-11, $CFCl_3$; F-12, CF_2Cl_2; F-13, CF_3Cl; F-21, $CFHCl_2$; F-22, CF_2HCl; F-113, CF_3CCl_3; F-114, $C_2F_4Cl_2$; F-115, C_2F_5Cl; F-116, C_2F_6; CCl_4; CH_3Cr; CH_3Cr; CH_3Cl; CF_4; CH_2Cl_2; C_2HCl_3; CH_3CCl_3; C_2Cl_4
6. *Nonmethane carbon compounds:* C_2H_2, C_2H_4, C_2H_6, C_3H_8, C_6H_6
7. *Peroxyacethylnitrate* (PAN)

Experts (73) discussed the contribution to the greenhouse effect, of the most important of the compounds listed above, as determined by their absorption bands in the 7 to 13-μm transparency window. Since some of these compounds are optically active both in the longwave (LW) and short-wave (SW) spectral regions (this pertains mainly to ozone and nitrogen dioxide), their effects can be manifested through both a warming and a cooling of climate, depending on the vertical concentration profile of these components.

Table 1.12 contains estimates of the mean global effect of increased con-centrations of different gases on climate. These estimates were obtained using 1-D models of the radiative-convective equilibrium (assuming a con-stant relative humidity of the atmosphere and, as a rule, fixed cloud-top height). In the case of ozone the dependence of its concentration $F(\psi, z)$ on latitude ψ and height z was considered, but a doubled O_3 content refers to the troposphere for all purposes.

TABLE 1.12 Summary of Climatic Effects Caused by Optically Active Minor Gaseous Components of the Atmosphere

Gas	Center of the Band (cm^{-1})	Background Mixing Ratio (ppb)	Disturbed Mixing Ratio (ppb)	Variations in the Outgoing Emission at the Level of the Tropopause (W/m^2)	Surface Air Temperature Variation (K)
CO_2	667	330×10^3	660×10^3	4.0	3.0
N_2O	589, 1168, 1285	300	600	0.65	0.3–0.4
CH_4	1306, 1534	1500	3000	0.61	0.3
O_3	1041, 1103	$F(\Psi, z)$	$2F(\Psi, z)$	1.24	0.9
$CFCl_3$	846, 1085, 2144	0	1	0.31	0.13
CF_2Cl_2	915, 1095, 1152	0	1	0.26	0.13
CH_4	632, 1241, 1261, 1283	0	1		0.07
CF_2HCl	1117, 1211	0	1		0.04
CCl_4	776	0	1		0.14
$CHCl_3$	774, 1220	0	1		0.1
CH_2Cl_2	714, 736, 1236	0	1		0.05
CH_3Cl	732, 1015, 1400	0	1		0.013
CH_3CCl_3	707, 1084	0	1		0.02
C_2H_4	949	0.2	0.4		0.01
SO_2	518, 1151, 1361	2	4		0.02
NH_3	950	6	12		0.09
HNO_3	1695, 1333, 850, 459	$F(\psi, z)$	$2F(\psi, z)$		0.06
H_2O	0–2000	3×10^3	6×10^3		0.6

With decreased O_3 content in the stratosphere, two effects of opposite sign occur: (i) heating of the troposphere and surface due to a growing income of the SW radiation, but decreasing of stratospheric temperature; (ii) the latter decreases the LW radiation flux coming from the stratosphere to the troposphere (hence, cooling). In the case of a homogeneous (vertical) decrease in the O_3 concentration in the stratosphere, the effect of LW cooling prevails. Another situation is observed in the troposphere, where the growth of the O_3 content causes a warming due to LW and SW radiation. Therefore, the climate of the troposphere is more sensitive to variations in the ozone content in the troposphere than the in the stratosphere.

Since the contribution of individual minor components to the greenhouse effect is additive, it follows from Table 1.12 that the total contribution is nearly equal to that due to a double CO_2. It is important to note that observational data do not permit a separation between the effects of CO_2 and those of minor components. The important role of minor components is determined both by their very intensive absorption bands and by linear relationships between their contributions to the greenhouse effect and an increase in their concentrations (in the case of CO_2 this dependence is logarithmic).

Estimates of the climatic effect of variations in the content of the optically active components are hindered by the necessity to consider the radiative-chemical and dynamic interactions among them, the results of which can be both additive and mutually balanced radiative effects. For example, CO oxidation (in the presence of NO) in the troposphere is a source of tropospheric ozone, and hence variations in the concentrations of these gases can indirectly affect climate, although CO and NO do not have a direct radiative effect. A similar situation is observed with respect to chlorofluorocarbons, which affect stratospheric ozone.

The radiative-dynamic interaction between the troposphere and stratosphere can play an important role. The effect of such pollutants as CO, NO, CH_4, among others, is determined by their influence on the chemically active hydroxyl, whose content, in turn, depends on the amount of water vapor in the troposphere, which increases with the warming caused by the intensified greenhouse effect.

Such a climate-chemical interaction is also substantial in the stratosphere: CO_2-induced cooling of the stratosphere can lead to a considerable increase in the ozone content of the upper stratosphere at the expense of the temperature dependence of the rates of chemical reactions determining the formation and destruction of ozone. Naturally, complicated relationships exist between the processes of the anthropogenic impact on the stratosphere, the radiative-dynamic interaction between the troposphere and stratosphere, and the climatic effect of the anthropogenically induced changes in the composition of the troposphere and stratosphere.

To resolve these difficult problems reliably, the following steps are necessary: (i) to continue the global monitoring of the optically active minor gaseous components, bearing in mind an identification of trends on the order of several tenths of 1% per year; (ii) to perform independent monitoring of changes in the ozone content of the troposphere and stratosphere and to reveal transformations of the vertical profile of O_3 concentrations; (iii) to develop 1-D and 2-D climate models with an interactive account of chemical and climate-forming processes; (iv) to apply 3-D GCMs to assess the climate sensitivity to atmospheric ozone variations; (v) to develop GCMs capable of reproducing the dynamics of water vapor in the stratosphere, taking into account chemical sources and sinks; (vi) to enlarge substantially a set of the spectroscopic parameters of minor gaseous components (there are still no data on the intensity of bands for some of them) and to make direct measurements of transmission functions (at a spectral resolution not less than 5 cm^{-1}) under both laboratory and field conditions; and (vii) to obtain more reliable information about variations in the extra-atmospheric insolation, especially in the spectral intervals 200 to 210 and 300 to 320 nm.

Bojkov (117) reviewed recent studies on tropospheric ozone stimulated by the discovery of anthropogenic effects that lead to an increase of the ozone content in the troposphere. His review revealed considerable effects of ozone on the atmospheric greenhouse effect and on climate. There is an urgent need for studies of tropospheric ozone, as determined by the substantial impact of tropospheric chemical processes on the processes in the stratosphere. Such impacts are caused, in particular, by the following factors: (i) the role of the troposphere as a source of numerous gaseous components for the stratosphere (NO_x, OH, CH_4, CO, CH_3Cl, CH_3CCl_3, and so forth), one of the most important sinks of which is their reaction with hydroxyl; (ii) changes in the content of such components as NO_x, CH_4, and CO which affect the tropospheric ozone content and hence the thermal regime of the troposphere; and (iii) a contribution by increased tropospheric ozone of several percent to variations in the total content of atmospheric ozone.

The absence of specific means for observations of tropospheric ozone precludes the sufficient reliability of available information and of preliminary conclusions based on any analysis of this information. Although the surface ozone concentration was measured at hundreds of locations, the data obtained are tainted by the effects of local sources of pollution and cannot be used as a reliable source of information about ozone in the free troposphere (except for background and alpine stations). Data that were more representative were obtained using the Umkehr technique and ozone sondes. However, a sufficiently long series of regular observations made with similar ozone sondes are available for only two locations (Höhenpeissenberg, FRG: Talville and Paiern, Switzerland).

An analysis of available data had made it possible to uncover basic features of the spatial and temporal variability of the troposphere's ozone content which can be characterized by an increase with height of the mixing ratio. A substantial annual change and meridional variability of tropospheric ozone with maxima near $33°$N and $55°$N as well as interhemispheric asymmetry (with the tropospheric ozone content less in the Southern Hemisphere than in the Northern one) have been observed.

From the Höhenpeissenberg observations, the annual changes of the surface air ozone concentration are characterized by a minimum in December–January, a subsequent rapid increase in February–March, a maximum in May–June, and a rapid decrease beginning in September. During the period 1971–1972 to 1981–1982, the total tropospheric ozone content (TTOC) increased by about 30%. This increase was confirmed by data from other stations.

A consideration of all the available data has led to the conclusion that during the last two decades of ozone-sonde observations, a trend was detected for a TTOC increase at a rate of 0.7 to 3% per year (at the time the TTOC fluctuated only slightly by $±2$%, from which it follows that TTOC variations do not reflect variations in the total ozone content). This positive TTOC trend should be considered to have been anthropogenically induced.

In this regard Bojkov (117) briefly reviewed the factors involved in the formation of tropospheric ozone from the viewpoint of the income of stratospheric ozone and of photochemical processes of ozone formation in the tropics. Also reviewed were the processes of ozone destruction. The tropospheric ozone balance is presented in Table 1.13.

Calculations made using a 1-D radiative-convective model have shown that an increase in TTOC in Höhenpeissenberg from 1967 to 1981 should have led to a $0.3°$C warming of the troposphere, an increase comparable to a CO_2-induced warming for the same period.

**TABLE 1.13   Preliminary Assessments of the Tropospheric Ozone Balance
(10^6 tons/year)**

Sources and Sinks	NH	SH	Globe
Income from the stratosphere	400	250	650
Destruction at surface level	450	180	630
Photochemical destruction	1300	570	1870
Photochemical formation	1350	500	1850

1.5 DETECTION OF THE CO_2 SIGNAL

Madden and Ramanathan (46) have noted that the CO_2 climatic effect predicted by numerical models should have been detectable using data on the interannual variability of surface air temperatures at 60°N latitude. Since, however, an analysis of data for a 72-year series of observations did not reveal such changes (e.g., mean temperature for 1958–1977 did not prove to the higher than that for the 1906–1925 period), the following alternatives can be suggested: (i) there are factors that compensate for the warming caused by an intensification of the greenhouse effect; (ii) the manifestation of a warming is delayed because of the thermal inertia of the ocean; and (iii) numerical modeling overestimates the climatic effect of an increased CO_2 concentration.

Estimates by Cess and Goldenberg (118) have led to a conclusion that the thermal energy of the ocean causes a two-decade delay in the manifestation of the effect of a warming. Thus the lack of convincing evidence for the existence of warming can be attributed to three circumstances: (i) the ocean-induced delay in the manifestation of warming; (ii) difficulty in identifying the CO_2 signal against the effects of other factors (e.g., changes in extra-atmospheric insolation, aerosol content, and so forth); and (iii) the limited time series of observations. Of great importance is the realization of an extensive long-term climate monitoring program and the identification of parameters most sensitive to the CO_2 signal. Apparently, the latter includes mean zonal summer air temperature (except for high latitudes, where the signal-to-noise ratio deteriorates); mean zonal summer air temperatures in the stratosphere and mesosphere (where a strong anthropogenic cooling must be pronounced) with a relatively small natural variability; the temperature of the surface and deep layers of the ocean; and the extent of the polar ice cover.

MacCracken (43) noted that in order to reveal with some degree of reliability the occurrence of a climate warming, one that has either already taken place or might occur in the future, the following conditions need to be met: (i) the representativeness and adequacy of climatic data that permit a reliable statistical analysis of climate variability; and (ii) the availability of information that enables us to identify and distinguish between the influences of different factors on climate change.

An analysis of observational data showed that these requirements cannot now be met. For example, the values of preindustrial CO_2 concentrations ranged between 250 and 290 ppm, and the data on CO_2 concentrations for 1850–1950 were not sufficiently reliable in those periods when the role of the biosphere as a factor in CO_2 concentration variability was more substantial than the role of industrial activities (70). It means that the CO_2 concentration from 1850 to 1980 could vary between 40 and 90 ppm, and therefore the estimates of a climate warming could be obtained only to an accuracy of a factor of 2.

As Luther (105) noted, a most natural possibility for detection of a CO_2 signal involves the monitoring of changes in the spectral distribution of the short- and longwave radiation fluxes, which reflect radiative signatures of CO_2. The outgoing longwave radiation (ORL) spectrum, for example, can serve as a signature. Calculations have shown, however, that even with a doubling of CO_2, a maximum decrease of outgoing longwave radiation (OLR) will be only about 12% in the wavelength interval 13 to 17 μm. Consequently, early detection of the CO_2 signal is impossible.

In this regard, Luther (105) calculated the spectral distribution of the downward thermal emission (DTE) flux in the atmosphere in the midlatitudes in a search for a spectral interval where the DTE was most sensitive to variations in CO_2 concentrations. Calculations have shown that DTE is most sensitive in the interval 700 to 800 cm^{-1} at a height of several kilometers. The height of maximum sensitivity of DTE to CO_2 increases depends on latitude [it decreases with increased latitude (caused by an accompanying decrease of the atmospheric water vapor content)]. The sensitivity of DTE to CO_2 variations in high latitudes increased by three to four times. Maximum absolute DTE variation with CO_2 increases of 20 to 60 ppm is reached in the interval 755 to 760 cm^{-1} and takes place at a height lower than that for the maximum relative variation constituting 9.63% at a height of 8 km, with the mixing ratio increasing from 340 to 400 ppm.

Luther's (105) results suggest that high-mountain observations of the variations in the spectral distribution of atmospheric emissions are more promising from the viewpoint of detecting the CO_2 signal as compared to satellite measurements of the OLR spectral composition. However, even in these relatively favorable conditions, the solution of this problem will be possible only in several decades. Early detection of the CO_2 signal can be improved by using a spectral resolution better than 5 cm^{-1}. In view of the long-term process of the CO_2-induced climatic impact, periodic observations will probably be sufficient.

Kiehl (104) assessed the possibilities of detecting the CO_2-induced climate changes from satellite spectral measurements of OLR in the wavenumber interval 500 to 800 cm^{-1} (the CO_2 15-μm band), where OLR is generated mainly by the stratosphere. The choice of OLR in the 15-μm region as an indicator of climate changes is determined by two factors: (i) CO_2 doubling-induced temperature changes are greater in the stratosphere than in the troposphere; and (ii) OLR formed in the stratosphere is independent of cloud conditions.

The outgoing longwave radiation was calculated with CO_2 mixing ratios 320 and 640 ppm at a spectral resolution of 5 cm^{-1}. With a doubling of CO_2,

OLR decreases by 2.7 W/m^2 over the entire interval under study (provided that all the vertical profiles of temperature and humidity remain constant) and by 2.2 W/m^2 if the changes taking place with a CO_2 doubling are considered. This decrease constitutes 0.96% with respect to the total OLR (230 W/m^2).

In some cases relative variations of OLR in individual spectral intervals 5 cm^{-1} wide exceed 10%. In particular, OLR variations at a wavenumber of 667 cm^{-1}, corresponding to one of the channels of the TIROS-N IR spectrometer, constitute about 5% and can be detected, especially with time averaging of the observational data, to reduce the instrumental noise. In view of the OLR natural variability, however, only data for latitudes below 60° must be considered, and only daily means for 6 warm months must be analyzed. Since stratospheric temperature variations can be caused not only by CO_2 but also by other optically active components (primarily ozone), it is necessary to monitor OLR variations in various spectral intervals.

From the viewpoint of other studies, five aspects of the problem of CO_2-induced climatic impacts are of paramount importance: (i) estimation of the possible fossil fuel use during the next century; (ii) probable changes of the global biosphere as a result of human activity; (iii) additional studies of the carbon cycle and the redistribution of carbon among basic reservoirs (of particular importance are the monitoring of CO_2 in the atmosphere and ocean, the estimation of biomass changes, an analysis of the interaction between the cycles of various components); (iv) more reliable and adequate assessments of the climate sensitivity to increase in atmospheric CO_2; and (v) studies of the impact of climate on natural ecosystems and on human activities.

REFERENCES

1. J. Adem and R. Garduño, 1982. Preliminary experiments on the climatic effect of an increase of the atmospheric CO_2 using a thermodynamic model, *Geofis. Int.* **21,** 309–324.

2. W. Bach, 1984. Carbon dioxide and climatic change: an update, *Prog. Phys. Geogr.* **8,** 83–93.

3. H. W. Bernard, Jr., 1980. *The Greenhouse Effect,* Cambridge, Mass.: Ballinger.

4. B. Bolin, 1981. Increase of atmospheric CO_2 and possible climatic changes, *Proc. Tech. Conf. on Climate—Asia and Western Pacific, Dec. 15–20, 1980, Guangzhou,* WMO Publication 578. Geneva: WMO, pp. 227–238.

5. K. Bryan, F. G. Konro, S. Manabe, and M. J. Spelman, 1982. Transient climate response to increasing atmospheric carbon dioxide, *Science* **215,** 56–58.

6. M. I. Budyko, K. Ya. Vinnikov, and N. A. Efimova, 1983. The dependence of

air temperature and precipitation on the CO_2 content in the atmosphere, *Meteorol. Gidrol.* **4**, 5–13.

7. L. B. Callis, M. Natarajan, and R. E. Boughner, 1983. On the realtionship between the greenhouse effect, atmospheric photochemistry, and species distribution, *J. Geophys. Res.* **C88**, 1401–1426.

8. G. I. Pearman (Ed.), 1980. *Carbon Dioxide and Climate: Australian Research,* Canberra: Australian Academy of Sciences.

9. National Research Council (U.S.), CO_2/Climate Review Panel, 1982. *Carbon Dioxide and Climate: A Second Assessment.* Washington, D.C.: National Academy Press.

10. W. C. Clark (Ed.), 1982. *Carbon Dioxide Review,* New York: Oxford University Press.

11. W. Bach et al. (Eds.), 1983. *Carbon Dioxide: Current Views and Developments in Energy/Climate Research.* Dordrecht, The Netherlands: D. Reidel.

12. W. C. Clark, G. Marland, C. Hojvat, V. R. Padgett, and S. Cody, 1984. The CO_2 question, *Science* **223**, 1014–1018.

13. A. J. Crane and H. W. Ellsaesser, 1984. The climatic effect of CO_2: a different view, *Atmos. Environ.* **18**, 1494–1496.

14. H. W. Ellsaesser, M. C. MacCracken, and J. J. Walton, 1984. Global climatic trends as revealed by the recorded data, *Preprint UCRL-90950,* Lawrence Livermore National Laboratories, Livermore, Calif.

15. H. W. Ellsaesser, 1984. The climatic effect of CO_2: a different view, *Atmos. Environ.* **18**, 431–434.

16. H. Flohn, 1984. Das CO_2-klima problem, *Nachr. Chem. Tech. Lab.* **32**, 305–309.

17. H. Flohn and Fantechi (Eds.), 1984. *The Climate of Europe: Past, Present, and Future,* Dordrecht, The Netherlands: D. Reidel.

18. W. L. Gates, K. H. Cook, and M. E. Schlesinger, 1981. Preliminary analysis of experiments on the climatic effects of increased CO_2 with an atmospheric general circulation model and a climatological ocean, *J. Geophys. Res.* **C88**, 6385–6393.

19. A. Gilchrist, 1983. Increased carbon dioxide concentrations and climate: the equilibrium response, in W. Bach et al. (Eds.), *Carbon Dioxide: Current Views and Developments in Energy/Climate Research.* Dordrecht, The Netherlands: D. Reidel, pp. 219–258.

20. J. Gribbin, 1983. *Future Weather: Carbon Dioxide, Climate and the Greenhouse Effect.* New York: Penguin Books.

21. J. Hansen, D. Johnson, A. Lacis, S. Lebedeff, P. Lee, D. Rind, and G. Russel, 1983. Climate impact of increasing carbon dioxide, *Science* **213**, 957–966.

22. A. Henderson-Sellers and A. J. Meadows, 1979. A simplified model for deriving planetary surface temperatures as a function of atmospheric chemical composition, *Planet. Space Sci.* **27**, 1095–1099.

23. E. O. Holopainen, 1983. Detection and monitoring of climate change: some remarks and recommendations, *Preprint*. Helsinki.

24. J. T. Houghton, 1979. Greenhouse effect of some atmospheric constituents, *Philos. Trans. R. Soc. London* **A290**, 515–521.

25. J. R. Hummel and R. A. Reck, 1981. Carbon dioxide and climate: the effects of water transport in radiative-convective models. *J. Geophys. Res.* **C86**, 12,035–12,038.

26. B. G. Hunt and N. C. Wells, 1979. An assessment of the possible future climatic impact of carbon dioxide increases based on a coupled one-dimensional atmospheric-oceanic model, *J. Geophys. Res.* **C84**, 787.

27. B. G. Hunt, 1981. An examination of some feedback mechanisms in the carbon dioxide climate problem. *Tellus* **33**, 78–88.

28. S. B. Idso, 1982. *Carbon Dioxide: Friend or Foe?* Tempe, Ariz.: Institute for Biospheric Research Press.

29. S. B. Idso, 1984. What if increases in atmospheric CO_2 have an inverse greenhouse effect? 1. Energy balance considerations related to surface albedo, *J. Climatol.* **4**, 399–410.

30. W. Bach, J. Pankrath, and J. Williams (Eds.), 1980. *Interactions of Energy and Climate, Proc. Int. Workshop, Mar. 3–6, Münster.* Dordrecht, The Netherlands: D. Reidel.

31. International Workshop on Climate Issues: Climate Research Board, 1978. *International Perspectives on the Study of Climate and Society.* Washington, D.C.: Assembly of Mathematicians and Physical Scientists.

32. J. Jäger, 1983. *Climate and Energy Systems: A Review of Their Interaction,* Chichester, West Sussex, England: Wiley.

33. W. W. Kellogg and R. Schware, 1981. *Climate Change and Society: Consequences of Increasing Atmospheric Carbon Dioxide.* Boulder, Colo.: Westview Press.

34. K. Ya. Kondratyev and L. I. Nedovesova, 1958. On the thermal emission of carbon dioxide in the atmosphere, *Izv. Akad. Nauk SSSR, Geofis.* **12.**

35. K. Ya. Kondratyev and H. Nijlisk, 1963. *On the Thermal Emission of Carbon Dioxide in the Atmosphere,* Problem fisiki atmosfery. LSU Publ. House, issue 2.

36. K. Ya. Kondratyev and A. M. Bunakova, 1978. Factors of the greenhouse effect of the atmosphere and their influence on climate, in J. Willaims (Ed.), *Carbon Dioxide, Climate and Society.* Oxford: Pergamon Press.

37. K. Ya. Kondratyev and N. I. Mosakalenko, 1980. Carbon dioxide and other optically-active components as factors of the atmospheric greenhouse effect, in *Problems of Atmospheric Carbon Dioxide, Proc. American-Soviet Symp.* Leningrad: Gidrometeorizdat.

38. K. Ya. Kondratyev, 1982. *Impact of Carbon Dioxide on Climate.* Obninsk: VNIIGMI-MCD.

39. K. Ya. Kondratyev and N. I. Moskalenko, 1984. *The Greenhouse Effect of the*

Planetary Atmospheres, Uspekhi nauki i tekhniki, Meteorologia i klimatologia. Moscow: VINITI.

40. K. Ya. Kondratyev and N. I. Moskalenko, 1984. *The Greenhouse Effect of the Atmosphere and Climate,* Uspekhi nauki i tekhniki, Meteorologia i klimatologia. Moscow: VINITI.

41. K. Ya. Kondratyev and N. I. Moskalenko, 1984. The role of carbon dioxide and other minor gaseous components and aerosols in the radiation budget, in J. T. Houghton (Ed.), *The Global Climate.* Cambridge: Cambridge University Press.

42. A. Lacis, J. Hansen, P. Lee, T. Mitchell, and S. Lebedeff, 1981. Greenhouse effect of trace gases, 1970–1980, *Geophys. Res. Lett.* **8,** 1035–1038.

43. M. C. MacCracken, 1983. *Is There Climatic Evidence Now for Carbon Dioxide Effects?* ARPA Paper 83–62.3. Livermore, Calif.: Lawrence Livermore National Laboratories.

44. M. C. MacCracken and H. W. Ellsaesser, 1983. Climatic data bases for detecting CO_2-induced climate change. *Preprint UCRL-Draft,* Lawrence Livermore National Laboratories, Livermore, Calif.

45. M. C. MacCracken, S. B. Idso, J. Hansen, D. Johnson, A. Lacis, S. Lebedeff, P. Lee, D. Rind, and G. Russel, 1983. Climatic effects of atmospheric carbon dioxide, *Science.* **220,** 873–875.

46. R. A. Madden and V. Ramanathan, 1980. Detecting climate change due to increasing carbon dioxide, *Science* **209,** 763–767.

47. S. Manabe and R. T. Wetherald, 1975. The effects of doubling the CO_2 concentration on the climate of a general circulation model, *J. Atmos. Sci.* **32,** 3–15.

48. S. Manabe and R. J. Stouffer, 1979. A CO_2-climate sensitivity study with a mathematical model of the global climate, *Nature* **282,** 491–493.

49. S. Manabe and R. T. Wetherald, 1980. On the distribution of climate change resulting from an increase in CO_2 content of the atmosphere, *J. Atmos. Sci.* **37,** 99–118.

50. S. Manabe, R. T. Wetherald, and R. J. Stouffer, 1981. Summer dryness due to an increase of atmospheric CO_2 concentration, *Clim. Change* 3, 347–386.

51. S. Manabe, 1983. Carbon dioxide and climatic change, *Adv. Geophys.* **25,** 39–82.

52. C. S. Mitchell, G. L. Potter, H. W. Ellsaesser, and J. J. Walton, 1981. Case study of feedbacks and synergisms in a doubled CO_2 experiment, *J. Atmos. Sci.* **38,** 1906–1910.

53. J. F. B. Mitchell, 1983. The seasonal response of a general circulation model to changes in CO_2 and sea temperatures, *Quart. J. R. Meteorol. Soc.* **109,** 113–152.

54. R. E. Newell and T. G. Dopplick, 1979. Questions concerning the possible influence of anthropogenic CO_2 on atmospheric temperature, *J. Appl. Meteorol.* **18,** 822–825.

55. R. E. Newell and T. G. Dopplick, 1981. A reply to R. G. Watts Discussion of

questions concerning the possible influence of anthropogenic CO_2 on atmospheric temperature, *J. Appl. Meteorol.* **20,** 114–117.

56. World Meteorological Organization, 1981. On the assessment of the role of CO_2 on climate variations and their impact, presented at the Joint WMO/ICSU/UNEP Meeting of Experts, Nov. 1980, Villach, Austria. Geneva: WMO.

57. A. H. Perry, 1984. Recent climatic change—is there a signal amongst the noise? *Prog. Phys. Geogr.* **8,** 111–117.

58. A. B. Pittock and M. J. Salinger, 1982. Towards regional scenarios for a CO_2-warmed Earth, *Clim. Change* **4,** 23–40.

59. A. B. Pittock, 1983. Recent climatic change in Australia: implications for a CO_2-warmed Earth, *Clim. Change* **5,** 321–340.

60. G. L. Potter and H. W. Ellsaesser, 1982. Understanding why models agree (or disagree), *Am. Inst. Phys. Proc.* **82,** 15–22.

61. U.S. Department of Energy, 1982. *Proceedings of the Workshop on First Detection of Carbon Dioxide Effects, Harpers Ferry, W. Va.* Contr. 8106214. Washington, D.C.: USDE.

62. U.S. Department of Energy, 1983. *Proceedings: Carbon Dioxide Research Conference: Carbon Dioxide, Science, and Consensus, Sept. 19–23, 1982, Berkeley Springs, W. Va.,* Contr. DE-ACO5-760 R 00033, Washington, D.C.: USDE.

63. L. R. Rakipova, 1980. Evaluation of stratospheric climate sensitivity to variations in the CO_2 content, *Tr. Gl. Geofiz. Obs.* **438,** 37–41.

64. V. Ramanathan, M. S. Lian, and R. D. Cess, 1979. Increased atmospheric CO_2: zonal and seasonal estimates of the effect on the radiation energy balance and surface temperature, *J. Geophys. Res.* **C84,** 4949–4958.

65. V. Ramanathan, 1980. Climatic effects of anthropogenic trace gases, *Interaction of Energy and Climate Proc., Int. Workshop, Münster,* Dordrecht, The Netherlands: D. Reidel.

66. V. Ramanathan, 1982. The role of ocean-atmosphere interactions in the CO_2 problem, *Am. Inst. Phys. Proc.* **82,** 275–291.

67. S. I. Rasool and S. H. Schneider, 1971. Atmospheric carbon dioxide and aerosols: effects of large increases on global climate, *Science* **173,** 138–141.

68. R. A. Reck, 1980. Carbon dioxide and climate: comparison of one- and three-dimensional models, *Environ. Int.* **2,** 387–391.

69. R. A. Reck, 1982. Introduction to the proceedings of the workshop, *Am. Inst. Phys. Proc.* **82,** 1–2.

72. W. W. Kellogg and R. D. Bojkov (Eds.), 1983. *Report of the JSC/CAS Meeting of Experts on Detection of Possible Climate Change, Moscow, Oct. 3–6, 1982,* Geneva: WMO.

71. World Meteorological Organization, 1981. *Report of CAS Global Climate Group, Second Session, April 27–May 1, 1981, Potsdam.* Geneva: CAS/WMO.

72. World Meteorological Organization, 1983. *Report of CAS Group of Rappor-*

teurs on Climate, Third Session, Nov. 29–Dec. 3, 1982, Toronto. Geneva: WMO.

73. World Meteorological Organization, 1983. *Report of the Meeting of Experts on Potential Climatic Effects of Ozone and Other Minor Trace Gases, Sept. 13–17, 1981, Boulder, Colo.* WMO Global Ozone Research and Monitoring Project Report 14. Geneva: WMO.

74. A. Robock, 1984. Detection of volcanic, CO_2, and ENSO signals in surface air temperature, *Preprint,* Department of Meteorology, University of Maryland, College Park, Md.

75. H. I. Schiff, 1981. A review of the carbon dioxide greenhouse problem, *Planet. Space Sci.* **29,** 935–950.

76. M. E. Schlesinger, 1983. A review of climate models and their simulation of CO_2-induced warming, *Int. J. Environ. Stud.* **20,** 103–114.

77. M. E. Schlesinger, 1983. *A Review of Climate Model Simulations of CO_2-Induced Climatic Change,* Climatic Research Institute Report 41. Corvallis, Ore.: Oregon State University.

78. S. H. Schneider, 1981. Climatic impact assessment in the CO_2 context—an editorial, *Clim. Change* **3,** 345–346.

79. S. H. Schneider and S. L. Thompson, 1981. Atmosphere CO_2 and climate: importance of the transient response, *J. Geophys. Res.* **C86,** 3135–3147.

80. J. Smagorinsky, 1983. Climatic changes due to CO_2, *Ambio* **12,** 83–86.

81. J. Smagorinsky, 1984. *The Problem of Climate and Climate Variations,* World Climate Paper 72.

82. J. T. Houghton (Ed.), 1984. *The Global Climate.* Geneva: WMO. Cambridge: Cambridge University Press.

83. S. L. Thompson and S. H. Schneider, 1982. Carbon dioxide and climate: has a signal been observed yet? *Nature* **295,** 645–646.

84. S. L. Thompson and S. H. Schneider, 1982. Carbon dioxide and climate: the importance of realistic geography in estimating the transient temperature response, *Science* **217,** 1031–1033.

85. G. B. Tucker, 1981. *The CO_2-Climate Connection: A Global Problem from an Australian Perspective.* Canberra: Australian Academy of Sciences.

86. W. C. Wang, Y. L. Young, A. A. Lacis, T. Mo, and J. E. Hansen, 1976. Greenhouse effects due to manmade perturbations of trace gases, *Science* **194,** 685–691.

87. W. C. Wang, J. P. Pinto, and Y. L. Young, 1980. Climate effects due to halogenated compounds in the Earth's atmosphere. *J. Atmos. Sci.* **37,** 333–338.

88. W. A. Washington and G. A. Meehl, 1983. General circulation model experiments on the climatic effects due to a doubling and quadrupling of carbon dioxide concentration, *J. Geophys. Res.* **C88,** 6600–6610.

89. L. B. Callis and M. Natarajan, 1981. Atmospheric carbon dioxide and chloro-

fluoromethanes: combined effects on stratospheric ozone, temperature, and surface temperature, *Geophys. Res. Lett.* **8**, 587–590.

90. J. R. Hummel and R. A. Reck, 1981. The direct thermal effects of $CHClF_2$ and CH_3CCl_3 and CH_2Cl_2 on atmospheric surface temperatures, *Atmos. Environ.* **15**, 379–382.

91. Impact of increased carbon dioxide in the atmosphere on climate, *Materials of the American-Soviet Meeting on Studies of the Impact of Increased CO_2 in the Atmosphere on Climate.* Leningrad: Gidrometeoizdat.

92. M. A. K. Khalil and R. A. Rasmussen, 1984. Carbon monoxide in the Earth's atmosphere: increasing trend, *Science* **224**, 54–55.

93. J. T. Kiehl and V. Ramanathan, 1983. CO_2 radiative parameterization used in climate models: comparison with narrow-band models and with laboratory data, *J. Geophys. Res.* **C88**, 5191–5202.

94. K. Ya. Kondratyev, 1977. *Present-Day Climate Changes and Their Determining Factors,* Uspekhi nauki i tekhniki, Meteorologia i klimatologia. Moscow: VINITI.

95. K. Ya. Kondratyev and G. E. Hunt, 1982. *Weather and Climate on Planets.* Oxford: Pergamon Press.

96. K. Ya. Kondratyev, 1983. *Earth's Radiation Budget, Aerosol, and Cloudiness,* Uspekhi nauki i tekhniki, Meteorologia i klimatologia. Moscow: VINITI.

97. K. Ya. Kondratyev and V. V. Kozoderov, 1984. *Anomalies of the Earth's Radiation Budget and of the Heat Content of the Upper Oceanic Layer as Indicators of Energetically Active Zones,* Uspekhi nauki i tekhniki, Atmosfera, okean, kosmos, Program "Sections." Moscow: VINITI.

98. P. S. Liss and A. J. Crane, 1983. *Man-Made Carbon Dioxide and Climatic Change: A Review of Scientific Problems.* Norwich: (Geo Books).

99. G. Vogel, 1984. Zur Berechnung infraroter Abkühlungsraten im Bereich der 15 μm CO_2-Bande unter Verwendung empirisch approximierter Transmissionsfunktionen nach Smith, *Z. Meteorol.* **34**, 118–130.

100. V. G. Gorshkov, 1984. Possible role of ocean biota in the global carbon cycle. *Okeanografia,* **24**, 453–459.

101. K. Ya. Kondratyev and V. I. Binenko, 1984. *Impact of Cloudiness on Radiation and Climate,* Leningrad: Gidrometeoizdat.

102. W. L. Gates, 1980. Modeling the surface temperature changes due to increased atmospheric CO_2, in W. Bach et al. (Eds.), *Interaction of Energy and Climate.* Dordrecht, The Netherlands: D. Reidel, pp. 169–190.

103. E. S. Epstein, 1982. Detecting climate change, *J. Appl. Meteorol.* **21**, 1172–1182.

104. J. T. Kiehl, 1983. Satellite detection of effects due to increased atmospheric carbon dioxide, *Science* **222**, 504–506.

105. F. M. Luther, 1983. Detecting the radiative effects of carbon dioxide: developing a strategy, *Fifth Conf. on Atmospheric Radiation, Oct. 31–Nov. 4, 1980, Baltimore, Md.* Boston: American Meteorological Society, pp. 166–169.

106. M. C. MacCracken and H. Moses, 1982. *Proc. Workshop on First Detection of Carbon Dioxide Effects, June 8–10, 1981, Harpers Ferry, W. Va.* Washington, D.C.: U.S. Department of Energy.

107. D. Rind and S. Lebedeff, 1984. *Potential Climatic Impacts of Increasing Atmospheric CO_2 with Emphasis on Water Availability and Hydrology in the United States,* Report prepared for the Environmental Protection Agency, EPA-230-04-84-006.

108. J. Hansen, A. Lacis, D. Rind, G. Russel, P. Stone, I. Fung, R. Ruedy, and J. Lerner, 1984. Climate sensitivity: analysis of feedback mechanisms, in *Climate Processes and Climate Sensitivity,* Geophysics Monograph 29, Maurice Ewing, Vol. 5, Boston: American Meteorological Society, pp. 130–163.

109. R. C. J. Somerville and L. A. Remer, 1984. Cloud optical thickness feedbacks in the CO_2 climate problem, *J. Geophys. Res.* **89**(D6), 9668–9672.

110. R. E. Newell and J. Hsiung, 1984. Sea surface temperature, atmospheric CO_2 and the global energy budget: some comparisons between the past and present, in N.-A. Mörner and W. Karlen, (Eds.), *Climatic Changes on a Yearly to Millenial Basis.* Dordrecht, The Netherlands: D. Reidel, pp. 533–561.

111. S. B. Idso, 1982. A surface air temperature response function for Earth's atmosphere, *Boundary-Layer Meteorol.* **22**, 227–232.

112. S. B. Idso, 1982. An empirical evaluation of earth's surface air temperature response to an increase in atmospheric carbon dioxide concentration, *Am. Inst. Phys. Proc.* **82**, 119–134.

113. S. B. Idso, 1983. Carbon dioxide and global temperature: what the data show, *J. Environ. Qual.* **12**, 159–163.

114. R. D. Cess and G. L. Potter, 1984. A commentary on the recent CO_2-climate controversy, *Clim. Change,* **6**(4), 365–376.

115. J. Adem and R. Garduño, 1984. Sensitivity studies of the climatic effect of an increase of atmospheric CO_2, *Geofis. Int.* **23**(1), 17–35.

116. I. M. Held, D. I. Linder, and M. L. Suarez, 1981. Albedo feedback, the meridional structure of the effective heat diffusivity, and climate sensitivity: results from dynamic and diffusive models, *J. Atmos. Sci.* **38**, 1911–1927.

117. R. D. Bojkov, 1984. *Tropospheric Ozone, Its Changes, and Possible Radiative Effects,* WMO Special Environmental Report 16. Geneva: WMO.

118. R. D. Cess and F. D. Goldenberg, 1981. The effect of ocean heat capacity upon global warming due to increasing atmospheric carbon dioxide, *J. Geophys. Res.* **C86**, 498–502.

119. R. L. Gilliland, 1982. Solar, volcanic, and CO_2 forcing of recent climatic changes, *Clim. Change* **4**, 111–132.

120. A. Robock, 1985. Detection of volcanic, CO_2 and ENSO signals in surface air temperature, *Advances in Space Research,* **5**(6), 53–56.

121. V. Ramanathan, R. J. Cicerone, H. B. Singh, and J. T. Kiehl, 1985. Trace gas trends and their potential role in climate change, *J. Geophys. Res.* **90**(D3), 5547–5562.

Chapter 2

The Climatic Impact of Volcanic Eruptions

Studies of volcanic eruptions are very interesting from the point of view of understanding both the laws of paleoclimate and present-day climate changes (1 – 127). Volcanic eruptions have contributed mainly to the formation of the atmosphere and oceans. The products of explosive volcanic eruptions reach the stratosphere and remain there for a year or longer, changing the chemical composition of the stratosphere and thereby affecting the Earth's radiation budget. It is these variations in the Earth's radiation budget caused by volcanic aerosols that may be responsible for posteruption climate changes.

Volcanic eruptions are well known to be a most powerful "natural experiment" on the environmental impact on climate. From the viewpoint of the tropospheric and stratospheric dynamics, of great interest is the monitoring of the global-scale transport of eruption products (128).

Benjamin Franklin had made a supposition that an eruption of the Hekla volcano in Iceland could have been a reason for the severe winter of 1783 – 1784 (120). Since then, numerous studies have been undertaken in search for correlations between individual volcanic eruptions and adverse weather conditions or between serial eruptions and climate anomalies. These studies revealed that volcanic eruptions may be one of the important factors of climate changes.

Of all the possible causes of climate changes, volcanic eruptions are the most adequately documented and understood. For example, a stratospheric temperature increase trend was observed after the 1963 Agung eruption in

Indonesia. This trend was superimposed on a growing-in-amplitude quasibiennial stratospheric temperature variation and could be explained by the Agung eruption. This increase was clearly pronounced in the Southern Hemisphere, where the variability of the amplitude of quasibiennial oscillations is small, whereas in the Northern Hemisphere (with a small input from the eruptive cloud) there was no stratospheric temperature increase. These observational data agree with theoretical estimates, from which it follows that a stratospheric warming is mainly caused by absorption of the upward thermal emission flux and, to a smaller degree, by the solar radiation absorption. In the troposphere and at the surface there was a global-scale temperature decrease (in the Northern Hemisphere it was the largest, 0.6°C). Hence the decrease could not have been caused totally by the eruption.

Following the 1815 Tambora volcanic eruption in Indonesia "a year without the sun" was observed, and a summertime temperature decrease in New England and West Europe (with respect to the climatic mean) reached 1 to 2.5°C.

Mass and Schneider (75) analyzed the consequences of numerous eruptions causing an increase to several tenths of the aerosol optical thickness in the stratosphere in the visible. They found that a statistically reliable temperature decrease a year after the eruption must constitute 0.3°C; and after 2 years, 0.1°C and more.

Although volcanoes have long been considered a cause of glaciations, this hypothesis remains unsubstantiated. Analysis of the aerosol particle content in the Greenland ice cores has not revealed any increase in the aerosol content during the early stages of the Wisconsin glaciation cycle.

The role of volcanic eruptions in the present temperature trend in the Northern Hemisphere (120) is not clear, since the post-1940 cooling was not preceded by an intensification of volcanic activity. Theoretical estimates revealed a strong effect of the composition and size distribution of volcanic aerosol on climate changes. For example, even with a small amount of an absorbing silicate aerosol, a considerable increase of solar radiation absorption may take place, and the presence of silicate particles larger than 0.5 μm can cause a warming at the surface as a result of the thermal emission of such particles. Calculations of temperature variations at the surface and in the stratosphere made by Hansen et al. (59) after the Agung eruption revealed good agreement with observations.

Bryson and Goodman (40) analyzed the "modulation" of the incoming direct solar radiation under the influence of volcanically induced changes in atmospheric transparency, using data of the longest series of actinometric observations. To estimate the temporal dynamics of the volcanic aerosol content in the stratosphere, it is assumed that the characteristic reaction time for the formation of the H_2SO_4, aerosol is 115 days, and the time for deposition of the particles about 400 days. In view of a rapid gravitational settling of ash particles, the mean annual content of the submicron H_2SO_4 aerosol is assumed to be twice as large as that of ashes directly ejected into the stratosphere. With these assumptions, one can describe the temporal dynamics of the ash and H_2SO_4 components of eruptive aerosols, typical of an explosive eruption (Fig. 2.1).

Until recently, observations of atmospheric transparency that were used to estimate the climatic impact of eruptions (3,12,13,40) have been a major source of information on the atmospheric effect of volcanic eruptions. Clear-sky solar radiation data for 1880–1980, used to identify the secular trend of the atmospheric optical thickness, revealed a distinct correlation between transparency decrease and large-scale volcanic eruptions (Fig. 2.2). These include the following eruptions among others: Pelée and Soufrière (1902), Katmai (1912), Bezymiannaya (1955), and Agung (1963). A comparison of the secular trend of the optical thickness and a "synthesized" chronology of powerful volcanic eruptions, taking into account the moment of eruption and the subsequent decrease of the concentration of erupted aerosols for low, middle, and subpolar latitudes, yielded a correlation coefficient of +0.87. An additional consideration of moderate eruptions raised the correlation coefficient to +0.95. This permits one to believe that variations in atmospheric transparency are practically completely governed by volcanic eruptions.

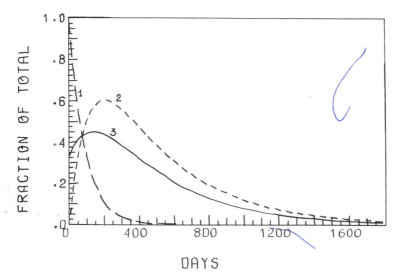

FIGURE 2.1. Temporal evolution of volcanic erupted aerosols. 1, Ash particles; 2, sulfuric acid aerosol resulting from gas-to-particle conversions; 3, total content of particles.

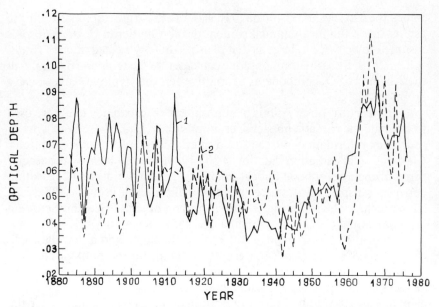

FIGURE 2.2. Secular trend of the atmospheric optical thickness. Data for the year 1900 should be considered unreliable. 1, Observations; 2, theoretical. (From Ref. 42.)

The residual variability of the optical thickness (about 5 to 10%) reveals an increasing trend that may be ascribed to an anthropogenic impact. Attempts to find 11- or 22-year cycles in the residual variability have not been successful. This attests to the absence of the impact of solar activity on incoming solar radiation. An analysis of data on clear-sky global radiation and mean annual surface air temperature after the year 1920 yielded a correlation coefficient of 0.91. Although this correlation should not be considered as proof of the respective cause-and-effect connection, a volcanically induced modulation of the mean hemispherical temperature secular trend can nevertheless be supposed to take place (Fig. 2.3).

Examination of this assumption, by calculating the temperature variations in the twentieth century with the use of an energy-balance climate model (the mean hemispheric cloud cover is assumed to be fixed) showed that with a prescribed observed trend of the optical thickness, temperature variations close to those observed are obtained. Using this as a basis, Bryson and Goodman (40) drew the conclusion that the observed variability of the mean surface air temperature in the Northern Hemisphere is determined primarily by volcanic eruptions (Fig. 2.4). On the other hand, it follows that with the volcanically induced variability excluded, one can filter out the

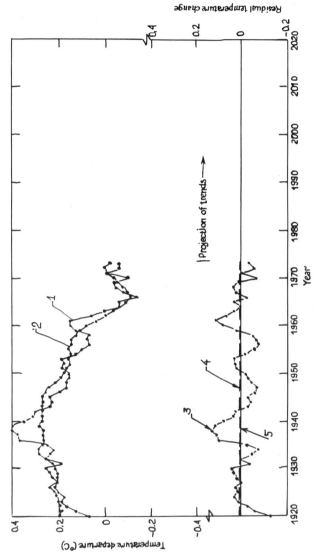

FIGURE 2.3. Variability of the NH mean surface air temperature and optical thickness of the atmosphere. 1, Observed NH surface air temperature variations from data of Borzenkova et al. (172); 2, secular trend of the optical thickness anomalies; 3, residual temperature variations; 4, linear trend (dashed line); 5, exponential trend (solid line).

77

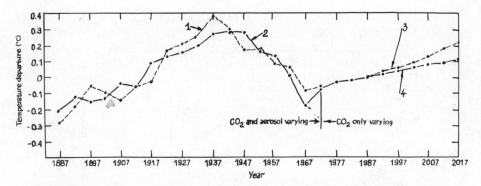

FIGURE 2.4. Secular trend of the 5-year averages of the NH mean surface air temperature: 1, observed; 2, calculated. Calculated temperature variations at a fixed optical thickness of the atmosphere (at the level of the year 1975): 3, due to an exponential increase of the CO_2 concentration; 4, due to an increase calculated from data on possible dynamics of fossil fuel consumption.

trend due to an increase in the atmospheric CO_2 concentration. This attempt, however, failed: the residual variability shows no trend of temperature increase. Even the forecast taking into account a CO_2 concentration increases until the year 2020 (with the optical thickness of the atmosphere fixed at the level of 1975) revealed a future warming during the next 50 years less than the warming observed during the recent years. In Bryson's opinion (40), in forecasting possible future climate changes the problem of possible prediction of volcanic eruptions rather than the runaway CO_2-induced greenhouse effect must be of primary importance. It must be emphasized, however, that this conclusion is of questionable validity (10,12,13). Assessment of posteruption variations in the Earth's radiation budget is a key aspect of the climatic impact of volcanic eruptions. Using the two-stream approximation and a three-layer model of the surface-troposphere-stratosphere system, Lenoble et al. (69) estimated the effect of a stratospheric aerosol layer constituting a 75% water solution of sulfuric acid or background aerosol particles (Table 2.1) on the system's radiation budget and its

TABLE 2.1   The Size Distribution and Optical Properties of Stratospheric Aerosols[a]

Aerosol model	Parameters					
	α	b	γ	ω_s^{eff}	b_s^{eff}	$c_s = \tau_{IR}^{eff}/\tau_s^{eff}$
Background aerosol	1.0	18	1.0	0.998	0.189	0.0383
Volcanic aerosol	1.0	16	0.5	0.994	0.169	0.0217

[a]ω_s^{eff}, b_s^{eff} are effective values of the single scattering albedo and backscattering; τ_{IR}^{eff}, τ_s^{eff} are effective values of the optical thicknesses for the shortwave and longwave radiation, respectively; $\tau_a = 0.03$.

components, as well as on the surface temperature (the surface is considered to be an isotropic reflector). The distribution of the particles number density (n) by radius (r) is prescribed by a modified gamma function:

$$n(r) = A_r^\alpha \exp\left(-br^\gamma\right),$$

where A, α, b, and γ are parameters. A comparison of the applied two-stream approximation to an accurate solution has shown that the errors of an approximate calculation of the diffuse transmission do not exceed 10%, with the single scattering albedo $\omega \geq 0.6$, optical thickness $\tau \leq 0.5$, and normally incident solar rays. Here the system's albedo is calculated accurately, and the error in calculations of albedo variations, due to changed properties of the stratospheric aerosol layer, does not exceed 10%. The Table 2.2 data illustrate variations in the system's radiation budget (ΔB) at $\tau_s = 0.03$, testifying to the fact that in both cases the radiation budget decreases since the contribution by backscattering prevails. An estimation of the mean global surface temperature decrease (at $\tau_s = 0.03$) gave $-0.45\,°C$ (volcanic aerosol) and $-1.1\,°C$ (background aerosol). An account of the albedo feedback resulted in an intensified cooling down to $-0.8\,°C$ and $-2.15\,°C$, respectively. A weaker effect of volcanic aerosols on climate would exist if there were a lower single-scattering albedo (e.g., stronger absorption).

When discussing the results of observations of volcanic aerosols, it is important to assess possibilities of the quantitative characteristics of the eruption intensity. In this regard, Kelly and Sear (129) analyzed the adequacy of Lamb's dust veil index (DVI), which was introduced to characterize quantitatively the impact of volcanic aerosols on the Earth's radiation budget (ERB) and climate during the several years following an eruption. An estimate of the DVI taking into account observational data and empirical and theoretical studies of the possible climatic implications of the dust veil is based on the nature of such implications as understood in 1960. The DVI is calculated by averaging over the maximum possible number of DVI estimates determined with the use of various techniques.

TABLE 2.2 Variations in the System's Radiation Budget (RB) $\Delta B(W/m^2)$

	RB Variations Due to:			
Aerosol Model	Backscattering	Shortwave Radiation Absorption	Greenhouse Effect	Total
Background aerosol	−28.5	0.0923	1.56	−26.8
Volcanic aerosol	−26.4	2.28	0.887	−23.2

Five available techniques were mutually calibrated (normalized) to give the DVI value 1000 for the 1883 Krakatoa eruption. According to technique B, the mean global DVI is

$$DVI = 0.97R_{max}E_{max}t_{mo},$$

where R_{max} is the posteruption maximum extinction of direct solar radiation determined as a monthly mean for the midlatitudes of the given hemisphere (maximum extinction can sometimes be reached only two years after the eruption); E_{max} is the geographical extent of the dust veil in conditional units as a function of the volcano's latitude ($E_{max} = 1.0$ for the band 20°N to 20°S and 0.3 for latitudes 40°); t_{mo} is the lifetime of the dust veil (in months).

Technique C gives

$$DVI = 52.5T_{D_{max}}E_{max}t_{mo},$$

where $T_{D_{max}}$ is the mean temperature decrease (°C) in the midlatitudes of the given hemisphere during the year when the effect of an eruption is at a maximum. Approximate estimates of the DVI, which may be considered as showing only an order of magnitude, are explained by different available observational data and hence by different techniques chosen for the estimation.

Recently specified data on the chronology of eruptions and a more adequate understanding of the nature of their climatic impact (e.g., the role of the secondary H_2SO_4 aerosol, and so forth) necessitate a revision of the understanding of the DVI.

The climatic implications of volcanic eruptions have recently been of great concern. For example, an extensive complex of observations of volcanic effluents and their propagation in the atmosphere have been carried out with the use of conventional means (e.g., ground-based, aircraft, balloon) as well as satellites. Substantial progress has been reached in the development of the techniques for numerical modeling of climate and large-scale transport (128,130,131). Of great importance was the successful numerical modeling of the processes of stratospheric aerosol formation.

Recent powerful eruptions of the Mount St. Helens (1980) and El Chichón (1982) volcanoes have become new "experiments" that have made it possible, for the first time, to document adequately posteruption changes in the atmosphere. Some relevant results are considered below, with emphasis on an analysis of El Chichón data.

2.1 THE NATURE OF VOLCANIC AEROSOL

The principal component of volcanic aerosol (e.g., background stratospheric aerosol) are droplets of the concentrated H_2SO_4 solution resulting from a cycle of reactions of sulfur dioxide oxidation. Gigantic injections of SO_2 into the stratosphere, like those occurring during the 1963 Agung eruption, lead to a considerable and long-term (for many months) decrease of the hydroxyl content in the stratosphere according to the reaction

$$SO_2 + OH \rightarrow HSO_3.$$

H_2SO_4 droplets formed from HSO_3 have a considerable effect on radiative transfer in the atmosphere.

The H_2SO_4 droplets and particles of volcanic ashes can start heterogeneous reactions, and the volcanic HCl can seriously affect the ozone content in the stratosphere. The aerosol of eruptive clouds appearing after eruptions consists mainly of volcanic ash particles (powdered lava) and droplets of dirty H_2SO_4 (Fig. 2.1). Since the processes of the stratospheric aerosol formation have already been characterized (13,16), we shall confine ourselves to discussing the results of some of the more recent studies.

Aircraft measurements have shown that during a moderate volcanic eruption, ash particles with diameters between 1 and 0.25 μm and more are ejected to the atmosphere. Murrow et al. (132) analyzed the size distribution of the Fuego volcano (Guatemala) eruptive products that were deposited onto the Earth's surface. An extrapolation of the size distribution function obtained toward its tails (it is nearly lognormal for an average diameter of about 0.6 μm) enabled estimation of the unconsidered portion of small and large particles.

An application of two techniques to estimate the portion of unconsidered small silicate particles with diameters smaller than 2 μm gave their relative and absolute volumes 0.8% (6×10^{-4} km^3) and 0.01% (3.4×10^{-4} km^3), respectively, for the whole of the erupted material, with a mass of about 2.2×10^9 g. These estimates characterize the possible amount of material ejected to the stratosphere (the eruptive cloud reached a height of 18 to 23 km, greatly exceeding the height of the tropopause). The actual amount of material injected into the stratosphere is apparently much smaller but despite a relatively small total volume, the total number of particles is very large and can therefore play an important role as condensation nuclei in the formation of stratospheric aerosol droplets.

Estimations of the mass of H_2SO_4 droplets and ejections of sulfur dioxide (a potential source of the H_2SO_4 aerosol) gave, respectively, 1.4×10^{10} g and 1.6×10^{12} g, for example, one to two orders of magnitude higher than for silicate particles (emissions of gaseous HCl reach 6.2×10^{10} g). Thus the formation of the eruptive stratospheric aerosol is mainly determined by H_2SO_4 aerosol droplets and not by the directly ejected small-sized silicate ash particles.

The contribution by volcanoes into sulfur gases production averages about one-third, and sulfur dioxide is its prevailing component, but volcanic ejections also contain hydrogen sulfide, carbonylsulfide, and carbonyldisulfide. Stewart (133) summarized (Table 2.3) the characteristics of the gases ejected by the Mount St. Helens volcano (the S/Cl molar ratio was obtained taking aerosols into account). Being ejected into the atmosphere, all the gaseous components are oxidized, producing sulfates; this process consists either of homogeneous gas-phase reactions or heterogeneous reactions with the liquid phase (the total set of these reactions is still inadequately studied).

In the case of sulfur dioxide, the initial stages of oxidation are the following reactions:

$$SO_2 + OH \rightarrow HSO_3; \qquad SO_2 + HO_2 \rightarrow SO_3 + OH.$$

TABLE 2.3   The Composition of the Mount St. Helens Eruptive Products

Component	Basic role	Relative Concentration		
		Environmental Air	Fumarole Release	Stratospheric Veil
H_2O	Cloud formation; the HO_x chemistry; SO_4, Cl^-, NO_3 washout	$(3-5) \times 10^{-6}$	$\sim 10^{-2}$	$\sim 10^{-4}$
SO_2	Precursors of sulfate aerosol, consume hydroxyl	$\sim 5 \times 10^{-11}$		$\sim 10^{-7}$
H_2S		10^{-9}	10^{-5}	$\sim 10^{-9}$
COS		$(1-5) \times 10^{-10}$		$\sim 10^{-9}$
CS_2		$\lesssim 10^{-12}$		$\lesssim 10^{-10}$
HCl	Catalysis of ozone	$\sim 10^{-10}$		$\lesssim 10^{-9}$
CH_3Cl		$\sim 10^{-10}$		$\sim 10^{-9}$
CH_3Br		$\sim 10^{-11}$	10^{-5}	
S/Cl		$\sim 0.1-1.0$	1.0	$1.0-100$
CO_2	Formation of the atmospheric thermal emission	3.6×10^{-4}	$\sim 3.6 \times 10^{-4}$	3.6×10^{-4}
CO	Tracer of hydroxyl	$\sim 10^{-8}$	?	$\sim 10^{-7}$
NH_3	Aerosol reactions	$\lesssim 10^{-10}$		—
N_2O	Precursors of the oxides of nitrogen	$\sim 10^{-7}$	?	$\sim 10^{-7}$
NO_x	Active in the ozone cycles	$\lesssim 10^{-8}$		$\sim 10^{-8}$

Source: Ref. 133.

Apparently, heterogeneous reactions play a decisive role in SO_2 oxidation. Of great importance in this connection is an understanding of the mechanism for catalysis by small admixtures of metals of the processes of oxidation of bisulfite (HSO_3^-) and sulfite (SO_3^{2-}), giving SO_4^{2-}. The contribution by bisulfite oxidation involving such strong oxidants as hydrogen peroxide and (to a lesser degree) ozone is still unreliably estimated. These, along with some other poorly studied factors, explain the difficulty of making a quantitative description of the processes of the gas-phase formation of eruptive aerosols.

Capone et al. (42) developed a 2-D simulation model of the formation of volcanic aerosols in the stratosphere, based on an interactive account of photochemical processes, transport, and size distribution of the aerosols. This model is a combination of the earlier used 2-D photochemical model of the atmospheric gas composition and a 1-D aerosol model; it covers the atmospheric layer 0 to 60 km at a vertical resolution of 2.5 km within $\pm 80°$ latitude (horizontal resolution 5°). The transport of gases and aerosol is described taking into account the vertical and horizontal turbulent mixing, advection, and gravitational settling.

A 25-level size distribution of the aerosol with radii ranging between 0.01 and 2.56 μm as well as an interactive evolution of six gaseous components (CS_2, COS, SO_2, SO_3, HSO_3, and H_2SO_4) are considered. The concentrations of O, O_2, O_3, CO_2, CH_3O_2, and H_2O that react with the components above are found with the use of a 2-D photochemical model of the stratosphere. The considered microphysical processes determining the formation and transformation of the aerosol include the following: H_2SO_4 condensation, H_2O evaporation, coagulation, and gravitational settling and washing-out of tropospheric aerosols.

Oxidation of the erupted sulfur dioxide is a most important mechanism for the formation of the H_2SO_4 aerosol. Under conditions of the volcanically disturbed atmosphere, it is important to take into account the effect of SO_2 on hydroxyl (instead of the prescribed OH concentration), which has been done with the use of a 1-D photochemical model that provides the parameterization of the dependence of the OH concentration on the H_2O and SO_2 releases.

Numerical modeling results (42) made it possible for the first time to reproduce within a closed model (describing the photochemistry, transport, and size distribution) the ozone concentration distribution and those of sulfur-containing gases, and also of the background and volcanic gas-phase aerosols. These results indicate that the residence time in the stratosphere of the volcanic aerosols exceeds 2 years, and that their effects on atmospheric radiation agrees with reality, provided that an initial total mass of the aerosol cloud is 10 megatons of sulfur dioxide. This value, however, may be specified as being smaller, provided that a more accurate description of the areal extent of the aerosol cloud and its movement can be made. It is assumed that the total water vapor release constitutes 100 megatons, and the total aerosol within radius of >0.01 μm release is on the order of 12 megatons. Immediately after the eruption, most of the SO_2 is transformed into H_2SO_4. In about 1 year the SO_2 concentration level returns to the background level.

The model's eruptive cloud is subject to a faster vertical and horizontal diffusion than is actually observed, although the calculated field of the ozone concentration agrees well with Nimbus-7 data. It follows that the model based on the use of averaged coefficients of diffusion cannot adequately describe in detail the evolution of a localized disturbance. Results of numerical modeling have shown that the use of various sets of the optical parameters of the volcanic aerosol obtained from data of lidar soundings and satellite measurements of extinction at the 6.8-μm wavelength provides very good consistency. This is especially promising in view of the preliminary character of the observational data and the substantial specificity of measurement techniques characterizing different optical properties of the erupted aerosol.

Of particular recent concern has been the monitoring of the evolution of the chemical composition of atmospheric aerosols from measured concentrations of trace metals in snow samples taken from different depths in the center of the Antarctic polar cap, since a correlation has been found between the concentration of trace metals and aerosol content in the lower atmosphere (apparently, in these conditions there is no marked chemical fractionation at the snow surface-atmosphere interface, as has been observed in other polar regions, such as Alaska).

Of particular interest are studies of such elements as Pb, Cd, Cu, Zn, Ag, and Hg, which belong to "anomalously enriched elements" (AEE). Their concentration in the current aerosol samples taken in various remote regions of the globe exceeds by many orders of magnitude the concentrations of the same elements in a sample of the soil or the oceans that are involved in the formation of aerosols. Although present observational data do not explain the cause of the above-mentioned enrichment (either anthropogenic or natural factors such as volcanic eruptions) a comparison of the results of measurements covering the whole century enables us to analyze the causes of the enrichment.

Boutron (134) discussed the analysis of 38 firn samples from ice sheet cores taken in January 1975 in Dome C (74°40'S; 125°10'E) from depths of 0.8 to 5.35 m, and later (January 1978) at depths down to 8.4 m. These data correspond to the period 1881 – 1977. The thickness of the analyzed layers of firn is 0.20 m and is equivalent to 2-year averaging. The analysis has shown that the concentration and factors of enrichment for all the elements considered for the year 1977 (Na, Mg, K, Ca, Fe, Al, Mn, Pb, Cd, Cu, Zn, Ag) agree to within 50% of those for the early 1880s, although a considerable variability of the concentration and factor of enrichment was observed during the century. This definitely testifies to a negligibly small effect of global-scale anthropogenic pollution on the content of the aforementioned elements in the aerosols of remote regions of the globe.

The presence of "anomalously enriched elements" in the aerosol is determined mainly by the effects of natural processes such as the following: low-temperature evaporation, emissions (through the propagation of aerosols) of vegetation and forest fires as well as of oceanic surface microlayers and volcanic eruptions. Although the relative contribution of these processes has not yet been estimated, it is most probable that volcanism plays the leading role. Variations in the factor of enrichment for lead and zinc for the last 100 years correlate well with variations in global scale volcanic activity, from which it may follow that the high present values of the enrichment factor for all the elements are also apparently determined by volcanic activity.

In the period 1912–1916 (±5 years) a very sharp maximum in the concentrations of most of the AEE was registered. This could have been caused by a large-scale volcanic eruption in the Antarctic, with its 13 active volcanoes. Traces of activity of these volcanoes have been detected from the analysis of ice samples (for the past 16,000 to 30,000 years) in the form of ash layers.

Until recently, it was a common view that the fate of ions in the stratosphere had been determined by a spontaneous neutralization that resulted from the recombination of charged particles of opposite sign. Arnold (135) showed that such a conclusion may be, however, incorrect with respect to complex ion clusters, since clustering reduces the mobility of ions and slows the neutralization. Thus if neutralization does not take place, stable pairs of ions can be observed. Such a possibility became more probable following the first simultaneous measurements of the composition of positive and negative ions in the stratosphere (at an altitude of about 36 km), which revealed the presence of the more stable components than those that had been theoretically predicted. This underscored the necessity to study the possible role of ion pairs in the formation of aerosols.

Arnold (135) assessed the possibilities of the formation of stable ion pairs in light of new data on stratospheric composition derived from the following criterion of stability:

$$E_D > E_N + E_C$$

where E_D is the effective energy needed to pull an electron from a negative cluster ion; E_N is the effective energy released in neutralization of a positive cluster ion by a free electron; E_C is the energy released in the formation of a chemical bond which may result from the interaction of ion nuclei.

An analysis of the characteristics of ions prevailing in the stratosphere has shown that the criterion in question is not fulfilled during the recombination of proton hydrates $H^+(H_2O)_4$, $H^+(H_2O)_5$, and $NO_3^-(HNO_3)_2$; that is, here the recombination leads to the formation not of stable ion pairs but to a spontaneous recombination. Stable pairs of ions are formed, however, in the case of recombination of more complex cluster ions. Due to a great dipole moment of such pairs, a process of the successive joining of free ions can occur leading to the formation of multi-ion complexes (MICs) similar, with respect to their nature, to ion crystals or salts particles. The mutual joining of ions is an alternative mechanism for MIC growth. The possibility of coagulation of MIC with aerosol particles should also be considered.

Thus the formation and growth of multi-ion complexes constitute a probable mechanism for gas-to-particle conversion, which does not require condensation and hence the necessary presence of oversaturated gaseous components. If, however, such components (e.g., H_2SO_4 vapor) do exist, then MIC, having reached a considerable size, can function as condensation nuclei. They can also substantially affect minor gaseous components of the stratosphere, causing not only condensation but also reactions on MIC surfaces.

In case the role of MIC in the formation of stratospheric aerosols becomes substantial, this will be of great importance from the viewpoint of determining the mechanisms of the impact of solar activity on the atmosphere, since the solar and galactic cosmic rays are the main source of ionization in the stratosphere and troposphere. In connection with the results discussed, of great importance is a search for MIC in the stratosphere by means of balloon measurements using ion mass spectrometers operating in a wide range of masses.

The oxides of nitrogen are known to be a stratospheric component. If they can react with sulfur and form the components of the stratospheric aerosol, this can indicate the existence of a mechanism for the nitrogen oxide sink and thus the purging of the stratosphere of pollutants. Although ammonium sulfate $(NH_4)_2SO_4$ has been discovered in stratospheric aerosols, it cannot be considered as providing a sink for the oxides of nitrogen, since ammonium sulfate is apparently produced from the reaction between gaseous ammonium and H_2SO_4 droplets.

Farlow et al. (136) pointed out, however, that recent studies had revealed the possibility of direct reactions between the oxides of nitrogen and sulfuric compounds.

$$2SO_2 + 3NO_3 + H_2O \rightarrow 2NOHSO_4 + NO,$$
$$2NO_2 + H_2SO_4 \rightarrow \quad NOHSO_4 + HNO_3,$$
$$NOHSO_4 + SO_3 \rightarrow \quad NOHS_2O_7.$$

In this connection, an attempt has been made to find these two forms of nitrososulfuric acid (e.g., $NOHSO_4$ and $NOHS_2O_7$). An X-ray analysis of 17 impactor samples of stratospheric aerosols obtained in 1976 using a U-2 aircraft flying at 15 to 20 km altitude revealed the presence of both forms of nitrososulfuric acid as well as $(NH_4)_2SO_4$ and $(NH_4)_2S_2O_8$. The estimates have shown that the maximum NO content which can be absorbed by stratospheric aerosols may range between $\frac{1}{3}$ and 2 with respect to the content of nitrogen monoxide in the environment. Proceeding from the fact that $NOHSO_4$ constitutes 18 to 19% of the precipitated aerosols, we obtain that from 1×10^7 to 5×10^7 kg per year of nitrogen monoxide can be removed at the expense of the formation of nitrososulfuric acid. Further efforts are needed, however, to verify the obtained results.

The problem of the origin and dynamics of the sulfate component of the stratospheric aerosol layer significantly affecting the Earth's radiation budget has evoked great interest in studies of the chemistry of sulfur compounds in the atmosphere. Carbonyl sulfide OCS had earlier been supposed to be an important source of sulfur for the stratosphere (apart from sulfur dioxide ejected during volcanic eruptions). As a result of OCS photolysis in the stratosphere, the atoms of sulfur had been released and then oxidized giving H_2SO_4. It had also been assumed that the reaction between OCS and OH could have been a source of the background tropospheric SO_2, and a similar reaction with participating CS_2 produced OCS and SO_2. Thus the reactions of OCS and CS_2 with hydroxyl can determine considerable sinks for these molecules in the troposphere.

In this connection, Cox and Sheppard (137) undertook measurements of the rates of reactions of hydroxyl with a number of sulfur compounds, including carbonyl disulfide at an atmospheric pressure of 10^5 mbar and a temperature of 297 ± 2 K. Results show that sulfur dioxide is a major product of reactions between hydroxyl and the sulfur compounds under investigation. The presence of hydroxyl in the troposphere determines the existence of a sink for all the gases reacting with it. At a rate of the OH + CS_2 reaction equal to 4.3×10^{-3} cm^3/mol·s, the lifetime of carbonyl disulfide in the troposphere constitutes 0.2 year. The reaction of CS_2 with OH also leads to the formation of OCS in the troposphere. Since the measured rate of the OH + OCS reaction is small ($<4 \times 10^{-14}$) and is still unreliably estimated, there is no basis for the conclusion that an intensive sink of carbonyl sulfide exists in the troposphere.

Crutzen and Schmailzl (49) analyzed new data on the budgets of odd oxygen, odd nitrogen, methane, and carbonyl sulfide obtained by numerical modeling with the use of the following: more reliable information about the extra-atmospheric spectral distribution of the UV solar radiation (wavelengths less than 300 nm), cross sections of the absorption by molecular oxygen (wavelength interval 200 to 240 nm), coefficients of the rates of reactions of the odd nitrogen destruction ($OH + HNO_3 \rightarrow H_2O + NO_3$: $OH + HNO_2 \rightarrow H_2O + NO_2$; $OH + HO_2 \rightarrow H_2O + O_2$), temperature dependence of the quantum yield of $O('D)$, resulting from ozone photolysis.

Calculations show that the rate of NO formation in the stratospheric air column by the reaction

$$N_2O + O(O'D) \rightarrow 2NO$$

constitutes $(1.1 - 1.9) \times 10^8$ molecules/(cm^2s), and the rates of losses for N_2O, CH_4, and COS are, respectively, $(0.9 - 1.4) \times 10^9$, 1×10^{10}, and 0.5×10^7 molecules/(cm^2s). An examination of the available observational data of the global-scale distribution of the N_2O and CH_4 content has led to the conclusion that at heights below 35 km the concentration of OH calculated using the latest data on reaction kinetics, is overestimated by a factor of 2. Most important, however, is the conclusion about the impossibility of conforming the available photochemical models with the observed ozone content, especially in the layer from 25 to 35 km (it is largely based on insufficient reliability and inadequacy of observational data). As a rule, calculations overestimate the rate of ozone formation in the very important layer from 25 to 75 km, greatly exceeding the values needed to explain the observed downward ozone flux to the troposphere.

It has been shown (49) that volcanic ejections of SO_2 to the stratosphere can cause a local increase in the ozone content and radiative heating of the atmosphere due to an intensified absorption of solar radiation by ozone, which can affect the dynamics of the clouds of erupted products, prior to their being subject to a large-scale diffusion and their propagation over a large stratospheric volume.

Therefore, studies of the field of the stratospheric ozone content after large-scale volcanic eruptions are of great interest from the viewpoint of testing the reliability of the photochemical theory for the stratosphere. Also, it is very important to study the impacts of the eruptive cloud on the chemistry of the stratosphere during the first days after the eruption, when it is still not subject to large-scale diffusion.

Sulfur dioxide strongly absorbs UV radiation, especially in the range 200 to 230 nm, where the absorption cross sections constitute 10^{-18} to 10^{-17} cm^2 per molecule. The SO_2 photolysis in this range triggers a set of reactions

$$SO_2 + hv \rightarrow SO + O; \qquad SO + O_2 \rightarrow SO_2 + O,$$

whose total result is the dissociation of oxygen: $O_2 + hv = 2O$ [the rate coefficient of the second reaction above is $k_{52} = 2.4 \times 10^{-13} \exp(-2370/T)$]. According to recent data on the kinetics of reactions, alternative reactions of SO with O_3 and NO_2 are less significant than the reaction with O_2. The considered succession of catalytic ozone producing reactions is, apparently, more important than the well-known reaction $O_2 + hv = O + O$ ($\lambda < 242$ nm), with a SO_2 mixing ratio exceeding 200 ppb (by volume).

Therefore, during the initial phase of the eruptive cloud (when there is a high concentration of SO_2) a large amount of ozone must be generated in the upper part of the cloud. On the other hand, it is important to consider the absorption of the UV radiation by O_3 and SO_2, as well as an accompanying intensification of a local radiative warming of the atmosphere, Q. With the assumed SO_2 mixing ratio in the eruptive cloud 10^{-4}, we obtain the values of ozone concentration and radiative heating given in Table 2.4

As shown, the fields of ozone concentration and radiative heating (temperature change) are greatly disturbed. Variations in the temperature field must affect the stratospheric dynamics and diffusion of the eruptive cloud. If during the eruptions, the stratospheric layer of sulfate aerosols is maintained (and intensified) at the expense of the eruptive SO_2, during normal periods this layer is maintained by COS from the troposphere.

In 1960 the U.S. Commission on Atomic Energy (later reorganized into the U.S. Department of Energy) initiated a program of air sampling at high altitudes. Subsequent analysis determined the rate of removal of radioactive pollutants from the stratosphere resulting from nuclear tests in the late 1950s and early 1960s. Beginning in 1971, some of the filters were washed after air sampling to reduce their sulfate content. However, some of the samples were analyzed for sulfate content, in support of studies of the stratospheric sulfate aerosol layer (the "Junge layer").

TABLE 2.4   Calculated Values of the Ozone Mixing Ratio and Radiative Heating (Q) in the Upper Part of the Eruptive Cloud, 6 Hours after Eruption

Height of Ejections (km)	O_3 (ppm)	Q (deg/h)
45	39.4	20
40	140	24
35	270	20
30	140	8
25	27	2
20	3.3	1

Sedlacek et al. (109) reproduced in detail and discussed the results of processing the entire data base accumulated during 10 years (errors in measurements of sulfate concentration constitute about ±25%). The concentration of sulfates is expressed as a mass mixing ratio in ppb (1 ng of sulfate per 1 g of air) and is determined for the layer from the tropopause up to 20 km. The air mass was sampled mainly during flights near the eastern coastlines of North and South America (75°N to 51°S).

The results obtained were compared with the chronology of volcanic eruptions. This comparison showed that during the decade several (7 to 12) eruptions took place either unnoticed or the erupted products reached much higher altitudes than originally thought. The e-fold decay time of the stratospheric aerosol content after the 1974 Fuego eruption was 11.2 ± 1.2 months. A 10-year-averaged contribution by the eruptive aerosol to the total background stratospheric aerosol content reached 58 to 62%. In addition, a 6 to 8% annual increase contributed by the anthropogenic aerosol is quite possible. No doubt, further continuous measurements of the stratospheric aerosol are needed.

2.2 VOLCANIC ACTIVITY ON OTHER PLANETS

In connection with the study of stratospheric aerosols in the Earth's atmosphere, the comparative planetological aspect of this problem is of great interest (138). A considerable amount of sulfur containing micron-sized particles (mainly as SO_4 sulfate) is observed in the atmospheres of the Earth and Venus. On Earth this aerosol is concentrated in a layer at an altitude of 18 to 20 km (in the lower stratosphere), with a typical number density of particles with diameter $D > 0.3 \ \mu m$, constituting 1 to 2 cm^{-3}. The stratospheric sulfate layer consists of droplets of the concentrated H_2SO_4 water solution and solid granules of sulfate salts (e.g., ammonium sulfate). In this case volcanic eruptions, industrial wastes, and biological processes serve as sources of sulfur.

The global cloud cover on Venus, located at an altitude of 45 to 75 km, contains particles with $D > 0.5 \ \mu m$ and a number density of about 100 to 300 cm^{-3}, which are mainly droplets of the concentrated H_2SO_4 water solution. Beneath the main cloud cover a haze layer (30 to 45 km) is located, with a particle number density of about 1 to 2 cm^{-3}.

As Settle (139) noted, the atmosphere of Mars lacks both the persistent cloudiness and the global aerosol layer of a long lifetime. There is convincing photogeological evidence, however, for volcanic activities on the Martian surface. Both the analysis of the cosmochemical models and data on volcanic processes on the Earth revealed the possibility of ejecting sulfur-containing gases into the Martian atmosphere as a result of volcanic activities occurring on the surface during earlier geological periods. (The ejected products should have been sulfur dioxide and hydrogen sulfide.)

Due to a stable wind convergence in the lower atmospheric layers observed in the Tharsis region, the sulfur-containing gases could have entered the upper layers of the Martian atmosphere (up to 20 to 30 km). Intensive circulation in the upper atmosphere had provided the global distribution of gases during a period of about 25 Earth days.

Surprisingly, similar photochemical processes of the sulfur dioxide conversion into the sulfate aerosol are observed in both the lower stratosphere of the Earth and lower layers of the Martian atmosphere (0 to 30 km). During the sulfur dioxide ejections to the present Martian atmosphere, the same chemical and microphysical processes should take place as in the lower stratosphere of the Earth. The lower layers of the Martian atmosphere in the equatorial and middle latitudes are somewhat cooler and exhibit comparable or higher humidity and higher concentration of condensation nuclei than those of the Earth's lower stratosphere. The latter two factors promote faster processes of conversion of SO_3 (gas) into SO_4^- (particles). However, a lower temperature slows down this process (168).

Mutual compensation of the two effects under consideration determines the comparability of the time scales of formation of sulfate aerosol in the lower Martian atmosphere and in the lower stratosphere of the Earth. Therefore, by analogy with the Earth, the principal factor limiting the formation of aerosols on Mars is, apparently, the oxidation of sulfur dioxide resulting from chemical reactions in which O, OH, and HO_2 take part.

Under conditions of the present Martian atmosphere, the formation of aerosols should take several thousand days. However, during persistent volcanic activities on the surface, the concentration of odd hydrogen (OH, HO_2) should increase considerably and, as a result, the time for aerosol formation shortens to several hundred days. The submicron aerosol particles formed in such a way can remain in the atmosphere for a long time, moving dozens of times around the planet before they are removed from the atmosphere by gravitational sedimentation—a major mechanism for removing the particles.

The sedimentation of aerosols onto the surface over a wide range of the equatorial and middle latitudes becomes possible because of their global distribution. The Martian volcanic sulfate aerosol should consist of the H_2SO_4 solution droplets, as well as the aggregates of droplets with inclusions of solid particles. The chemical activity of such aerosols deposited on the planetary surface should have stimulated the process of leaching the material of the planetary surface. Sulfate aerosol sedimentation can be considered a possible mechanism for the transport of sulfur to the regions of the Viking landing sites where sulfur compounds had been found in soil.

Numerical modeling of the general circulation of the Martian atmosphere has shown that the Chryse Planitia and Utopia Planitia regions are characterized by the presence of averaged downward vertical motions during most of the year, from which it follows that in these regions the prevailing sedimentation of the sulfate aerosol has taken place. A comparatively high concentration (by the Earth's standards) of sulfates in the surface layer on the Viking landing sites can be explained by the global or hemispheric propagation of aerosols.

Powerful volcanic ejections change not only the aerosol composition of the atmosphere but also its gas composition. It should be noted that with an intensification of volcanic activity, the amount of SO_2 injected into the stratosphere and troposphere grows. Sulfur dioxide is known to have strong absorption bands in the UV and IR spectral regions. Hence an increase in the SO_2 concentration decreases the planetary albedo, intensifies the shortwave radiation absorbed by the planet, and raises the effective temperature of the planet. Strong IR SO_2 bands lead to an intensification of the radiative cooling of the upper troposphere and to an enhancement of the greenhouse effect near the surface.

The H_2SO_4 droplets resulting from this conversion bring forth substantial changes in the content and optical properties of the total atmospheric aerosol: decreased absorption of shortwave radiation by aerosols and increased aerosol activity in the longwave spectral interval. The process of formation of sulfate aerosols is accompanied by a decrease in SO_2 concentration. This results in an increase in planetary albedo and a decrease in surface temperature.

Numerical modeling carried out by Kondratyev et al. (9) and Kondratyev and Moskalenko (18) showed that the radiative regime and surface temperature on the Earth and Mars could have changed largely as a result of variations in volcanic activities. Calculations of the evolution of climate and the greenhouse effect on the Earth with regard to the evolution of the atmospheric chemical composition have shown that in the initial stage of the Earth's evolution the volcanically caused greenhouse effect increases up to 7 K. In the period of maximum volcanic activity the optical thickness of the stratospheric aerosol cloud reached $\tau_a \approx 4.9$, variations in surface temperature being determined to a large extent by a correlation between the vertical temperature profile and the atmospheric moisture content, as well as by lower- and middle-layer cloudiness.

In the present atmosphere an intensification of volcanic activity leads to a decrease of the tropospheric temperature and an increase in the stratospheric temperature. The latter is partially caused by the absorption of the shortwave radiation by SO_2 and stratospheric aerosols, as well as by the absorption of thermal emission from the surface and the troposphere by stratospheric sulfate aerosols.

The stratospheric aerosol layer with an optical thickness $\tau_a = 0.3$ leads to an increase in the Earth's albedo of about 7%, an increase in the stratospheric temperature of about 6 K, a decrease of 10% in the global radiation, a decrease of 7% in the shortwave absorbed radiation, and a decrease of 3 to 7 K in the global surface temperature (depending on a manifestation of the accompanying feedbacks).

It is of interest to compare the climatic impacts of volcanic activities on the Earth and Mars in the course of their evolution. In contrast to the Earth, the atmosphere of Mars is more rarified and consists mainly of CO_2. However, during volcanic activity, the chemical composition of the Martian atmosphere had differed substantially from that of the present day, which could have led to substantial changes in the climate on Mars. The properties of the Martian lava resemble those of the basalt lava of the Earth, whose outgassing was followed by the release of a large amount of SO_2 and H_2S (the mixing ratio of sulfur compounds constitutes about 700 ppm). The concentration of sulfur compounds during the outgassing of the Martian lava should have been higher due to more intensive outgassing in conditions of low pressures near the Martian surface. We assume the mixing ratio for sulfur compounds to be 1000 ppm.

The lifetime of SO_2 for the present-day Martian atmosphere constitutes about 3.9×10^3 days (in the terrestrial atmosphere it is about 110 days). In still earlier periods the pressure and temperature near the Martian surface could have been higher. However, the lifetime of SO_2 had always been long, which appears to have been a cause of the accumulation of a considerable amount of SO_2 in the Martian atmosphere during volcanic eruptions.

During the eruptions, a large amount of volcanic ashes and water vapor had been ejected into the atmosphere. The latter had favored the formation of clouds of H_2SO_4 water solutions. The velocity of particle sedimentation depends on their size, and the time of sedimentation on the surface decreases rapidly with the growing size of particles. For example, particles of 5 μm fall out over a period of 30 days, and 1-μm particles over 4 years.

Since the time for the stratospheric sulfate aerosol formation on Mars constitutes about two Martian years, the long lifetime under conditions of the Martian atmosphere is possible only for submicron aerosols. The consideration of the Martian atmospheric circulation suggests that during the formation of submicron particles, the total mixing of aerosols took place within both the Northern and Southern Hemispheres on Mars. Thus the Martian sulfate aerosol during an eruption could have been a global-scale phenomenon, simultaneously affecting the climate on Mars.

The climatic impact of the greenhouse effect on Mars was calculated in approximation of the radiative-convective equilibrium. The mean global surface albedo was assumed to be constant and was estimated from the spectral albedo for Mars.

Major characteristic features of the radiative heat exchange in the Martian atmosphere for the period of volcanic activity are determined by the following processes: (i) strong absorption of the shortwave radiation by the atmospheric SO_2 in the wavelength interval 0.2 to 0.4 μm leads to a decrease in the planetary albedo and to the formation of a temperature inversion in the upper atmospheric layers; (ii) the formation of clouds from H_2SO_4 water solution particles promotes an increase in the planetary albedo and an intensification of the greenhouse effect at the expense of longwave radiation absorption in the bands of the H_2SO_4 water solution; and (iii) strong IR SO_2 bands intensify the greenhouse effect in the atmosphere and favor an increase in the atmospheric and surface temperatures.

Mean global Martian surface temperatures as a function of surface pressure P_s are listed in Table 2.5. The atmosphere of Mars is assumed to consist of CO_2, SO_2, and water vapor ($P_{SO_2} = P_{CO_2} \times 10^{-3}$). The surface albedo $A_s = 0.19$ and the atmospheric relative humidity does not exceed 0.5.

When the surface temperature of the planet increases, due to the SO_2 greenhouse effect, the moisture content in the atmosphere increases. As a result, the absorption of the thermal emission from the planetary surface and lower troposphere by its upper layers increases, and the greenhouse effect is intensified. Sulfur dioxide substantially influences the atmospheric greenhouse effect. Rainfall and snowfall become possible even at relatively low pressures near the planetary surface P_s 0.05 to 0.1 atm. Accumulation of SO_2 in large amounts shifts this level toward still lower pressures. During the period of volcanic activity, SO_2 had played a decisive role in the formation of the climate on Mars, governing the radiative regime of its atmosphere and regulating its climate.

TABLE 2.5 Mean Temperature T_s near the Martian Surface as a Function of P_s (CO_2)[a]

P_{CO_2}	A_s	$\omega(H_2O)$ (cm)	T_s (K)	T_s^* (K)
0.0065	0.205	1×10^{-2}	238	230
0.01	0.22	5×10^{-2}	245	233
0.025	0.235	0.15	254	238
0.05	0.25	0.25	263	244
0.1	0.27	0.4	272	252
0.25	0.29	1.0	285	268

[a] $\omega(H_2O)$ is the water vapor content in the vertical air column on Mars; T_s^* is the surface temperature in the absence of SO_2 in the atmosphere of Mars.

2.3 THE EL CHICHÓN VOLCANO ERUPTION

On 28 March 1982 at 23:30 local time an eruption of El Chichón volcano in Chiapas (a southeastern state of Mexico) took place. This stratovolcano of the late Pliocene or early Pleistocene, located at the east end of the Mexican neovolcanic zone in the current Chiapas volcanic arc, has exhibited only solfataric activity during a long period. Since Chiapas is located at a junction of the American, Cocos, and Caribbean tectonic plates, its volcanic activity is supposed to be caused by continuous shifting of the Cocos plate beneath the southeastern part of Mexico.

As Hoffer et al. (140) noted, during the first 10 days of volcanic activity, when three large-scale eruptions occurred, about 0.3 km^3 of andesite pyroclastic matter was ejected (without lava streams). The initial product erupted was tephra enriched with a crystalline component, with a larger amount of silicon and alkalines compared to the pyroclastic matter of the second and third eruptions. The initial tephra consisted mainly of juvenile substances, and then the eruptive product contained both the juvenile and lithoid fractions.

First eruptions were characterized by large amounts of ejected ash, moderate amounts of pumice, and smaller amounts of lithoid fragments. During the first phase of the eruption that continued until 2 April, large amounts of light gray ash were ejected which covered the adjacent area northeast of the volcano. The thickness of the ash layer reached 0.5 m at a distance of 15 km and decreased to 0.2 m at a distance of 75 km from the volcano. The ash layer in Villahermosa, the capital city of the adjacent state of Tabasco, was 0.1 m thick.

The second phase consisted of two large-scale eruptions (3 April, 19:33; 4 April, 05:36), when brown-gray ash with a large percentage of lithic thephra was ejected and propagated mainly to the east of the volcano. On 4 April at 10:30 the rate of settling for ash near Teapa reached 0.33 g/(m^2·s), which immersed this region in nearly total darkness, reducing visibility to 5 m. By 12:30 the rate of settling decreased to 0.5 g/(m^2·s). On this day, pyroclastic streams, consisting of hot ash and large blocks of pumice, moved down the slopes of El Chichón. The thickness of the tephra layer measured on 5 April near Paleco (125 km east of the volcano) exceeded 0.4 m.

The chemical analysis of 30 samples of tephra collected on 3 to 7 April at different locations revealed two types of tephra: (i) light gray matter with a high percentage of silicon (59%, on average) ejected on 28 March – 2 April, covered with tephra containing large amounts of lithic components (the April 3 and 4 eruptions); and (ii) basalt tephra with a high percentage of iron, the oxides of magnesium, and calcium.

As Robock (103) noted, the stratospheric post-El Chichón eruptive cloud on 4 April 1982 was apparently the most powerful cloud in the last century. This cloud caused a local temperature decrease of more than 5°C, and a further cooling down to 0.5°C was expected near the surface in 1984–1985. The interpretation of the unique data base of complex posteruption ground-based balloon, aircraft, and satellite observations opens up the possibilities of verifying the models of gas-to-particle conversion as a source of the stratospheric H_2SO_4 aerosol, as well as calculating the transport and the gravitational settling of particles and the impact of aerosols on the radiative regime and climate.

Robock (103) gave a brief review of the reports at two conferences held in late 1982, at which the following aspects of the problem were discussed: global-scale diffusion of the eruptive cloud, variations in its gas and aerosol composition, and possible impact of the eruption on weather and climate. The large size of the eruptive cloud permitted its continuous monitoring using satellite-derived imagery (following the Mount St. Helens eruption in 1980, the eruptive cloud could be identified only during the first 3 to 4 days).

The eruptive cloud from El Chichón made a turn around the globe for 21 days at an average speed of 22 m/s. Lidar soundings in Hawaii, over which the thickest part of the cloud passed on 9 April, gave values of the backscattering coefficient (determined with respect to the Rayleigh scattering) exceeding 200, which had never been observed before, with a maximum at the 26-km level. The eruptive cloud was stratified. Two months after the eruption the upper layer of the cloud covered the latitudinal belt 10°S–30°N. The lower layer (with a maximum concentration at a height of 20 km) propagated apparently over the entire globe.

Air and aerosol samplings from aircraft at altitudes up to 20 km permitted the chemical analysis of the cloud to be made. Aerosol sounding data contained information about the concentration, size distribution, and (indirectly) the composition of the aerosol, which turned out to contain 95% of H_2SO_4. From data of observations on the Hawaiian Islands an additional eruption-caused attenuation of the solar constant reached 20%, and of the net radiation 7%. An estimation of the cloud mass from its optical thickness gave 15 million tons.

The spatial propagation of the eruptive cloud was monitored in detail from satellite data (14). The first of four large-scale El Chichón eruptions happened on 28 March 1982 at 23:32 local time. An analysis of immediate GOES imagery in the visible (0.55 to 0.75 μm; spatial resolution 1 km) and infrared (10.5 to 12.5 μm; 8 km) spectral regions gave a minimum brightness temperature (BT) of -75.2°C of the eruptive cloud at 02:30 on 29 March. A comparison with radiosonde data showed that this temperature corresponded to the height 16.3 km (the height of the tropopause was 16.5 km;

hence the cloud top reached the tropopause level). By 08:00 the cloud started to dissipate, moving northeastward in the troposphere (with the westward component in the stratosphere); and the minimum BT rose to $-63.2°C$ (height 13.3 km).

By the end of 28 March the cloud had reached Cuba and covered an area of about 600,000 km^2. On 3 April at 02:30 the second eruption took place on a reduced scale from the first one. Minimum BT at 04:00 constituted $-71.2°C$ (15.1 km; the height of the tropopause 17.0 km). The ash cloud moved to the northeast and southwest, respectively, in the layers 13.7 to 16.3 km and 0 to 4.9 km. About 20:00 on 2 April the third eruption took place, and in about 10 hours (at 05:22 on 4 April) the fourth and most powerful one. According to NOAA-6 data, minimum BAT dropped to $-83.0°C$, which pointed to the penetration of the eruptive cloud into the tropopause.

The power of the 4 April 1982 El Chichón eruption made it possible to monitor the motion and dynamics of the eruptive cloud using the satellite-derived IR and visible imagery for at least 1 month (eruptive clouds formed after weaker eruptions of the volcanoes Mount St. Helens, Alaid, and Gallangung could be recognized only during 2 to 3 days). Robock and Matson (142) reproduced a set of subsequent images from GOES-E, GOES-W, and NOAA-7, starting from 5 April 1982, when the front part of the eruptive cloud made a full turn around the globe. The meridional motion was very weak, since the eruptive aerosol was concentrated in the latitudinal belt $10-20°N$. The average speed of the westward-moving front edge of the cloud during a period of 21 days constituted 22 m/s.

The presence of volcanic clouds was also manifested through an anomalous SST decrease, retrieved without account of the radiation attenuated by the eruptive aerosol, which also made it possible to monitor the motion of clouds. As Krueger (143) noted, of great importance for the analysis of the effect of eruptions on the atmosphere was the use of atmospheric remote sounding data obtained from the satellites Nimbus-7 and SME (Solar Mesospheric Explorer). Processing of these data revealed a strong SO_2-induced absorption in two shortwave channels of the ozonometric instruments.

Global mapping of the total ozone content without account of SO_2 absorption, using the 9 April data, revealed a vast zone of heightened values corresponding to an eruptive cloud. The SME visible spectrometer data showed a considerably increased atmospheric brightness at the horizon at the wavelength 0.5 μm in the region of the volcano. The volcanically induced disturbances of the brightness field were also observed at 1.27 and 1.87 μm from spectrometric data of the aurora observations and from IR radiometric data (6.3 and 9.6 μm). All of these data suggest that during the first 3 months, eruptive aerosol covered the $0-30°N$ latitudinal belt.

Although the upper boundary of the major eruptive cloud was about 6 km higher than the ceiling of the aircraft used for aerosol sampling, very important results were obtained, which characterized the properties of the secondary aerosol layers and the aerosol sediment from the major cloud. The following measurement instruments were used: the Knollenberg photoelectric counter, wire impactor, quartz cascade impactor, condensation nuclei counter, filter samplers, cryogenic cells for air sampling, and cells to measure the water vapor concentration from the Lyman-alpha radiation absorption. An important contribution was made by aircraft lidar soundings in July and October 1982. At Mauna Loa observatory, observations were made with the use of the airborne photometer, the Epply photometer and pyrheliometer, the Dobson spectrophotometer, and ruby lidar.

An analysis of the aerosol samples taken on 19 April over the Gulf of California showed a prevailing contribution by silicate particles covered with an H_2SO_4 layer, with a maximum number density at the diameter $2\,\mu m$, and a mass concentration of 1 to 4 mg/m^3. As a rule, ash particles were clusters of several particles. On 15 April the highest concentration of sulfates of about 160 ppb was recorded (compared with the background concentration of 1 to 2 ppb, by volume). The SO_2 concentration in April was low (50×10^{-12}), but in July at a height above 21.8 km the concentration exceeded 100×10^{-12}.

An unusually high COS concentration (350×10^{-12}) was registered on 5 May at a height of 18 km. The water vapor concentration at heights of 19 to 20 km is 5 ppm more than the background concentration (4.4 ± 0.4 ppm). The total SO_2 content in the whole eruptive cloud from the 4 April Nimbus-7 data was estimated at $(3-4) \times 10^6$ tons. During the period 9 to 20 April the cloud's optical thickness at $\lambda = 0.425\,\mu m$ varied between 0.6 and 0.75 and decreased to 0.25 in July.

The SME (Solar Mesosphere Explorer) satellite carried three devices to study the mesosphere: IR radiometer, and visible and IR spectrometers (both of them covering the wavelength interval 0.3 to 2.4 μm). They measured the concentration of various minor optically active components in the stratosphere and mesosphere. In particular, two IR scanning radiometers measured the distribution of atmospheric thermal emission at the limb in the bands of water vapor (6.3 μm) and ozone (9.6 μm).

After the 4 April 1982 El Chichón eruption, Earth's radiation field was disturbed by the eruptive aerosol cloud, which made it possible to use the SME data to monitor the process of gas-to-particle conversion and propagation of the eruptive aerosol. Most informative were data for channels 6.3 μm and 9.6 μm of the radiometer and 1.9 μm of the spectrometer, where there was practically no molecular scattering and therefore the aerosol scattering was clearly manifested. The vertical resolution of the observational data constituted 3.5 km, with a vertical reference error of 2 km.

Barth et al. (34) performed an analysis of data for three orbits for the latitudinal belt $30°W - 120°W$ (the shift of successive orbits was $24°$ longitude, and the latitudinal resolution $5°$). This analysis showed that the combined effect of the 21-day westward round-the-globe transport and mixing had determined the formation of an Earth-encircling aerosol cloud by the first week of June in the latitudinal band from the equator to $30°N$. According to the available data, about $(3-4) \times 10^{12}$ g of sulfur dioxide were ejected into the stratosphere (up to 30 km) on 4 April 1982. In the process of westward round-the-globe motion SO_2 reacted with odd hydrogen and odd oxygen residing in the stratosphere, giving H_2SO vapors and then H_2SO_4 water droplets (through homogeneous nucleation or condensation on nuclei). The growing size of droplets intensified the gravitational settling until its effect exceeded the rate of the growth of particles (at a height of 27 km this equilibrium state was reached 8 weeks after the eruption).

Apparently, an initial maximum of the eruptive aerosol concentration at a height of 27 km was determined by SO_2 ejected to this level, but it could also have been governed by a rapidly intensifying photodissociation of SO_2 above 27 km. A slower rate of the gas-to-particle conversion at lower altitudes could be connected both with slow formation of H_2SO_4 due to a lower concentration of odd hydrogen as well as with the specific initial vertical profile of the SO_2 concentration. From observational data, by the end of May, SO_2 ejected to the stratosphere was mainly used in the formation of H_2SO_4 aerosol, whose concentration started decreasing and the level of its maximum lowered, reaching a height of 20 km in 24 weeks.

Patterson et al. (92) performed laboratory measurements of the optical properties of three samples of El Chichón ash falling out at a distance of 12 to 80 km from the volcano. The ash consisted of pumice (in the form of nonmineral and glassy particles) and rock particles. The share of glass in pumice reached 80% (the phenocrysts that constitute glass are mainly andeside and hornblende with apatite, augite, and anhydrite). Rock particles are characterized by the following proportions: 20% glass, about 50% plagioclase, and 30% hornblende. Basic components of the insoluble part of ash are SiO_2 (59%) and Al_2O_3 (18%).

The chemical composition of ash ejected by different eruptions of El Chichón turned out to be similar. The samples examined had ions Ca^{2+}, SO_4^{2-}, Na^+, K^+, Mg^{2+}, HCO_3^-, and Cl^-. Table 2.6 illustrates the mineralogical composition of samples (values of the refractive index are given in parentheses). Each sample consisted of dark gray ash and some darker material.

Measurements of the imaginary part of the complex refraction index n_{IM} gave very small values of the order of 0.001 at a wavelength of about 0.5 μm and revealed a weak wavelength dependence. The optical parameters of sample AM 101 taken at a maximum distance from the volcano (80 km) happened to be close to the respective parameters of the stratospheric aero-

TABLE 2.6   Mineralogical Composition of Three Samples of Volcanic Ash (%)

Sample	Feldspar Minerals (~1.56)	Volcanic Glass (1.52–1.53)	Colored Minerals (1.7)	Opaque Particles (Magnetite?)
AM 125	50	20	25	5
AM 106B	45	30	20	5
AM 101	30	65	5	—

Source: Ref. 92.

sols. The results obtained show that the complex refraction index for stratospheric silicate ash is 1.53 to 0.001i.

Recent observations suggest that the erupted gases and particles are a major source of an abnormal enrichment of the background aerosol in remote regions (e.g., in the Antarctic). In this regard, on 3 November 1982, Kotra et al. (144) carried out measurements of concentrations of various gaseous components and aerosols in the eruptive plume from El Chichón using the NASA flying laboratory — the Electra-429. In the period of observations hydrogen sulfide was a major gaseous sulfur compound in the plume.

Using the neutron activation technique, 29 elements were found in the filter and impactor aerosol samples. The visually observed plume reached a height of 0.7 km, width of 3 km, and length of 1 km, extending in the wind direction.

These observational data were used to estimate the injections to the atmosphere of minor gaseous components, normalized against total injections of sulfur compounds. The results obtained revealed an enrichment with such volatiles as sulfur, chloride, arsenic selenium, bromine, antimony, iodine, tungsten, and mercury, with respect to the basic pyroclastic, by a factor of 60,000 to 20,000. Arsenic, antimony, and selenium were concentrated mainly in small particles less than 3 μm in diameter. Calcium and natrium occurred only in large particles, and the distribution of aluminum and magnesium was bimodal. The composition of ash particles ejected to the atmosphere and reaching the stratosphere during a large-scale eruption (the aircraft-derived May 1982 data for a height of 18 km) was characterized by a 10-to-30-fold increase of the content of some elements as compared to ash ejected to the atmosphere during the quiet period of volcanic activity.

In the periods before and after the eruptions of El Chichón (23 March – 17 December 1982) Wilson et al. (145) measured the size distribution of stratospheric aerosols at altitudes from 19.6 to 21.6 km with the condensation nuclei counter (an estimation of the number density of particles larger than 0.01 μm in diameter), wire impactor, and Knollenberg photoelectric counter (the diameter range of 0.1 to 3 μm), carried by the U-2 aircraft. The aerosol concentration field in the regions not affected by the eruption (the flights were made along the Pacific coastline of the United States) turned out to be homogeneous to a large extent. The largest contribution to the aerosol number density before the eruption (six to seven particles per cubic centimeter) had been made by particles less than 0.1 μm in diameter.

Since the characteristic time for coagulation of such particles constitutes about 9 months, they have probably been a product of the 1981 Nyamuragir volcano eruption, which is considered to be a major factor in the appearance of the "mysterious" volcanic cloud. A high concentration of posteruption submicron particles (April–May) reflected the effect of the posteruption process of gas-to-particle conversion. Measurements made in November and December revealed a decreased content of particles less than 0.1 μm in diameter compared to the preeruption conditions. Results of simultaneous measurements of the SO_2 volume concentration and aerosol near the lower boundary of the eruptive cloud (2 weeks after the eruption) made it possible to estimate the rate of transformation of SO_2 to the H_2SO_4 aerosol within the range 15 to 1200 molecules/cm^3.

During the summer, fall, and winter following the eruption of El Chichón, Gandrud et al. (146) obtained filter samples of sulfate aerosols using the high-altitude U-2 aircraft (in the region of the U.S. western coastline) and balloons (near 33°N) at altitudes of 15 to 28 km. An analysis of samples showed that the values of the posteruption sulfates mixing ratio is about two orders of magnitude greater than the background levels. The processing of aircraft filter samples gave results consistent with those of synchronous impactor samplings, and calculations of the total content of sulfates in an air column agreed with results of aircraft lidar soundings, Nimbus-7 data, and results of balloon-borne photoelectric counter samplings.

From data of the first series of the 23 July 1982 aircraft sounding (25–36°N, near 129°W) and 5 and 17 August balloon sounding, a maximum sulfate mixing ratio was observed at a height of about 23.5 km and constituted 78 ppb (by mass). The final series of aircraft (20–21 January, 35–50°N) and balloon (23 January) soundings gave a maximum of 500 ppb at a height of 20 km. The total content of sulfates in the layer from 15 to 29 km varied (from aircraft data) between 0.017 and 0.24 g/m^2, whereas the background value in 1976 was 0.00096 g/m^2; the posteruption content of sulfates increased 18 to 25 times. Taking into account available data, the content of sulfates in spring 1982 at 33°N latitude averaged 0.020 to 0.021 g/m^2.

Vedder et al. (147) discussed results of the gas chromatographic analysis of the SO_2 content in air samples taken from U-2 and ER-2 aircraft between 16 April and 13 December 1982 at altitudes of 15 to 22 km, whose flights had been planned to take into account the evolution of the eruptive cloud from

El Chichón, which covered the region $23-52°N$, $108-130°W$. Measured values of the SO_2 mixing ratio varied within 8 to 132 ppb (by volume), with an error of $\pm23\%$. Mixing ratios registered before the eruption constituted $(20-170) \times 10^{-12}$. The level 50×10^{-12} was considered as the background level. Data of aerosol measurements indicate that the results of the first flights refer to the periphery of the volcanic cloud, and later flights refer to the central part of the cloud, but in the period of its evolution, when most of the erupted SO_2 was transformed into the H_2SO_4 aerosol.

From 4 December 1982 until 4 January 1983, using the flying laboratory Convair-990, Dutton and DeLuisi (148) performed measurements of the spectral transparency of the atmosphere above the aircraft (against the sun) with two calibrated filter photometers operating at 0.368, 0.500, 0.675, and 0.778 μm (the first photometer), and 0.380 and 0.500 μm (the second photometer). The aircraft flew at a level of 250 to 200 mb along the 120°W meridian in the 5°S-55°N latitudinal belt.

The use of observational results to calculate the aerosol optical thickness (AOT), determined as a vertically integrated product of the mass coefficient of aerosol extinction by the mass aerosol concentration, revealed a relative maximum near 5°N and relative minimum near 25°N (in the northern latitudes, AOT started increasing after a relative constancy in low latitudes). A volcanically induced averaged (over wavelengths and in time) AOT increase reached 0.10 ± 0.01 (the background aerosol was practically absent, constituting only 0.005).

Estimates showed that such an AOT increase must cause an 18% drop of direct solar radiation extinction at the level of the tropopause (for the representative solar zenith angle 60°). Data on the AOT latitudinal and spectral variability point to the existence of two specific latitudinal zones characterized by different spectral signatures of aerosol, which is determined by different size distributions of aerosol in the southern and northern latitudinal bands caused mainly by the aerosol transformation in the process of its transport.

Wittenborn et al. (149) performed measurements of the IR atmospheric transparency (against the sun) in the wavelength interval 8 to 13.5 μm from an altitude of about 11 km using the Convair-990 spectrometer with a wedge interference filter. (The flights were made in December 1982 in the latitudinal band of 5°S-50°N.) An analysis of registered transmission spectra revealed not only the CO_2 and O_3 bands but also an absorption band near 8.5 μm that can be identified as that of the erupted H_2SO_4 aerosol. The share of absorbed radiation in the 10–41°N latitudinal band constituted 0.02 per unit air mass, about half this absorption being determined by a contribution from aerosols. An estimate of the ratio of optical thicknesses of the eruptive cloud at 8.5- and 0.5-μm wavelengths gave 0.14 at 20°N in December 1982.

This value corresponds to the modal radius of particles $0.4\,\mu$m, which agrees with data of direct measurements of the aerosol size distribution.

Dutton and DeLuisi (150) considered the results of a determination of the atmospheric aerosol optical thickness from the spectral atmospheric transparency observed at four surface stations located in various geographical regions, before and after the El Chichón eruption (as well as from some aircraft measurements). At Mauna Loa Observatory the eruption-caused evolution of AOT was detected 5 days after the eruption. At the South Pole and at Barrow (Alaska) stations the appearance of the volcanic cloud was observed in 8 to 12 months; continuous observations were impossible because of cloud and low sun elevation. In American Samoa it was impossible to detect an eruptive signal because of the large variability of background aerosol in the maritime atmospheric boundary layer.

Aircraft observations near the tropopause made it possible to observe directly the evolution of stratospheric aerosols. Data of AOT observations at 0.5-μm wavelength at Mauna Loa for the 2.5-month period after the eruption and for the next year revealed at the beginning of the period a large variability of AOT caused by spatial inhomogeneity of the eruptive cloud. Maximum values of AOT were 20 to 30 times greater than the background values. A large decrease of this variability and the AOT took place by December 1982.

Even at the South Pole, during the last 3 months of 1982, an increase of AOT by 0.025 to 0.030 was observed. A volcanically caused increase of AOT also took place in Barrow (Alaska), although observations there were possible only from mid-March to mid-May. Aircraft measurements in the interval $70°$N–$55°$S made between October 1982 and May 1983 revealed a maximum of AOT at the altitude of the volcano and a minimum near $20°$N.

After a series of El Chichón eruptions on 28 March and 3 and 4 April 1982, 18 aerosol soundings were made in Laramie, Wyoming ($41°$N) and in southern Texas (27–$29°$N) to study the mechanism of gas-to-particle conversions that determine the transformation of erupted SO_2 ejected to the atmosphere into the concentrated H_2SO_4 solution droplets (151–154). Previous observations made after the eruptions of Mount St. Helens (1980) and Alaid (1981) volcanoes revealed a continuous formation of both submicron droplets ($r \sim 0.02\,\mu$m) through gas-to-particle conversions during about 1.5 months after the eruption and of very large droplets ($r \sim 1\,\mu$m). The latter made the largest contribution to the formation of the backscattering signal during lidar soundings.

These observational results were obtained using various instruments that measured the number density of condensation nuclei ($r \sim 0.01\,\mu$m; devices are used beginning in 1973), particles with radii $r \geq 0.15\,\mu$m and $r \geq 0.25$ μm (aerosol radiosonde; used beginning in 1971), and large particles $r \geq$

0.25; 0.95; 1.2; 1.8 μm); this sensor, used from 1981, could measure the concentration of particles down to minimum values 1 cm^{-3}).

Since the post-El Chichón stratospheric aerosol at altitudes above 20 km during the first 4 to 5 months was located in the band 0–30°N, along with aerosol soundings in Laramie, soundings were made in southern Texas on 18–19 May in Laredo (27.3°N), 21 August in Sinton (27.8°N), and 23 October in Del Rio (29.2°N). In the latter case all four devices were launched in one balloon and an experiment was made with the samples warmed to determine the boiling point for the aerosol.

Analysis of observational data showed that there were always two major aerosol layers; the lower layer located between the tropopause (heights 12 to 16 km) and an approximate level of 21 km, and the center of the upper 5-km layer was near 25 km (Fig. 2.5). Typical values of the number density of particles in these layers constituted about 10 cm^{-3} ($r > 0.15\,\mu$m) during the entire period from May to October, but the concentration of large particles ($r \geq 1\,\mu$m) decreased at the time from ~1 to 0.1 cm^{-3}.

The aerosol layers were always divided by an unusually pure atmospheric layer, without any vertical mixing between them. The horizontal development of the upper layer was limited, but the lower layer had extended to Laramie by June. The 18–19 May sounding data revealed high concentrations of both condensation nuclei (CN) and large particles in both layers, which points to the process of continuous formation of new particles with a radius (r) of 0.01 μm as a result of gas-to-particle conversions, since the time for coagulation of such particles (their concentration constitutes about 750 cm^{-3}) is less than 5 days.

Ground-based measurements of the atmospheric spectral transparency made by Lockwood (155) in Flagstaff, Arizona, during a week in mid-May in the wavelength interval 330 to 850 nm gave maximum values of the optical thickness of the aerosol cloud in the visible of about 0.3 (this means that soundings in Laredo refer to conditions of thickest aerosol layers in the stratosphere). Forty-five days after the eruption the aerosol size distribution in the layer 24.5 to 25.5 km was bimodal with modes, to which radii of about 0.02 μm (new particles) correspond, as well as 0.7 μm (enlarged particles due to the condensation of H$_2$SO$_4$ vapors existing before the eruption, when their modal radii constituted about 0.08 μm).

Although the number density of submicron aerosols is much higher than that of large particles, the latter contribute most to the total mass concentration of aerosols (35 μm/cm^3 compared to 1μg/cm^3). The 21 August data of soundings in Texas revealed a relatively low concentration of the smallest particles in both cases, indicating that the process of formation of new particles had ceased by that time. These data revealed the existence of a thin aerosol layer at altitudes of 31 to 33 km.

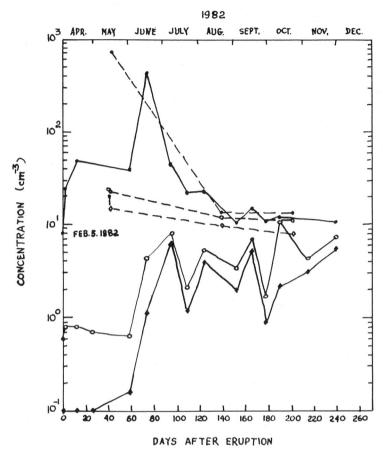

FIGURE 2.5. Dependence of aerosol number density near a level of maximum concentration above 20 km from observations at Laramie (solid lines) and in southern Texas (dashed lines) after the eruption of El Chichón. •, $r \geq 0.01$ μm; ○, $r \geq 0.15$ μm; ◇, $r \geq 0.25$ μm.

Results of the 23 October soundings in Texas were similar to those for 21 August. Since from the 25-km height large particles (~ 1 μm) drop by 1 km every 10 days, the persistent concentration of these particles points to continuing growth of particles due to the accretion of H_2SO_4 vapors, which requires a concentration of H_2SO_4 not lower than $(3-4) \times 10^{-7}$ cm^{-3} (about 0.05 ppb). According to data of soundings in Laramie, Wyoming, the formation of the gas-to-particle conversion aerosol ($r \sim 0.01$ μm) ceased after June, when a rapid decrease of particle concentration began, typical of the process of coagulation. By December 1982 the aerosol concentration observed in Laramie was approximately the same as in the latitudinal belt

27–29°N and was characterized by a relatively larger contribution by particles with a radius of $r \geq 0.25$ μm than by particles with $r \geq 0.15$ μm.

Calculations of the aerosol mass in an air column above the prescribed level, made on the supposition that particles are droplets of a 75% H_2SO_4 water solution (density 1.65 g/cm^3), revealed a rapid initial decrease of the aerosol mass in Texas due to the transport and gravitational settling of particles, and at Laramie a slow growth of the aerosol mass was observed as erupted material got into the atmosphere (Fig. 2.6).

In May, when according to satellite data the spreading of aerosols was confined to the latitude belt 0–30°N, the aerosol mass in air column above 15 km at 27.3° latitude constituted 0.18 g/cm^2 (about 100 times greater than before the eruption), and in the entire atmosphere within 0–30°N it reached

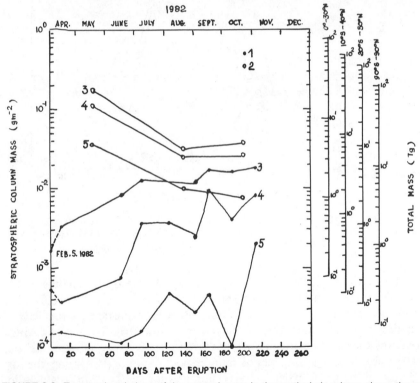

FIGURE 2.6. Temporal variations of the aerosol mass in the vertical air column above three altitudes, from observations at Laramie (•) and in southern Texas (○) after the eruption of El Chichón. The scale on the right makes it possible to estimate the aerosol mass (1 Tg = 10^{12} g) in different latitudinal belts. 3, $r > 15$ μm; 4, $r > 20$ μm; 5, $r > 25$ μm.

nearly 23 megatons (apparently, this was an upper limit, since aerosols were nonuniformly distributed).

Based on data of aircraft lidar soundings, by October the upper aerosol layer covered the latitudinal belt at least from $10°S$ to $40°N$, and the lower layer from $20°S$ to $50°N$, with the aerosol mass in the atmosphere above 20 km reaching 5.5 and 2.3 megatons, respectively (5.1 megatons for the lower layer at altitudes above 15 km). The total aerosol mass in both layers is 8.3 megatons, from which it follows that the initial mass apparently exceeded 10 megatons (but was below 20 megatons). By December the effect of gravitational settling had not still dominated the process of particle growth, making it possible to estimate the lifetime of erupted stratospheric aerosol.

The heating of the input aperture of the photoelectric counter up to $150°C$ (performed in October) at a slow descent of the balloon ($29°N$) showed that $\geq 99\%$ of the upper layer particles and 98% of the lower layer particles were volatile (or had a nonevaporating nucleus with $r < 0.15\ \mu m$). This suggests that the bulk of the upper layer is composed of large droplets of a 80% H_2SO_4 water solution, and that of the lower layer, droplets with an H_2SO_4 concentration of about 60 to 65% (different concentrations are explained by the different temperatures of the layers). These estimates are in agreement with the theory of stratospheric sulfuric acid aerosol formation (121–124).

From observations at Mauna Loa 2 days after the 4 April eruption the maximum SO_2 content in the atmosphere was equivalent to a mixing ratio of about 5 ppm, assuming that the SO_2 was concentrated in the 5-km layer near 25-km altitude. Faster growth of particles at 25 km, than at 18 km (although in the latter case less H_2SO_4 is required) can be explained by the accelerated SO_2 oxidation reaction resulting from a higher temperature at a 25-km height and (or) higher concentrations of H_2SO_4 in the upper layer. Probably, the existence of two aerosol layers is explained by the view that the lower one formed as a result of weaker ejections of sulfur on 28 March and 3 April, and the one resulted from the most powerful 4 April eruption. This conclusion does not contradict the results of analysis of heights and motion of eruptive clouds using satellite imagery.

Aerosol soundings were made by Hofmann and Rosen (156) in Laredo ($26.3°N$) and Del Rio ($29.2°N$) on 19 May and 23 October (1.5 and 7.5 months, respectively, after the 4 April El Chichón eruption). An analysis of the data by Hofmann and Rosen (156) showed that the distribution of aerosol by radius $n(r)$ in the layer of maximum concentration (24.5 to 25.5 km) is bimodal and well approximated by the lognormal function

$$n(r)\ dr = \sum_{i=0}^{1} (N_i/\sqrt{2\pi})\exp(-\alpha_i^2/2)\ d\alpha_i.$$

Here N_i is the total number density of the ith mode; $\alpha_i = -\ln(r/r_i)/\ln \sigma_i$, where r_i is the ith modal (median) radius; and σ_i the ith (nondimensional) modal width. The value of $\int_0^r n(r)\,dr$ is the total number of particle $N(r)$ with a radius exceeding r.

From the 19 May observations, the aerosol of fine (large) size is characterized by the following parameters: $N_0(N_1) = 150(4\ \text{cm}^{-3})$; $\sigma_0(\sigma_1) = 2.80$ (1.77); $r_0(r_1) = 0.02\ (0.72\ \mu\text{m})$; $m_0(m_1) = 1\ (35\ \mu\text{g/m}^3)$.

From observations at Laramie from February to April 1982, the following values correspond to the background aerosol (before the eruption): $N = 5$ cm^{-3}, $\sigma = 1.60$, $r = 0.08\ \mu\text{m}$, and $m = 0.05\ \mu\text{g/m}^3$ (where m is the mass concentration of aerosols).

On 23 October, 7.5 months after the eruption, the following values were registered: $N_0(N_1) = 10\ (0.2\ \text{cm}^{-3})$, $\sigma_0(\sigma_1) = 1.50\ (1.10)$, $r_0(r_1) = 0.27\ (1.0$ $\mu\text{m})$, and $m_0(m_1) = 2.9\ (1.5\ \mu\text{g/m}^3)$.

Thus modal radii of volcanic aerosol particles are 0.02 and 0.7 μm, the concentration of large-size particles being nearly the same as before the eruption. Hence the conclusion can be drawn that these particles resulted from the growth of old particles, whereas fine-size particles formed after the eruption is a product of gas-to-particle conversion. Regular aerosol soundings at Laramie showed that there was a gradual posteruption growth of the aerosol layer at a height of about 18 km, and the upper (25 km) aerosol layer started to appear sporadically in July and reached some stability by the end of 1982.

The 23 October sounding in Del Rio revealed an almost complete absence (ceased formation) of smallest particles ($r \sim 0.01\ \mu$m), which caused an increase of the modal radius of the small-size fraction up to 0.3 μm. Simultaneously, large-size particles grew, which manifested through an enlargement of the model radius from 0.7 to 1 μm and a 20-fold decrease of their number density (probably because of gravitational settling). Stratospheric aerosol mass concentration decreased eight times, with the main contribution to mass from the 0.2-μm mode particles.

Six to eight months after the Mount St. Helens (1980) and Alaid (1981) eruptions, a secondary 1-μm mode was also observed, with the main mode remaining consistent with the preeruption one of about $\sim$0.08 μm. This means that following the eruption of El Chichón, the aerosols formed consisted of much larger particles, and hence the total mass reached about 10 megatons in 6 months (as opposed to a total aerosol mass, following the Mount St. Helens eruption of 0.25 megatons).

Recent theoretical studies and observations have led to the conclusion that the atmosphere at altitudes of 25 to 30 km is characterized by an almost saturating concentration of water vapor of about 10^5 to 10^6 mol/cm^3, preserved in volcanically quiet periods due to photo-dissociation of carbonyl sulfide in the stratosphere. After eruptions, when the stratosphere gets large amounts of SO_2, a strong supersaturation occurs with respect to H_2SO_4, and favorable conditions are created for the formation of the sulfuric acid aerosol: (i) provided that the supersaturation is large enough (the sulfuric acid concentration at 25 km exceeds 10^7 molecules/cm^3), a homogeneous nucleation and formation of $H_2SO_4-H_2O$ droplets takes place without participation of condensation nuclei; and (ii) if there is an accretion of vapor on already existing particles, which stimulates their growth. An intermediate process (with respect to those mentioned) is associated with possible condensation on numerous ions and ion clusters.

Since the final product (small droplets of the sulfuric acid solution) does not suggest its mechanism of formation (either homogeneous or ion nucleation), the complex of aforementioned processes is usually considered as one process. The process of formation of new droplets in supersaturation conditions is very fast and intensive, which determined a high initial concentration of particles ($\geq 10^3$ cm^{-3}). On the other hand, however, coagulation of particles determined by their square-number density is also very intensive. The coagulation-induced growth of droplets causes a rapid decrease of their number density, but in this case the lifetime of droplets increases. The lognormal size distribution of aerosols is formed.

Consideration of the growth of particle radii observed from May to October (from 0.08 to 0.72 μm) gave the H_2SO_4 concentration 2×10^7 molecules/cm^2, which exceeds the background concentration at a 25-km height at least 100-fold. It follows that the rate of the H_2SO_4-H_2O homogeneous nucleation at a temperature of $-85\,°C$ varies within 10^3 to 10^{-3} cm^{-3}/s, with the water vapor mixing ratio growing from 3.5 to 10 ppm (by volume). This is equivalent to the formation of 10^2 to 10^6 particles/cm^3 per day and corresponds to the conserved maximum concentration of particles with $r \sim 0.01$ μm of 750 cm^{-3}.

The time constant that determines the loss of H_2SO_4 vapors spent on the formation of particles was estimated at 5 min. This is approximately the same time constant that characterizes the gas-phase reactions of continuous transformation of SO_2 into H_2SO_4 vapors. Estimates for previous eruptions gave a time constant for the eruptive aerosol dissipation (an e-fold decrease of concentration) of about 10 months. Since in December 1982 there was no marked extinction of eruptive aerosol layers after the eruption of El Chichón, one may expect that climatic consequences of this eruption would be more substantial than those observed earlier. It is not excluded that from the viewpoint of the long-term stratospheric effects of this eruption, this eruption is the largest for at least the last 100 years.

Using data of numerous balloon soundings at Laramie (41°N) and soundings on four occasions in southern Texas (27–32°N), Hofmann and Rosen (152) analyzed the mechanisms of SO_2 to sulfuric acid droplet conversion in the stratosphere following the eruption of El Chichón.

Data on condensation nuclei at Laramie show two periods of new aerosol nucleation during the first 100 days after the eruption and in early 1983. The first period is probably due to sulfuric acid vapor supersaturation. As for the other period, the mechanism of particle production is connected with thermal nucleation of aerosols in polar regions from eruptive H_2SO_4 vapors ejected to high latitudes and supersaturated, being rapidly cooled.

The initially pronounced bimodality of aerosol in the layer at altitudes of about 25 km (modes of 0.02 and 0.7 μm) can be considered to be a result of new particle nucleation from the gas phase and the intensive growth of preeruption stratospheric particles.

2.4 THE OPTICAL PROPERTIES OF VOLCANIC AEROSOLS

The need to estimate the effect of stratospheric aerosols on the radiation budget of the surface-atmosphere system and of a priori information about aerosols in interpreting data of remote optical soundings determines the urgency of reliable complex data on the optical properties of aerosols (1,11,16,17,44,48). In this connection Grams and Rosen (157) made a brief review of information about the concentration and size distribution of aerosols, the shape of particles, and the complex refraction index. They gave estimates of phase functions for $m = 1.5 + 0.005i$ for different values of the size parameter $x = 2\pi r/\lambda$ (r is the particle's radius, λ the wavelength). Comparison of the phase functions for the monodisperse and polydisperse aerosols illustrates the smoothing out of the angular fine structure of phase functions due to polydispersiveness. Calculations reveal a very strong decrease in the scattering cross section at angles exceeding 15° with the increasing imaginary part of m (the absorption index).

Clarke et al. (158) performed laboratory measurements of the aerosol absorption coefficient at wavelengths 450, 550, 560, and 800 nm using samples on a glass plate covered with silicon rubber, with subsequent application of the integrating plate technique (the absorption coefficient b_a is determined from the transmission of light by a sample as compared to that by a pure plate). Results of measurements for the samples taken by the U-2 aircraft on routes about 1000 km long (to accumulate a sufficient mass of aerosol) before, during, and after the eruptions of Mount St. Helens and El Chichón showed the presence of an absorbing component in aerosols in all cases.

Typical values of b_a at 550 nm vary from 10^{-9} to 10^{-8} m^{-1}, except for cases when the samples were taken in the most dense parts of the eruptive clouds, immediately following the eruptions. Despite a large variability of the absorption coefficient, its post-eruption values exceed those for early 1979. Data of the 5 May 1982 flight near the U.S.-Mexican border at a height of 21 km gave 8 μg/m^3 for the aerosol mass concentration, about half the mass falling on particles smaller than 5 μm in diameter.

The scattering coefficient was estimated at 1.2×10^{-5} m^{-1}, and the single-scattering albedo varied within 0.875 to 0.958 (apparently, ash is characterized by a value greater than 0.988). The imaginary part of the complex refraction index was about 0.0034. Estimates for the period following the eruption of El Chichón have led to the conclusion about a volcanically induced increase of the global albedo (if the single-scattering albedo for the aerosol layer exceeds 0.94), and a decrease of tropospheric temperature (if the single-scattering albedo exceeds 0.98).

Rao and Bradley (98) discussed results of observations of the total (0.29 to 2.9 μm) direct solar radiation and global radiation at Corvallis, Oregon, in late November and December 1982, when the stratospheric dust cloud from the eruption of El Chichón reached midlatitudes, manifested itself through the anomalous coloration of skies at dusk on 30 December. Data of observations (instantaneous and mean annual values) in clear skies and without local pollution at near-noon time were considered.

The results obtained show a considerable effect of volcanic aerosol on the shortwave radiation field near the surface. For example, the atmospheric transmission coefficient for the global (direct solar) radiation decreased by about 13% (27%). The scattered-to-global radiation flux ratio increased by a factor 2.25. Results of observations of direct solar radiation at several adjacent locations showed its decrease varying between 18 and 25%. Similar transformation of the incoming shortwave radiation was observed after the Agung eruption in 1963.

Shortly after the 18 May 1980 Mount St. Helens eruption, aircraft lidar soundings were made in the eastern United States, and the results were compared with data of ground-based lidar soundings. Results considered by McCormick (159) made it possible to analyze the global distribution of eruptive aerosols. During the first flight (21 May), when an Nd laser was used (1.06 μm), a value of 107 was registered for the backscattering ratio (compared to the Rayleigh backscattering signal) from a 1-km aerosol layer at a 13.6-km height. Similar aerosol layers at altitudes of 12 to 14 km, characterized by large horizontal inhomogeneity, were observed on 16 and 27 May.

Apparently, the composition of the initial eruptive cloud in the lower stratosphere was mainly silicate. The ratios of backscattering signals at two wavelengths (ruby and Nd lasers) and depolarization values for the lower stratospheric aerosol (12 to 18 km) were reduced to background levels 1 month after the eruption, which points to a rapid conversion of particles into spheres with a usual refraction index. At heights above 20 km the aerosol moved slowly westward, whereas the bulk of erupted material moved eastward with the speed and direction of motion, depending very much on height. The westward-moving aerosol was discovered by lidar stations in Japan and later by European and American stations. The circumglobal motion of the eruptive cloud took 60 days.

Data of aircraft lidar soundings in the United States in early September showed a smoothed spatial distribution of aerosols manifested in the formation of a vertically extended layer (14 to 21 km) with a maximum near 18 to 19 km and a moving layer at altitudes 21 to 22 km. In the case of ruby lidar, maximum values of the backscattering ratio were 1.3 to 1.5 Apparently, by September the formation of the backscattering signal had almost completely been determined by sulfuric acid aerosols. Total production of sulfate aerosols due to gas-to-particle conversion was estimated at 5×10^5 tons, which corresponds to a doubled mass of background aerosols and is much below the aerosol production that followed other eruptions: 3×10^7 tons (Agung, 1963) and 5×10^7 tons (Fuego, 1974).

From observations of the outgoing short- and longwave radiance (OSR and OLR) using the Nimbus-7 AVHRR, Schwedfeger et al. (107) retrieved the spectral albedo of system A (channel 0.58 to 0.68 μm) and total OLR flux (channel 11.5 to 12.5 μm) under clear-sky conditions for February to September 1982 (a total of 44 days) in the Hawaiian region, where ground-based observations of atmospheric transparency were made at the Mauna Loa Observatory. Data for the channel 0.58 to 0.68 μm were also used to retrieve the aerosol optical thickness of the atmosphere (τ) at a wavelength of 0.5 μm.

Comparison with results of ground-based observations has shown that errors in retrieving τ do not exceed ± 0.05. The linear regression of A values gave the estimate of the derivative $\partial A/\partial \tau = 7.8\%$ (with the correlation coefficient 0.615 and MSD 1.3%), which characterizes the effect of volcanic aerosols on albedo. In the case of the total albedo (obtained from data of channels 0.58 to 0.68 and 0.725 to 1.1 μm) and the use of a more precise technique for retrieval, $\partial A/\partial \tau = 5.7\%$. In the case of OLR the derivative $\partial F/\partial \tau = 5.1$ W/m^2, but the correlation coefficient, constituting only -0.16, points to a prevailing contribution by other factors of the OLR variability. For the Earth's radiation budget, $\partial R/\partial \tau = -9.3$ W/m^2. The results are consistent with calculations by Harshvardhan in 1979 (61), taking into account nonequivalent observation conditions and calculations.

DeLuisi et al. (52) discussed results of lidar soundings (at 694-nm wavelength), as well as observations of the atmospheric spectral transparency, direct solar and global radiation carried out at Mauna Loa Observatory in 1982. These results showed the effect on the atmospheric radiative regime of stratospheric aerosol of unknown origin (beginning 28 January) and of the eruptive stratospheric cloud from El Chichón (9 April). Since cloudiness on the days prior to those mentioned had prevented observations, it is possible that the considered disturbances could have appeared some days earlier.

The effect of El Chichón on the radiative characteristics of the atmosphere was the largest of all similar effects observed earlier at Mauna Loa. Lidar soundings revealed a fine vertical structure of the eruptive cloud with an initial maximum of the aerosol concentration above 22 km, where the concentration of the background aerosol is also maximum, and revealed the subsequent descent of aerosol particles. However, later the layer near 22 km became more intensive than that at 26 km, although the latter was observed more frequently. In 3 months the vertical mixing determined the formation of a homogeneous sun photometer data, the background AOT at 425-nm wavelength was 0.02; on 14 May it was 0.49, reaching 0.7 by the end of that day. The characteristic feature of the stratospheric aerosols was only a small variation of the wavelength dependence of AOT in June and July 1982, which reflected the efficiency of the gas-to-particle conversion mechanism for aerosol formation, as well as the effect of the mechanisms of coagulation and sedimentation.

Retrieving the aerosol size distribution from the spectral transparency suggested that it may well be approximated as lognormal with a broad maximum in the range of radii 0.2 to 0.6 μm. Estimation of the total aerosol content in the air column gave 0.06 g/m^2, which determines the aerosol mass (in the latitudinal belt 15°S–35°N) to be about 10^7 tons (this estimate does not take into account the aerosols at altitudes below 21 km).

Actinometric observations registered a substantial posteruption reduction of direct solar and global radiation. A decrease of the global radiation at near-noon constituted 5.6% (from observations on 6 June 1982 as compared to 5 June 1981), and in the case of direct solar radiation it reached 21.3%. A decrease of the daily sum of global radiation for 6 June 1982 (as compared to 5 June 1981) constituted 2490 kJ/m^2 (7.7 $\pm$ 1.5%). It is probable, however, that the estimates obtained should be considered maximum, since the thickest part of the eruptive cloud was spreading over Mauna Loa at the time.

The scale of the eruption of El Chichón can only be compared with the Agung eruption in March 1973 in Indonesia, but the effect of Agung was confined mainly to the Southern Hemisphere: AOT at Mauna Loa was only ~0.02, but in Aspendale, Australia, it reached 0.15. Previous eruptions had also been characterized by a smaller vertical extent of the eruptive cloud. Only the AOT estimates for the Krakatoa eruption (1883) gave maximum AOT of about 0.6.

As has been mentioned, the post-El Chichón observations revealed a serious disturbing effect of the eruptive aerosols on results of the remote sounding of the atmosphere and surface (e.g., SST retrieval). This effect can be particularly substantial with the use of limb or occultation techniques for sounding, and it is governed by such aerosol characteristics as the optical thickness, the phase function, and the complex refraction index.

King (160) processed data of the atmospheric spectral transparency measured at Mauna Loa in July. He obtained mean monthly optical thicknesses of the stratospheric aerosol layer in the visible ranging from 0.5 to 0.25. These were 10 times greater than the background values. Retrieving the aerosol size distribution averaged over the atmosphere (with the prescribed refraction index 1.45) revealed the presence of a narrow maximum in the interval of radii 0.3 to 0.4 μm.

With the density of aerosol matter 2 g/cm^3, the aerosol mass in the air column will be 0.0579 g/cm^2 (June) and 0.0602 g/cm^2 (July). These estimates were found taking into account the size distribution in the range of radii $0.1 < r < 4.0\,\mu$m, which was found by extrapolation of the values over the measured sizes. The size distribution can be approximated by a modified gamma distribution:

$$dN/d(\log r) = Cr^{\alpha} \exp(-br).$$

Estimation of the parameters for June (July) gave the following results: $c = 1.549 \times 10^{15}$ (9.897×10^{19}), $\alpha = 8.26$ (13.65), $b = 24.4$ (39.3 μm^{-1}), and $r = 0.338$ (0.347 μm).

On the assumption that the stratosphere aerosol consists of a 75% sulfuric acid–water mixture, King (160) calculated the spectral dependence of the aerosol optical thickness, the single scattering albedo, and the factor of the indicatrix asymmetry for the wavelength region 0.25 to 25.0 μm. The use of results for 3.7, 10.8, and 12.0 μm has shown that the SST retrieval from data of measurements at these wavelengths, without taking into account the eruptive aerosols, underestimates SST by 0.6 K at night and by 1.2 K in the day. Calculations for 0.7 μm gave 0.668 for the factor of indicatrix asymmetry, 1000 for the single-scattering albedo, and 73.3 for the ratio between the total extinction coefficient and backscattering. The angular distribution of the polarization degree is characterized by an unusually large number of scattering angles, to which a neutral polarization corresponds.

Analysis of data of regular observations of direct solar radiation at Mauna Loa beginning in 1958 has shown that the total atmospheric transparency is a very reliable indicator of the volcanically caused variations in the stratospheric aerosol content. This was illustrated by actinometric observations by DeLuisi et al. (161) after the 18 May 1980 Mount St. Helens eruption. From data of 3 June observations, when the eruptive cloud was again over Mauna Loa, having made a circle around the globe, the values of the ratio between global radiation variations and those of direct solar radiation were 0.256 (61.6°), 0.239 (59.7°), and 0.227 (57.8°). (The solar zenith distance is given in parentheses). Since the mean weighted solar zenith distance for the illuminated side of the Earth is 60°, these results can be considered representative from the viewpoint of characteristics of the global scale effect of the stratospheric aerosol layer. The respective post-Agung values were 0.2 (1963), 0.25 (1964), and 0.30 (1965).

Data on the ratio between spectral scattered and direct solar radiation made it possible to estimate the size distribution and absorbing properties of the aerosols. If $\Delta \tau$ is the change in the optical thickness between 3 and 4 July, then with the Junge size distribution,

$$dN/d(\log r) \sim r^{-\nu*},$$

where N is the number density and r *the radius of particles, we obtain*

$$\Delta \tau(\lambda) \sim \lambda^{-\gamma},$$

where $\nu* = \gamma + 2$ and $\gamma = 0.47$, according to observations at Boulder, Colorado.

Estimates of the imaginary part of the complex refraction index ranged between 0 and 0.002. The post-Agung aerosol was also weakly absorbing. Analysis of the secular trend of transparency made at Mauna Loa revealed reductions of transparency after different eruptions: 2% (Agung), 0.75% (Fuego), 0.57% (Souffrier), 0.55% (Sierra Negra), and 0.58% (Mount St. Helens). Processing of a small number of lidar soundings at Boulder gave a maximum optical thickness of 1.6, but the effects of cirrus clouds is not excluded here.

If the effect of stratospheric aerosol layer on the shortwave radiation transfer depends weakly on the variability of the absorption coefficient, an opposite situation is observed with respect to the longwave radiation. The role of the aerosol size distribution is also pronounced (at least, in the range of radii 0.1 to 2.0 μm), depending on whose variability the effect can appear of either warming or cooling of the climate. Although sulfuric acid droplets are mainly spherical, an account of the effect of nonspherical particles on the phase function is also essential.

To estimate the climatic impact of stratospheric aerosol, of great importance are studies of the heat balance of particles residing in the stratosphere. In this connection, Fiocco et al. (56) calculated the temperature of an aerosol particle determined by its absorption of solar radiation and radiative and conductive heat exchange with the atmosphere, as well as by phase conversions. Previous calculations had shown that at altitudes above 60 km the daytime difference between the temperatures of ambient air and particles (ΔT) can be above 100 K.

New calculations have shown that at altitudes below 50 km, maximum noontime $\Delta T < 1$ K, and below 10 km $\Delta T < 0.01$ K. Near noon, ΔT is always above zero; that is, aerosol particles heat the atmosphere. At night the situation is the opposite. From data of conductive gas-to-particle heat exchange, information was obtained about rates of air heating (or cooling) at different altitudes, depending on particle radii.

Analysis of daily means points to a strong variability of the heating rate, depending on the altitude and radius of the particles. The $r \sim 0.5 \mu m$ particles contribute most to air heating. Near the stratopause (50-km altitude), the presence of aerosols almost always causes a cooling. In the wintertime lower troposphere the aerosol, as a rule, also promotes a cooling. The cooling effect can also be observed in summer in regions with a low surface albedo. However, daily mean aerosol heating of the lower stratosphere (the layer 15 to 25 km) constitutes about 0.05 to 0.1 k/day, and after the eruptions it can reach 1 deg/day in the equatorial zone.

2.5   THEORETICAL ESTIMATES OF THE CLIMATIC IMPACT OF VOLCANIC ERUPTIONS

Different opinions can be found in literature on climatic implications of volcanic eruptions (12,27,40,55,67,68). It was assumed, for example, that volcanic activity had been the cause of ice ages. In a number of studies the effect of stratospheric volcanic dust on weather and climate has been analyzed. Examination of the observational data revealed that as a rule, during the first or second year after powerful eruptions the surface air cooled by 0.5 to 1.0 K, but in some cases such an event was not observed. Figure 2.7 shows the results obtained by Oliver (89).

2.5.1   Observational Data and Semiempirical Estimates of the Climatic Impact of Volcanic Eruptions

Ellsaesser (54) questioned the conclusions drawn by Angell and Korshover (28,162) about the possible effect of the 1963 Agung eruption on the change in the mean surface temperature manifested in the following (Figs. 2.8 through 2.13):

1. In 1963 the surface temperature dropped by 0.6°C in the northern extratropical latitudes, 0.2°C in the tropics, and 0.4°C in the southern extratropical latitudes.

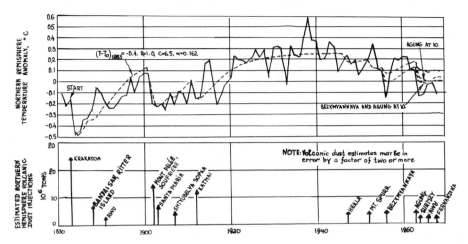

FIGURE 2.7. NH surface temperature anomalies and volcanic eruptions. (Data from Ref. 89.)

2. The temperature of the layer from 850 to 300 mb in the southern extratropical latitudes dropped by 0.7°C, while in the layer from 300 to 100 mb it increased by 0.7°C. Therefore, an assumption was made about possible changes in the previously observed climate cooling toward warming in the early 1970s — due not only to the Agung eruption, of course.

Doubts concerning the justifications for these conclusions were caused by the following considerations:

1. Since the stratospheric dust cloud resulting from the Agung eruption was tenfold denser in the Southern Hemisphere than in the Northern Hemisphere, one cannot understand why the eruption has caused surface temperature changes in the extratropical latitudes of the Northern Hemisphere which were one and a half times greater and lasting twice as long as in the Southern Hemisphere.

2. The warming of the layer from 300 to 100 mb in the southern extratropical latitudes by 0.7°C occurred simultaneously with similar warming in 1962–1963 in the Northern Hemisphere, but the latter cannot be ascribed to the eruption effect, since the dust cloud reached the southern boundary of the United States only in September 1963 (probably, both these phenomena were driven by the same, still unknown mechanism, which has nothing to do with the eruption).

3. If the eruption was the cause of temperature variations, it is not clear why the surface temperature in the extratropical latitudes of the Northern

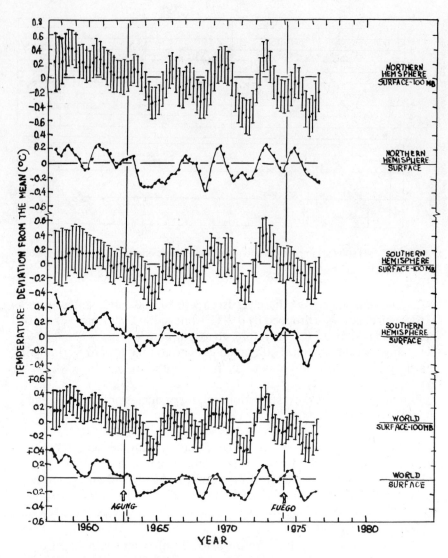

FIGURE 2.8. Deviations of mean seasonal surface temperature and atmospheric temperature averaged over geographical regions. (After Ref. 162).

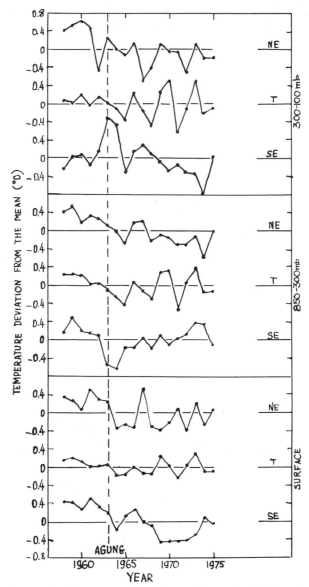

FIGURE 2.9. Nonsmoothed mean annual deviations of temperature for the extratropical latitudes of the Northern and Southern Hemispheres and in the tropics. The vertical dashed line marks the moment of the Agung eruption. (After Ref. 28.)

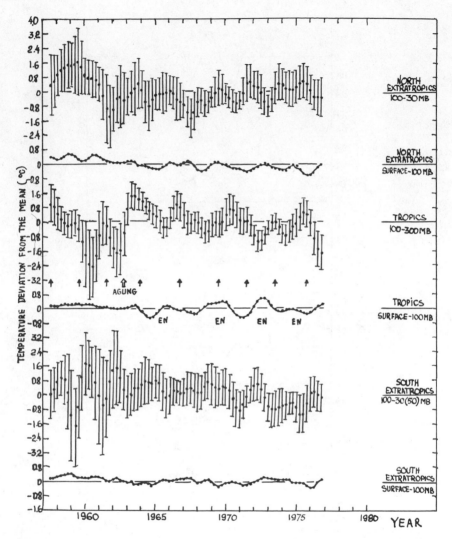

FIGURE 2.10. Same as in Fig. 2.8, but for different atmospheric layers. EN, El Niño. (After Ref. 168.)

Hemisphere had not dropped before 1963–1964, in contrast to the cooling of the layer from 850 to 300 mb and warming of the layer from 300 to 100 mb.

4. Also, it is not clear in what way the volcanic dust with residence time in the stratosphere not more than 30 days could have raised the mean annual temperature of the upper troposphere which continued until 1964.

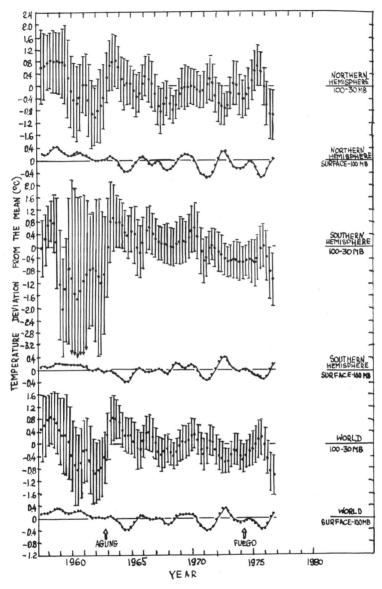

FIGURE 2.11. Same as in Fig. 2.10, but for both hemispheres and the globe. (After Ref. 168.)

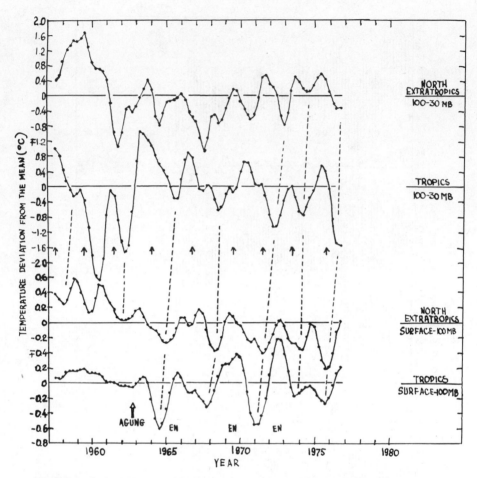

FIGURE 2.12. Smoothed deviations of mean seasonal temperature values from seasonal averages, which reflect the upward and northward spreading of temperature waves due to positive SST anomalies in the eastern tropical Pacific. EN, El Niño. (After Ref. 168.)

In view of these controversies, Taylor et al. (116) performed an analysis of data of surface air temperature observations for the 1815–1963 period using the technique of "superimposed epochs," making it possible to detect small-temperature signals caused by discrete events in data series with noise characterized by natural random temperature variations. These data were obtained from observations at 42 stations located in various parts of the globe.

Only those posteruption temperature data were analyzed to which, after Lamb, a dust veil index (DVI) of not less than 100 corresponded (DVI characterizes an additional attenuation of the solar radiation caused by the loading of the atmosphere with the volcanic dust). During the period under consideration, 25 similar eruptions took place during 18 years (sometimes two eruptions per year occurred). Separately, 13 eruptions for 8 years were considered to which a DVI > 500 corresponded. Most powerful were the following eruptions (DVI in parentheses): Tambora, Indonesia, 1846 (1000); Cosigüina, Nicaragua, 1875 (4000); Amargura, Fiji, 1815 (3000); Krakatoa, Indonesia, 1883 (1000); Agung, Indonesia, 1963 (800).

Application of the superimposed epochs technique revealed a weak but statistically significant volcanic signal in the long time series of temperature observations. Being caused by powerful volcanic eruptions, the temperature drop constituted about 0.5 K and usually continued into the first or second year after the eruption. The time of the temperature's return to its normal level is 2 to 5 years. The statistical significance level for the volcanic signal (an identical drop of temperature that can be attributed to random causes) varies within 0 to 5.4%.

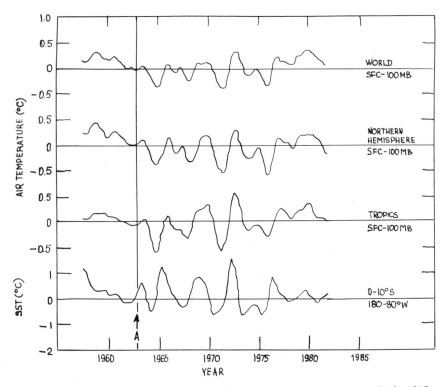

FIGURE 2.13. Variations in SST and surface air temperature in the eastern Pacific (0–10°S; 180–80°W). A, Agung eruption. (After Ref. 30.)

The effect of eruptions shows itself more strongly in high latitudes, but in this case it is combined with a much greater variability of the temperature field. Therefore, the statistical significance of a high-latitude signal is practically lower, and the duration of the signal is shorter than in low latitudes. Data on the annual change of temperature show that as a rule, in summer the signal turns out to be stronger and lower. However, the reliability of this conclusion is limited because of the absence of information about the spatial and temporal distribution of the ejected products as well as the lack of a sufficiently homogeneous and representative series of temperature observations. Therefore, an important objective of further studies is a repeated analysis of data with the use of more homogeneous (although shorter) series as well as observational results for the last decades which are more representative both in space and in time. Therefore, an objective analysis aimed at constructing global temperature fields may be useful.

Analyzing the data of Taylor et al. (116), Ellsaesser (55) emphasized the controversy in the interpretation of observational data from the viewpoint of the climatic effect of volcanic activity. A consideration of the chronology of eruptions has led to the conclusion (55) about nonrepresentativeness (i.e., fragmentary character) of observations before 1960.

After the 1963 Agung eruption, Newell (84) detected a 6 K warming of the stratospheric layer from 50 to 80 mb (20 to 18 km) over Australia, but from data by McInturff et al. the temperature increase near the 50-mb level in the tropics did not exceed 2 K. This showed the great difficulty of identifying a volcanically caused disturbance against variations due to other factors. Later studies by Angell and Korshover (29,30) pointed to a 3 K warming of the layer from 30 to 100 mb (24 to 16 km) in 1963, with a subsequent 2 K temperature decrease, but did not reveal the occurrence of any stratospheric warming at extratropical latitudes.

Using the NOAA-6 temperature sounding data and aerological soundings, Parker and Brownscombe (90) analyzed changes in the stratospheric temperature field following the eruption of El Chichón (the vertical resolution of satellite data was only 15 km). Data from the four tropical aerological stations point to a temperature increase in the lower stratosphere (at 30- and 50-mb levels) up to 6 K at 30 mb in July 1982 and 5 K at 50 mb in August. The latter date of maximum warming at the 50-mb level suggests that the temperature disturbance propagated downward (as after the Agung eruption).

An analysis of satellite indirect sounding data on the annual change of mean monthly temperatures for the latitude band of 15°N–15°S and for the entire globe in 1980, 1981, and 1982 at levels 80, 50, 15, and 2 mb (the "sharpest" weighting function and clearest observations correspond to the

50-mb level) showed that in 1982 a temperature increase took place in the tropics at 50 mb (as compared to 1980–1981), with a maximum of 2 K in July and August. Maximum warming at 15 mb (28 km) was observed in May, and near 2 and 5 mb cooling took place in July and August. Temperature variations at 80 mb were uncertain.

Variations in the mean global temperature whose day-to-day MSDs were only 0.03 K revealed a considerable warming in 1982 at 80, 50, and 15 mb, but probably at the two lower levels (80 and 50 mb) it had begun before the eruptions. The latter circumstance, as well as an asymmetry of warming with respect to the equator, suggests the supposition that it was partly determined by quasibiennial variability. This assumption is favored by the view that the eruptive cloud covered only the latitudinal band of 60°N–20°S, and by the probability of the downward-spreading disturbance connected with the quasibiennial periodicity.

However, since the amplitude of the observed warming exceeded that corresponding to the quasibiennial periodicity, there is no doubt that a volcanically caused stratospheric temperature increase was also manifested.

Although there are causes for the possible substantial warming of the stratosphere following large volcanic eruptions, even the analysis of data on the stratospheric temperature field after the Agung eruption (8°S, 115°E) in March 1963 was not sufficiently convincing from the viewpoint of filtering out all the factors, except the eruption. Therefore, Quiroz (95) undertook a filtering out of the stratospheric temperature change due to the volcanic eruption of El Chichón, taking into account the following factors: (i) the annual, semiannual change, and quasibiennial oscillations (QBO); (ii) sudden warmings of the high-latitude stratosphere followed by cooling in the low latitudes; and (iii) SST anomalies, especially those connected with the El Niño event.

Since the concentration of the eruptive stratospheric aerosols reaches its maximum several months after the eruption, emphasis was placed on variations in the temperature field in mid- and late 1982. As for the contribution by the annual and semiannual change, it can easily be excluded by extracting climatological monthly means from the observed temperature values. It is much more difficult with other factors. Analysis of observational data of the mean zonal temperature at 15°N at 30 mb showed that between April and June 1982 (after the eruption) a 2°C temperature increase took place, but it is difficult to identify this warming as having been caused by the eruption, because it had begun in January, long before the eruption.

No doubt, the stratospheric warming in low latitudes observed until mid-April (immediately after the eruption) is a "false volcanic signal." This conclusion can be drawn on the basis that this warming spread to middle latitudes of both hemispheres, whereas the distribution of the volcanic aerosol even in the fall of 1982 was strongly asymmetrical with respect to the equator, with the maximum aerosol content near 10°N.

Previous studies have led to the conclusion that about 6 months after a strong SST increase in the eastern equatorial Pacific (El Niño), a warming by 1 °C of the tropical troposphere takes place, followed by a cooling of the lower stratosphere. Since the 1982 SST increase started only in the spring and could be pronounced only in the fall and winter, no doubt the effect of the summer SST anomaly was negligible, but it must later be taken into account. Thus of major importance is an account of QBO.

Based on an analysis of the spatial and temporal variability of QBO for the previous years, Quiroz (95) showed that an altitude-dependent phase shift is typical of QBO at 30 mb. Since the trend of stratospheric warming is manifested simultaneously, it makes it possible to consider it to be volcanically induced. The exclusion of the QBO contribution gave values of the eruptive residual variability varying within 1.0 to 3.0°C in the 0–30°N band (with an error of 0.5 to 1.0°C), whereas the total observed temperature anomaly at 30 mb reached 6°C at several locations, including Singapore (1°N).

Within the considered latitudinal belt a slightly increased warming took place south of 15°N, although the volcano's latitude is 17°N. This was probably connected to the maximum outgoing longwave radiation near the equator, the absorption of which by the eruptive aerosol is a major energy source for a stratospheric warming. Bearing in mind substantial errors in estimates, it is in fact possible that a maximum warming took place in the latitudinal belt of 10–15°N. Although the volcanic signal is pronounced at other altitudes (e.g., at the 10-mb level), the reliability of the respective data of observations must be thoroughly analyzed.

An interesting analysis of the effect of volcanic eruptions on the temperature field of the tropical troposphere was performed by Parker (91). Having considered data of aerological soundings for the latitudinal belt of 20°S–20°N for the period of 1950–1984 with seasonal averaging and the filtering-out of the contribution by the Southern Oscillation to the formation of the temperature field, Parker concluded that the mean temperature of the levels 850 to 500 and 500 mb during the year following the March 1963 Agung eruption was higher than before the eruption. A maximum temperature increase, reaching only 1°C, was observed 2 years after the eruption. No cooling was observed after the April 1982 El Chichón eruption (until the end of 1982). The cooling in the Northern Hemisphere in the winter of 1982–1984 did not differ from the coolings in 1961 and 1971, neither of which was caused by an eruption. The results of observational data analysis which revealed the impact of eruptions are given in the monograph by Loginov (23).

Definite correlations between climate variability and volcanic eruptions were found from an analysis of paleoclimatic data. It was shown, for example, in recent studies by Stothers and Rampino (113,114) that between 1500 B.C. and A.D. 1250 there had been a high correlation between volcanic eruptions and such characteristics as dry fogs in western Europe and acid rains in Greenland. Volcanic eruptions in the Mediterranean Sea and in Iceland are a reason of at least five of the nine layers of increased acidity detected in the Greenland ice cores for this period.

A complete analysis of the present-day global temperature field variations between 1958 and 1982 was made by Angell and Korshover (29,30). The analysis revealed a global-scale cooling of about 0.5°C between 1958 and 1970 and a subsequent weak warming. From spring 1981 to spring 1982 a substantial cooling was, however, registered in the Northern Hemisphere. Apparently, after the 1963 Agung eruption a 0.3°C decrease of the surface air temperature in the Northern Hemisphere took place.

As Mitchell (81) noted, the impact of volcanic eruptions on global climate manifests itself mainly through an increased posteruption dust loading of the stratosphere, which leads to a cooling of the lower atmosphere and warming of the stratosphere. According to Bray (38,39), from the viewpoint of paleoclimatology, a change in the albedo of the ash-covered surface could have been of some importance. An analysis of observational data showed that starting from 1880, variations of the mean temperature of the atmosphere could have been strongly related to volcanic eruptions, and a relative warming between 1920 and 1945 could have resulted from an almost complete absence of large-scale volcanic eruptions during the three decades following the 1912 Katmai eruption.

In the event that conclusions about the effect of volcanic eruptions on climate are confirmed, the problem of forecasting the volcanic activities will be of special interest (119). Bryson and Goodman (40) believe that eruptions should be considered as one of the important factors of climate change for time scales from several years to several thousand of years.

The periods of considerable change in the Earth's climate are marked by surprisingly small variations in the mean temperature of the Earth's surface T_s. During the last 10,000 years — in the Holocene period — T_s had changed by 1 K. Such variations had, however, substantially affected the development of humankind because they were followed by variations in sea ice and precipitation. Of great importance also is the high sensitivity to temperature variations of the vegetation season in high latitudes. For the last 2 million years (the Quaternary period) quasicyclic variations in T_s (by 5 K) had taken place during several thousands of years, in whose cooling phases glaciations had occurred over vast continental territories in the high and middle latitudes. There had been no such glacial-interglacial oscillations for a period of hundreds of million years prior to the Quaternary period. It is very important to reveal the factors of these climate changes and climate forecasts because of the vital importance of climate changes for the future evolution of humankind.

In this connection, Pollack et al. (93,94,163) and Angell and Korshover (162) discussed probable climatic implications of natural and anthropogenic stratospheric aerosols (volcanic eruptions and high-altitude aviation, respectively). This effect can manifest itself through variations in the global heat budget—both the solar radiation absorbed by the planet and the outgoing thermal radiation.

During the first several months after a volcanic eruption, the silicate dust, the optical constants of which can be assumed to be identical to the constants of the obsidian and basalt glass, is the prevailing component of the aerosols. Decreasing over time, the mean radius of particles is assumed to be $1.0\,\mu m\,\frac{1}{2}$ month after an eruption, $0.5\,\mu m$ in 2 months, $0.25\,\mu m$ in 4 months, and $0.1\,\mu m$ in 15 months. Several months after the eruption the sulfur-containing gases are transformed into H_2SO_4 droplets which become a dominating component of aerosols.

Aerosols produced by the repeated use of transport spacecraft (shuttles) are Al_2O_3 particles, while those produced by supersonic aircraft (SSA) and other high-altitude aircraft are H_2SO_4 droplets. In all cases the aerosol size distribution can be considered identical to that of the usual H_2SO_4 aerosol.

Table 2.7 shows estimates of probable variations in the stratospheric optical thickness $\Delta\tau$ at $\lambda = 0.55\,\mu m$ under the influence of the factors above. These data refer to the Northern Hemisphere. To obtain the $\Delta\tau$ values for the Southern Hemisphere, multiply the entries in Table 2.7 by $\frac{2}{3}$. The tabulated data show that $\Delta\tau$ variations that could be produced by supersonic aircraft and shuttles are much less, even for a mean decadal increase of $\Delta\tau$, than those

TABLE 2.7   Probable Variations in the Stratospheric Optical Thickness in the Northern Hemisphere under the Influence of Aerosols of Different Origin

Source of Aerosol	Time Scale	$\Delta\tau$
SSA and other high-altitude aircraft	1990, probable level	4.9×10^{-4}
	1990, maximum level	8.9×10^{-4}
	2000, probable level	1.1×10^{-3}
	2000, maximum level	3.3×10^{-3}
Shuttle	1990	3.3×10^{-5}
Individual powerful volcanic eruptions	Maximum level during 1 month	3×10^{-1}
	Mean level during 1 year	1×10^{-1}
Repeated volcanic eruptions	Mean decadal for small-scale volcanic activity	2×10^{-1}
	Maximum mean for 10 years	8×10^{-2}

due to small-scale volcanic activities. With due regard to the foregoing data on aerosols, calculations were made of variations in the solar radiation absorbed by the surface-atmosphere system, and in the outgoing thermal emission.

Stratospheric warming due to increasing $\Delta\tau$ is determined by the absorption of the upward longwave radiation flux: the stratospheric aerosol causes stronger cooling of the stratosphere at the expense of radiative heat exchange as compared to warming due to solar radiation absorption. Solar radiation absorbed by the Earth's surface-troposphere system decreases when the aerosol content in the stratosphere grows. To calculate radiative fluxes, the doubling technique was used, taking into account multiple scattering in the case of the shortwave radiation. The longwave (thermal) emission was calculated taking into account molecular and aerosol absorption. Calculations of surface temperature variations revealed the important role of the longwave radiation contribution that causes (due to attenuation of the radiant heat release) a substantial decrease in surface cooling due to decreased input of shortwave radiation.

The calculated results agree with the air cooling near the surface and the warming of the stratosphere observed after powerful eruptions. It is important that the main contribution to the stratospheric warming is determined by the absorption of the upwelling flux of the longwave emission, but not the solar radiation.

Estimates of the climatic effects of the shuttle-produced aerosols show that they are negligible. The same can be said for the effect of the SSA: in this case T_s variations should not exceed 0.1 K. This effect, however, requires reestimation with the availability of new data.

Probable trends in volcanic activity and in the development of anthropogenic aerosol sources show that these factors are not likely to surpass the effect of increasing amounts of carbon dioxide in the atmosphere. Provided that the CO_2 effect prevails, one can expect climate warmings in the near future.

The development of supersonic aircraft and shuttles still entails in future decades an additional source of pollution of the stratosphere with H_2SO_4 droplets (from SSA) and aluminum oxide particles (from shuttles). To estimate the possible effect of these pollutants, Pollack et al. (93,94) undertook detailed calculations of the resulting changes in the constituents of the mean global heat budget and the accompanying variations of mean surface temperature. The temperature was calculated based on the assumed equality of the solar radiation absorbed by the planet and the outgoing emission computed by selecting a vertical profile of temperature determined by the radiative-convective balance, which provides for the equality of the Earth's heat-budget components.

An analysis of calculated results showed (Fig. 2.14) that the aluminum oxide particles, with a refraction index of 1.77, cause a greater increase in the Earth's albedo with the growing optical thickness $\Delta\tau$ of the stratospheric aerosol layer than does H_2SO_4 aerosol, with a refraction index 1.43. Naturally, as $\Delta\tau$ grows, the portion of radiation absorbed by the surface-troposphere system (Fig. 2.15) decreases. The effect of aerosols on ozone photochemistry is weak.

In a continuation of the previous studies, Whitten et al. (164) obtained new estimates of the effect of effluents of supersonic aircraft and planned launches of shuttles on the stratospheric aerosol layer based on the detailed theory of formation of the sulfate aerosol layer in the stratosphere (121 – 124). The estimates obtained were used to calculate the effect of the anthropogenically caused changes in the stratospheric aerosol layer on climate (e.g., mean temperature of the Earth's surface).

The development of the theory of formation of the stratospheric sulfate aerosol from the gas phase made it possible to take into account both gaseous and aerosol effluents from SSA and shuttles which can affect natural aerosols. Continuous pollution of the stratosphere is assumed to have already continued for 5 years (the stratospheric model under consideration is stationary).

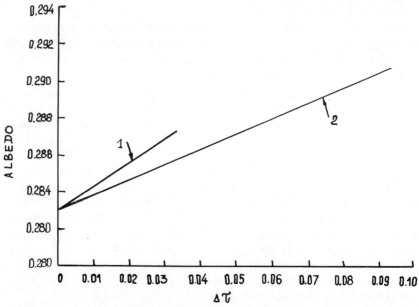

FIGURE 2.14. Dependence of the global albedo on variations in the stratospheric optical thickness caused by particles of aluminum oxide (1) and sulfuric acid aerosol (2).

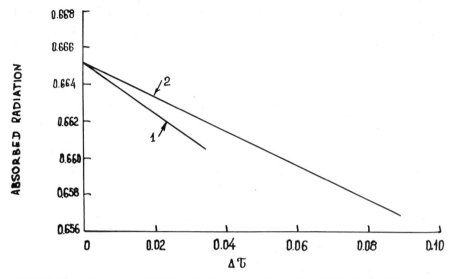

FIGURE 2.15. Shortwave radiation (with respect to the extra-atmospheric insolation) absorbed by the surface-troposphere system, depending on variations in the stratospheric optical thickness $\Delta\tau$ caused by aluminum oxide particles (1) and sulfuric acid aerosol (2).

From the point of view of the possible effect on climate, the main conclusion is that this effect is negligible. If one proceeds from the assumption that 300 SSA are flying daily at an altitude of 20 km, or that 50 shuttles are launched every year, this will increase the concentration of large aerosol particles in the stratosphere by about 20%, causing a decrease of mean surface temperature by less than 0.01 K (this estimate is correct to an accuracy of a factor of 5).

These estimates, obtained with the use of a much more complete model, agree well with previous approximate estimates. The new model, however, has made it possible to solve a number of problems concerning the effect on the stratosphere of large-scale ejections of small particles. In particular, it was shown that the number of small particles that can grow large is very small.

The total concentration of large particles is limited by the rate of input of sulfate aerosols to the stratospheric layer. This rate of input determined by natural factors constitutes about $1-2 \times 10^5$ tons of sulfur dioxide per year. Effluents from several hundred SSA constitute only 0.29×10^5 tons per year. The prevailing effect of natural processes of formation of the sulfate layer determines a relative constancy of the total mass of sulfate aerosols and its determining number of large particles in conditions of affecting the stratosphere.

As far as small particles of soot (SSA fuel wastes) and Al_2O_3 (shuttle wastes) entering the stratosphere are concerned, they coagulate rapidly with large particles. Therefore, of the SSA fuel wastes, not soot particles but sulfur dioxide affects the stratospheric aerosol most, and hence it is sulfur dioxide injections that must be controlled, if required.

Calculations show that if the soot wastes constitute only 0.3 g of soot per kilogram of fuel, then SO_2 is responsible for the formation of about 2 g per kilogram of H_2SO_4 aerosol. Probably the most important aspect of the anthropogenic effect of the stratospheric aerosol is associated with the indirect impact of industrial activity and not with the direct influence of material ejected to the stratosphere. So, for example, an increase in the background concentrations of SO_2 and carbonyl sulfide caused by growing levels of fuel combustion and coal gasification may lead to an increase in the concentration of the sulfate aerosol in the stratosphere. Possible effects of the SSA-caused changes in the concentration of condensation nuclei on the processes of formation of cirrus clouds in the upper troposphere and lower stratosphere should be analyzed.

Although the effect of stratospheric aerosols on the global albedo is much stronger than on the outgoing emission, an intensification of the greenhouse effect, caused by increasing $\Delta \tau$, provides a 40% compensation of the Earth's surface temperature drop, which results from an increase of the planetary albedo. Sometimes a slight increase of surface temperature is also observed (Table 2.8).

With 0.1 K as a detection threshold for the Earth's surface mean global temperature change, an analysis of available data on possible stratospheric pollution from SSA and shuttles in the decades to come shows that tempera-

TABLE 2.8 Calculations of the Optical Thickness $\Delta \tau$ at $\lambda = 0.55$ μm and Surface Temperature ΔT_s Determined by Silicate Dust, with the Relative Humidity Assumed to Be Constant

		Basalt Glass				
Modal radius of particles (μm)	0.1	0.25	0.50	0.50	0.50	1.0
Optical thickness	0.195	0.195	0.045	0.195	0.495	0.195
Surface temperature T_s	−2.305	−2.109	+0.031	+0.023	0.027	+2.985

		Obsidian			
Modal radius of particles (μm)	0.1	0.1	0.25	0.50	0.50
Optical thickness	0.045	0.195	0.195	0.045	0.195
Surface temperature T_s	−0.417	−1.617	−1.482	−0.012	−0.186

ture changes caused by these factors lie below the detection threshold. The pollution-caused variations are much smaller than those caused by small-scale volcanic eruptions in the mid-1960s (see Table 2.7).

Variations in the Earth's surface temperature can reach a threshold value if the number of SSA is as high as 7500 (at 5.5 flight-hours per day) or daily launches of shuttles totaling 70. Since the development of SSA can be sufficiently fast, further estimates will be needed with more accurate data.

The contribution of aerosols to the decrease of atmospheric transparency for the longwave radiation is comparatively small. Harshvardhan (61) undertook new calculations of the sensitivity of the Earth's surface-atmosphere system albedos and solar zenith angles.

In the absence of the stratospheric aerosol layer the system's albedo is written as

$$\alpha_p = \alpha_c C + \alpha_s(1 - C),$$

where α_s and α_c are the system's albedoes for clear-sky and cloudy areas, respectively, and C is the cloud amount.

The surface albedo is calculated for $10°$ latitudinal belts with due regard to observational data. The albedo of the cloudless system is determined using results of calculations made by Braslau and Dave (169).

Calculations resulted in mean monthly mean zonal α_s profiles, as well as a profile averaged over a year, taking into account monthly values of insolation (as weighting factors). The mean annual values of the system albedo thus obtained correspond closely to those of minimum albedo found by Vonder Haar and Suomi (170) from satellite observations (substantial discrepancies occur only near the equator and at high latitudes, because of cloud effects not completely excluded in processing satellite data).

The system albedo for overcast cloudiness α_c has also been obtained using the calculations of Braslau and Dave (169), specified with regard to satellite observations. The technique described above was used to calculate variations in the mean global albedo caused by stratospheric aerosols for a cloud amount taken from climatological data, as well as to calculate the variations of mean zonal albedo for individual months. Agreement was obtained between calculated (0.308) and observed (0.306) values of the mean annual albedo for the Northern Hemisphere.

A consideration of the effect of stratospheric aerosols on the outgoing longwave radiation allowed the estimation of changes in the radiation budget of the Earth's surface-atmosphere system. The sensitivity to stratospheric aerosol of the system albedo A for a hemisphere is determined by the derivative $dA/d\tau$, where τ is the optical thickness of the aerosol layer at $\lambda = 0.55$ μm. Since for the stratospheric aerosol layer $\tau \ll 1$, the albedo is directly proportional to the optical thickness and the sensitivity of the albedo is $\Delta A/\Delta\tau$.

Harshvardhan (61) assumed the aerosol to consist of droplets of a 75% solution of sulfuric acid, its size distribution being described by modified gamma distribution and phase function by the Henye-Greenstein function. Here the extinction coefficient at $\lambda = 0.55$ μm, identical to the scattering coefficient, constitutes 1.1×10^{-4} km^{-1}, the single-scattering albedo is 1, and asymmetry of the phase function is 0.73. The effect of stratospheric aerosols was calculated in approximation of an optically thin layer, and values of the system albedo were assumed from satellite data as standard (as initial data for a sensitivity estimate).

Results show that the prevailing influence of an optically thin strato-spheric aerosol layer shows itself in growing reflected solar radiation over the globe, except for winter polar regions. Analysis of mean monthly values of $\Delta A/\Delta\tau$ reveals a substantial dependence on the solar zenith angle, manifest-ing itself as a latitudinal-temporal variability within 0.08 to 1.0 (in percent at $\Delta\tau = 0.01$). In high latitudes the effect of the surface albedo is also great.

Numerical modeling results confirm the inadequacy of the assumption (used in energy-balance climate models) of equivalent climatic effects of solar constant changes and increasing stratospheric aerosol concentrations, since even in the presence of a globally homogeneous aerosol layer its effect on the distribution of solar radiation extinction turns out to be inhomogeneous.

In the latitude zone below 50°, variations in the radiation budget are sufficiently homogeneous and small. This is determined by a compensating contribution of the intensified greenhouse effect of the atmosphere. Calcula-tions of the greenhouse effect sensitivity [expressed in $(W/m^2)/\tau_{vis}$] showed that it is maximum in the tropics (13.9 and 7.7 at clear sky and overcast cloudiness, respectively) and minimum in the winter sub-Arctic (5.3 and 4.0, respectively). Thus an increase in optical thickness by 0.1 in the clear-sky tropics leads to an intensification of the greenhouse effect by 1.4 W/m^2.

An increase in the temperature of the aerosol layer due to the absorption of the upward longwave radiation flux makes a noticeable contribution to variations in the greenhouse effect. The strongest pertubations occur in the spring and fall in polar regions, when the equator-to-pole difference in the radiation budget increases by 6.5 to 7.0 W/m^2 for the aerosol optical thick-ness 0.1 at $\tau = 0.55$ αm.

Cess et al. (43) was concerned about the fact that various authors had obtained different values of the derivative $d\alpha_p/d\tau_{vis}$, the difference being about 0.2, according to Harshvardhan and Cess (60) and Cadle and Grams (165), while Angell and Korshover (162) had obtained a value of 0.082. Cess et al. (43) showed that the first value had been overestimated, because it had been obtained by using approximations of an effective wavelength.

Calculations, taking into consideration the spectral dependence of albedo (determined by Rayleigh scattering, in particular), showed that the selection of an effective wavelength 0.53 or 0.55 μm usually leads to a near-maximum value of $d\bar{\alpha}\lambda/d\tau_{vis}$ (where $\bar{\alpha}\lambda$ is the spectrally averaged albedo of the system). A consideration of the dependence of albedo on the solar zenith angle is also of great importance. Calculations of $d\alpha_p/d\tau_{vis}$ performed by Cess et al. (43) taking these factors into account suggested the unacceptability of the "one-wavelength" approximation and underscored the necessity to bear in mind the dependence of the solar zenith angle. A mean global value of $d\alpha_p/d\tau_{vis}$ for average cloud conditions was found to be 0.092; that is, it agreed well with the data of Pollack et al. (163). The effect of stratospheric aerosol on α_s was approximately five times stronger than on α_c, which points to a substantial specific character of the regional effect of stratospheric aerosols on the system albedo.

There is an opinion that the effect of stratospheric aerosols on longwave radiation transfer is negligible. As Coakley and Grams (46) noted, this opinion is erroneous, because in some cases the effect of aerosols on longwave radiation may be more significant than their effect on the shortwave radiation transfer. This is particularly the case when the aerosol layer consists of tiny but strongly absorbing particles ($x = 2\pi r/\lambda \ll 1$, where r is the particle's radius and λ is the wavelength).

To evaluate the effect of stratospheric aerosols on the global climate, Coakley and Grams (46) proposed a simple energy-balance model of a two-layer atmosphere to calculate variations in the mean global surface temperature caused by aerosols. The stratosphere and the surface-troposphere system were assumed to be in radiative equilibrium.

For a two-layer model of an atmosphere, consisting of troposphere and stratosphere in radiative equilibrium, an approximate formula was obtained expressing the dependence of outgoing longwave radiation $F\uparrow$ on the optical properties of stratospheric aerosols:

$$\Delta F\uparrow/F\uparrow = \tau'(1 - \omega' + 2\omega'\beta') - \tau[(1 - \omega)(1 - 3A/1 - A) + 2\omega\beta(1 - A)]$$

where τ is the mean weighted (considering the solar spectrum) aerosol thickness for the shortwave radiation, ω the effective aerosol albedo of single scattering, β the parameter of phase function asymmetry which characterizes the forward scattering, and $A = 0.3$ the surface albedo.

Stroke marks in the preceding equation identify the parameters that refer to the longwave radiation. With the use of an empirical relationship between outgoing radiation and surface temperature, one can apply this formula to evaluate the effect of stratospheric aerosols on surface temperature. Previous approximate calculations have shown that a 1% change of the outgoing radiation corresponds to a change of the mean global surface temperature by 1 K with the same sign. Based on these data, one can assess the effect of the stratospheric aerosol-induced variations in $F\uparrow$ on the surface temperature, assuming that the optical properties of the troposphere (such as the albedos of clouds and surface) are independent of surface temperature.

Using the Mie formula, calculations were made of the absorption and scattering cross sections at different values of the complex refraction index (the refraction index was assumed to be 1.5, the absorption index 0, 0.001, 0.01, and 0.1) and particle size (parameter x varied within 0.001 to 100). The results were used to calculate $\Delta F\uparrow/F\uparrow$ depending on the aerosol size distribution (at the aerosol concentration 1 μm/m^3).

These calculations suggested that both small particles ($r < 0.05\ \mu$m) and large ones ($r > 1\ \mu$m) have a stronger effect on the longwave radiation transfer than that of the shortwave radiation, and hence they cause an increase in the surface temperature. An alternative holds for particles of intermediate sizes which are responsible for surface cooling. Surface temperature variations are shown to be most substantial in the case of monomodal aerosols with radii 0.2 μm.

Actual temperature changes should be less, since aerosols cannot be monomodal, and the thermal inertial of the surface-atmosphere system should smooth out the aerosol effect. A maximum possible change of the surface temperature resulting from volcanic activity (like that of Agung) is a cooling by about 0.8 K. Calculations of $\Delta F\uparrow/F\uparrow$ had led to the conclusion that most likely the climatic effect of present aerosols is a lowering of the surface temperature, because particles with radii 0.1 to 1 μm contribute most to the mass of stratospheric aerosols.

Coakley and Grams (46) showed that a consideration of the stratospheric aerosol effect by way of equivalent changes in the solar constant may lead to large errors, particularly from the viewpoint of the effect of particles with radii beyond 0.1 to 1 μm. Such an equivalency may only take place in cases where the effect of aerosols on the longwave radiation transfer is negligible and the aerosols do not absorb in the shortwave spectral interval.

All these calculations were performed using the mean weighted spectral characteristics of aerosols. A comparison with calculation results for monochromatic characteristics at wavelengths 0.5 and 10 μm revealed substantial discrepancies and, consequently, the importance of considering the selective character of aerosol optical characteristics.

Large-scale volcanic eruptions are an excellent natural model to study the response of the global climate system to radiative perturbations caused by the formation of posteruption stratosphere aerosols. The powerful eruption of Agung in 1963 on Bali (8°S, 115°E) was, apparently, the next, by power, after Krakatoa (1883) and turned out to be the best documented when compared to previous eruptions. The global posteruption distribution of aerosols can be traced by observing the aerosol optical effects.

To estimate the effect of increasing concentration of stratospheric aerosols on the atmospheric temperature regime in low latitudes where this effect is maximum, Hansen et al. (59) used a 1-D model of radiative-convective equilibrium. The case of a "normal" atmosphere at the stratospheric aerosol optical thickness 0.005 was taken as the control background level.

Onto this background value was superimposed the dynamics of the posteruption aerosol on the assumption that 20%, 60%, and 20% of the aerosols were located, respectively, in the layers 20 to 26, 19 to 22, and 16 to 19 km. It was assumed that the aerosol was sulfuric acid and that the surface heat capacity was equivalent to a 70-m layer of the ocean.

A comparison of calculated and observed (in Port Hedland, Australia) changes of temperature at the 60- and 100-mb levels revealed a reasonable agreement, especially if one remembers that part of the observed temperature increase could have been caused by quasibiennial variability. The calculated course of the tropospheric mean temperature in the 30°N–30°S latitudinal belt, beginning with the eruption until early 1966, also agrees with observations: with a temperature drop of about several tenths of a degree, and with a time constant of about a year.

Table 2.9 shows calculation results for the cases of sulfate aerosols, weakly absorbing solar radiation, and silicate aerosols strongly absorbing solar radiation.

Considerable differences between the results point to a necessity for an adequate consideration of aerosol optical properties. Apparently, the prop-

TABLE 2.9   Temperature Changes in the Stratosphere and Troposphere Caused by Sulfate and Silicate Aerosols

Aerosol[a]	Time (Days)							
	30	60	120	180	360	540	720	1000
Sulfate								
T_{55mb}	0.63	2.15	4.77	5.34	3.75	1.90	0.84	0.02
T_{tr}	−0.01	−0.03	−0.12	−0.23	−0.48	−0.54	−0.51	−0.44
Silicate								
T_{55mb}	3.8	8.1	12.8	13.1	10.5	7.4	5.2	2.8
T_{tr}	−0.01	0.03	0.08	−0.13	−0.22	−0.23	−0.19	−0.13

[a]T_{55mb}, stratospheric temperature; T_{tr}, mean tropospheric temperature.

erties of real aerosols are intermediate compared to the characteristics of sulfates and silicates.

Results show that the value, sign, and phase shift of temperature changes in the troposphere and stratosphere (calculated using a 1-D climate model) agree with those observed after the Agung eruption. They also testify to the possibility of using such an approximate model to evaluate the climatic effects caused by the changed atmospheric composition.

It would be very important to undertake a complex program of observations and numerical modeling following a new large-scale eruption. The urgency of such a problem is determined, in particular, because a powerful eruption in the Northern Hemisphere can seriously disrupt agricultural productivity.

Oliver (89) suggested a semiempirical technique for estimating mean temperature changes in the Northern Hemisphere due to stratospheric dust loading based on the use of data on volcanic eruptions and subsequent coolings. Comparison of data on secular variations in the mean annual temperature in the Northern Hemisphere for 1880–1968 with chronological records of volcanic eruptions and assessments of dust outbreaks reveals the great power of these outbreaks and the existence of short periods of cooling after large-scale eruptions. So, for example, after the 1883 Krakatoa eruption, about 50 to 100 megatons of dust got into the global stratosphere, which attenuated solar radiation by 20 to 30%. Between 1880 and 1930 a general trend was observed toward a climate warming, which occurred even at the beginning of this period when very powerful eruptions took place.

During several decades (about 1915–1945) volcanic activity was rather weak, and air temperature was higher than in the preceding and subsequent (present) periods. Having analyzed various factors of climate change, Oliver (89) based his considerations on Mitchell's conclusion that neither increasing the CO_2 content nor variations in the tropospheric aerosol content, considered separately or together, could explain the temperature changes observed during the last century.

To estimate the eruption-induced variations in the Earth's surface mean temperature T_s, the following equation is proposed:

$$dT_s/dt = -\lambda(T_s - T_R),$$

where t is the time, λ the relaxation parameter (years^{-1}), and T_R an equilibrium temperature determined from the relationship

$$T_R = T_0 - \alpha M(t),$$

where T_0 is the mean equilibrium temperature of the dust-free atmosphere, α

the cooling coefficient (°C/megaton of dust), and $M(t)$ the dust mass as a function of time expressed as

$$M(t) = M_0 e^{-k(t-t_0)}.$$

It is assumed that T_0 is constant for a period of 10 to 20 years or 80 to 100 years and can be evaluated from data for a period of 5 to 10 years or, respectively, several decades after a weak volcanic activity. The first of these assumptions was used for the period late nineteenth-early twentieth centuries, and the second for the entire 1883–1968 period.

Introducing temperature anomalies

$$\theta(t) - \beta = T(t) - T_0$$

and bearing in mind the relationships above, one can easily obtain the formula

$$\theta(t) = (\theta - \beta)e^{-\lambda(t-t_0)} + \beta - \lambda\alpha M_0/(\lambda - k)[e^{-k(t-t_0)} - e^{-\lambda(t-t_0)}], \quad (2.1)$$

which determines $\theta(t)$ depending on time and parameters of the problem.

The value of αM_0 ($M_0 = m|_{t=t_0}$) can be inferred from actinometric measurements, with comparison with the heat budget of the surface-atmosphere system.

Calculations with formula (2.1) were made with a time step of 1 year, with M_0 as a sum of all the residual dust and the dust that appeared during this year. The constants λ and α were found by fitting $\theta(t)$ to observational data. Calculations were made starting from the year 1883.

An examination of calculation results showed that powerful volcanic eruptions lead to short-term climatic coolings which can be prolonged with frequent subsequent eruptions. Values of the cooling coefficient and time of reaction to a stratospheric dust loading thus obtained are consistent with the values obtained by other scientists.

An interesting feature is that in the case of an initial negative anomaly ($\theta_0 - \beta < 0$) corresponding to the preeruption warming, the post-eruption cooling effect is minimal and manifests itself almost immediately. Total cooling (from $t = t_0$ to ∞) caused by a certain eruption is $\alpha M^0/k$.

Calculations for short periods (20 years) and the entire period in question gave values of α(°C/megaton) of 0.07 to 0.10 and 0.16 to 0.21, and $c = 4\lambda$ (years) constituting 2.5 to 5.0 and 6 to 9 years, respectively. According to numerical modeling data performed within the Climate Impact Assessment Program (CIAP), the most probable α range is 0.27 to 0.36. Oliver (89) emphasized that even reliably estimated values of α, k, and λ parameters are inadequate to forecast the effects of a given volcanic eruption, since the consideration of the background trend is of great importance. This makes research on the long-term variability of stratospheric dust loading and other climate-determining parameters most urgent.

Harshvardhan and Cess (60) published approximate estimates of the climatic effects of stratospheric aerosols obtained on the assumption that it consisted of supercooled droplets of a 75% water solution of sulfuric acid condensed as a layer at an altitude of 20 km. The size distribution of number density N is determined from the formula

$$dN/d(\log r) = ar^b.$$

Here $a = 10^{-0.474}$ and $b = -382$ in the interval $0.3 \leq r \leq 1$ μm; $a = 10^{3.72}$, $b = 4.19$ at $0.1 \leq r \leq 0.3$ μm; $a = 10^4$, $b = 0$ at $0.03 \leq r \leq 0.05$ μm.

In the solar spectrum range a constant value of 1.43 was assumed for the refraction index (with the imaginary part of the complex refraction index about 10^{-7}). In this case the extinction coefficient is 10^{-6} and therefore the shortwave radiation absorption by stratospheric aerosols is four orders of magnitude weaker than the longwave radiation absorption. For the background optical thickness $\tau_{vis} = 0.02$, which corresponds to the geometric thickness of a homogeneous aerosol layer of 2.3 km.

An approximate estimate of the aerosol layer reflectivity gives $R = 0.37\tau_{vis}$ at a single-scattering albedo of $\omega_0 = 1.0$. Since $\omega_0(\lambda) < 0.1$ in the wavelength region $\lambda > 5$ μm, one can neglect the scattering when calculating the longwave radiation fluxes. An approximate consideration permits the following simple formulas to be derived for the emissivity and absorptivity of aerosol, respectively:

$$\epsilon = 0.0816\tau_{vis} \qquad \text{(for } T = 217 \text{ K)}$$

and

$$\alpha = 0.1029\tau_{vis}.$$

The value $\alpha/\epsilon = 1.26$ (other than unity) points to the effect of aerosol selectivity. Leaving out an account of the dynamic factors and estimating the aerosol-layer temperature proceeding from the radiative equilibrium condition (the radiative heat exchange is approximated by Newton approximation), we obtain the air temperature T_A:

$$T_A = 214 \left[(1 + 13.90\tau_{vis})/(1 + 13.22\tau_{vis})\right]$$

That is, in the absence of aerosols the air temperature at 20 km is 214 K, and in the case of the aerosol background concentration ($\tau_{vis} = 0.02$) it rises to 216.3 K. For a purely aerosol layer (not taking into consideration the molecular absorption effect) the temperature reaches 225 K.

One can, therefore, believe that the maximum possible warming of the stratosphere caused by a strongly increased aerosol concentration will reach 9 K. This estimate fits the observational data obtained after the 1963 Agung eruption (8°S), when the air temperature at 19.5 km altitude (from observations in Australia) rose by about 6 K and lasted at this level during a year, remaining at 2 to 3 K higher than the pre-Agung temperature for the next several years. The temperature increases at lower altitudes were much weaker, seldom exceeding 2 K at 12.4 km altitude.

The post-Agung value of τ_{vis} increased in the Northern Hemisphere by a factor of 2 (47°N) or 3 (19°N). According to Mauna Loa Observatory data, a lowered atmospheric transparency had been observed for 7 years after the eruption. The optical thickness in the Southern Hemisphere (25 to 28°S) increased by a factor of 8, remaining the same for a year. Although such a heavy dust loading of the stratosphere could not have caused a climate cooling equivalent to calculation data for the equilibrium state, one can assume that the 1960s volcanic activity had contributed somewhat to maintaining the temperature decrease that started in the 1940s. It should be emphasized that the aerosol-induced stratospheric temperature increase is the consequence of aerosol selectivity ($\alpha < \epsilon$).

At $\alpha = \epsilon$, the temperature would have remained practically constant. Consideration of the equation for the heat (radiation) budget for the Earth's surface-atmosphere system and the use of the empirical relationship between the outgoing longwave emission, surface temperature, and cloud amount make it possible to obtain the following approximate formula for a surface temperature change caused by stratospheric aerosol:

$$\Delta T_s = \frac{S}{dF/dT_s}\left[1 - \frac{(1-R)/(1-AR) - \epsilon\sigma T_A^4/S}{1-\alpha}\right],$$

where S is the solar constant, F the outgoing longwave emission, and $A = 0.3$ the planetary albedo in the absence of aerosols.

With $dF/dT_s = 15$ W/(m²-K) for a condition of 50% cloudiness, the normal (background) stratospheric aerosols ($\tau_{vis} = 0.02$) cause a decrease in the global surface temperature by 0.7 K. Doubling the aerosol content should double this temperature decrease. This is determined because the effect of an aerosol-induced increase of the planetary albedo outweighs the impact of the aerosol greenhouse effect.

These estimates showed that the solar radiation extinction caused by stratospheric aerosols leads to Earth surface cooling, whereas an intensification of the greenhouse effect entails its warming. The aerosol solar radiation absorption causes a stratospheric warming. Calculations of these effects associated with volcanic aerosols agree with observational data. It turned out, however, that the effect of warming or cooling of the Earth's surface depends critically on the size of aerosol particles.

Since there is no observational evidence on the post-eruption stratospheric aerosol size distribution, the evolutionary path of the aerosol size distribution can be described based on the use of a physicochemical model of the H_2SO_4 aerosol formation. Toon et al. (119) applied this approach and came to the conclusion that the stratospheric aerosol of volcanic origin causes a stratospheric warming and surface cooling, the calculated values for warming and cooling corresponding closely to those observed.

2.5.2 Numerical Modeling of the Climatic Impact of Stratospheric Aerosols

The most reliable way of revealing the climatic impact of volcanic eruptions consists, naturally, of the numerical modeling of climate based on the use of sufficiently complete (2-D or 3-D) climate models with maximum regard to observational data.

The lower 50-km layer of the atmosphere can be considered as a two-layer model consisting of the troposphere in the state of convective equilibrium and the stratosphere, where the conditions of approximate radiative equilibrium are observed. The general circulation of the troposphere is determined by heat sources located in its lower layers in equatorial and subtropical latitudes. Heat sinks are located in the middle and upper stratosphere and in polar latitudes. Stratospheric circulation is caused by the energy transfer from beneath, but of great importance also is the radiative heat flux divergence in the layer from 40 to 50 km, which is positive in the summer hemisphere and negative in the winter hemisphere.

Characteristic features of the vertical profiles of heat flux divergences and atmospheric motions determine the specific character of the vertical temperature profiles in the troposphere and stratosphere, and in this connection the difference in relative contributions by vertical and horizontal motions into the heat transfer, momentum, and various pollutants. One of the important specific features is a great difference between characteristic lifetimes for trace gases in the troposphere (1 to 2 weeks) and stratosphere (1 to 2 years). Another difference is in the role of minor gaseous components as the factors that determine the processes in the troposphere and stratosphere.

An analysis of general laws of the tropospheric-stratospheric circulation

and climate permitted numerical modeling of the climatic impact of stratospheric aerosol. Kondratyev and Moskalenko (19) considered specific features of the radiative heat exchange in the atmosphere disturbed by an
evolving gas-particle cloud following a volcanic eruption.

Air masses disturbed by the products of volcanic eruptions are characterized by increased concentrations of CO_2, SO_2, and CH_4, as well as by considerable pollution by volcanic aerosol. An increased concentration of the
oxides of nitrogen is also possible, but there is no reliable information about
their concentration in the erupted products.

A thick gas-dust cloud of volcanic origin moves rapidly, driven by air
diffusion and circulation, covering a broad band of moving air masses, and
through gaps in the tropopause made by strong upward fluxes gets to the
stratosphere, forming a stratospheric aerosol layer with the optical and structural characteristics evolving in time. The stratospheric cloud also has an
increased volume concentration of SO_2 (as compared to the background
one), which reaches 5×10^{-6}.

Measurements revealed a bimodal size distribution of stratospheric aerosol represented by modes of "newly generated" and "old" stratospheric
aerosols. At the initial state the mode of new aerosols has a modal radius
$r = 0.003$ to 0.01 μm. With time, as a result of coagulation, this mode
integrates and upon reaching a critical size, particles fall out of the aerosol
cloud, due to sedimentation. The old aerosols evolve independently. The
time of sinking for the stratospheric aerosol cloud is about 0.8 to 1.5 years
(depending on the environmental conditions and the power of volcanic
ejection). A long lifetime of the stratospheric volcanic cloud determines its
great planetary-scale significance for the radiative heat exchange and
climate.

Models of the optical characteristics of the stratospheric aerosol layer,
taking into account the evolution of its chemical composition and size
distribution, were constructed (19) based on measured size distribution and
calculated coefficients of absorption, scattering, and on phase functions for
various types of stratospheric aerosols. Most substantial changes in the radiative regime of the stratosphere at heights of 20 to 30 km are caused by
volcanic aerosols. Numerical modeling showed that the albedo of the sur-
face-aerosol cloud system A_s depends on the value of surface albedo and on
the optical characteristics of aerosol and the concentration of the aerosol
cloud. With the surface albedo $A_0 > 0.39$, the aerosols cause a decrease in the
system albedo A_s. If $A_0 < 0.39$, stratospheric aerosols raise the system albedo.
With the optical thickness of the stratospheric cloud $\tau_a = 0.3$ (at $\lambda = 0.55$
μm) the global radiation decreases by almost 10%, the planetary albedo
increases by 7% and the stratospheric temperature increases by 6 K (mainly
at the expense of the absorption of the thermal radiation from the surface

and troposphere). The effect of increasing the planetary albedo at the expense of the shortwave radiation reflection is partly compensated for by the absorption of the UV radiation by erupted SO_2.

A consideration of the volcanic effect on the radiative heat exchange of the atmosphere suggests a conclusion about the cooling of the surface and troposphere. This cooling effect is amplified by increasing the planetary albedo due to the growing zone of glaciation and the shifting of the ice cover toward lower latitudes. Naturally, the reaction of the climate to an intensification of volcanic activity shows itself better on the continents than over the oceans. The thermal inertia determines weaker changes in the climate of the Southern Hemisphere than of the Northern Hemisphere, since the oceanic area in the Northern Hemisphere is less. There is an opinion that volcanism in present conditions increases the water exchange in the oceans-continents-atmosphere system, promoting an accumulation of ice and a broadening of the area under the ice cover. Thus a correct account of various feedbacks in modeling the climatic effect of volcanic activity is an important problem.

At the initial stage, prior to a volcanic eruption, the troposphere and stratosphere get an abundance of gaseous products. Powerful gas ejections to the atmosphere often take place prior to volcanic eruptions, and therefore the processes of warming and cooling can substantially be phase-shifted in time with respect to periods of strong volcanic eruptions. These raise the concentration of the stratospheric water vapor, atmospheric carbon dioxide, and sulfur dioxide. Pollution of the troposphere and stratosphere by the oxides of nitrogen is also possible. The presence of the latter in the atmosphere can share a strong climatic effect due to variations in the planetary albedo and manifestations of the greenhouse effect. If the strength of volcanism is enough to compensate for the sink of the gaseous components with fresh gas ejections or even to increase their concentration, an effect of warming is observed which is intensified, due to feedback. With sporadic and infrequent volcanic eruptions, a warming will first take place, followed by a slow cooling of the troposphere and surface, caused by a decrease in the concentration of volcanic gas ejections and by the evolution of the stratospheric aerosol cloud.

Calculations of climatic impacts taking into account feedback (and without such considerations) gave a decrease in the mean global temperature by 0.3 and 0.7 K, at a thickness of the stratospheric aerosol layer $\tau = 0.3$ and a $\frac{1}{10}$ coverage of skies. Pulsations of volcanic activity may be one of major reasons for the thermal modulations of climate and for the growth or melting of the ice cover in past epochs.

To estimate the possible climatic impact of the eruptive cloud from El Chichón, Pollack and Ackerman (93) performed calculations of the vertical profiles of radiative fluxes and temperature (using a 1-D radiative-convective equilibrium model) during the first 6 months after the eruption. It is assumed that there are water clouds in the atmosphere amounting to 50% (separate calculations were made for a clear and cloudy atmosphere). The optical thickness of clouds is taken to be 7.5, which corresponds to a system albedo of 0.6 The volcanic aerosol is prescribed as a polydisperse ensemble of silicate nulei (their share is an adjusting parameter) covered with the sulfuric acid shell. The optical thickness of the volcanic aerosol at 0.55 μm is taken as 0.3, and the lognormal size distribution for the model radius 0.6 μm and $n = 1.5$. The aerosol consists (by volume) of a 90% H_2SO_4 water mixture (concentration 75%) and 10% silicate ashes. The aerosol layer is located in the altitude range 16 to 31 km with a maximum concentration at 28-km height.

Calculations showed that the formation of the eruptive cloud causes an increase of the surface-atmosphere system albedo at $\mu_0 = \cos\theta_0 = 2/\pi$ (θ_0 is the solar zenith angle) by 0.027 (by about 10%); the albedo increases in proportion to the aerosol optical thickness, increasing from 0 to 0.5. It is, however, practically insensitive to the share of ash f (varying from 0 to 0.9) and to the modal radius of particles. Calculations of the volcanically caused reduction of the global radiation in clear-sky conditions at Mauna Loa ($2.2 \pm 0.4\%$) gave results consistent with observations.

A stratospheric temperature increase caused by the absorption of the upward longwave radiation flux by the eruptive aerosol was maximum at 24-km height (30-mb level), reaching 3.5 K, which agrees well with the summer 1982 observation data (these calculations contain an adjusting element consisting in setting $\mu_0 = 0.60$, which provides better agreement with the observed temperature profile in the tropics before the eruption). Maximum warming at 24-km height (but not 28 km, where the aerosol concentration is maximum) is connected with an additional contribution to the warming at the expense of the absorption of the downward longwave radiation emitted by the aerosol layer above. If f (the share of ashes) > 0.2, a considerable contribution to the solar radiation absorption is made by volcanic ash. The warming of the eruptive aerosol layer is much greater over the ocean than over land, since the land surface temperature markedly decreases.

MacCracken (72) of the California University Lawrence Livermore National Laboratory applied a 2-D zonal model of general atmospheric circulation (ZMA-2) to estimate the climate's sensitivity to variations of the total content of ozone, to stratospheric aerosols, and to water vapor as well as to variations in the solar constant. The first stage of numerical modeling consisted of test calculations aimed at comparing calculated fields of meteorological elements with observations.

Calculations made with insolation assumed to be constant during a year and also with due regard to annual and diurnal changes showed that in the first case a persistent climatic regime sets in much faster. Calculated (at a fixed mean annual insolation) and observed temperature fields fit rather well, except for the lower stratosphere in the polar latitudes, where calculated values of temperature are overestimated. An agreement between the specific humidity fields is also satisfactory. The calculated distribution of precipitation is characterized by an underestimated minimum of precipitation in the subtropical belt. The fields of the total atmospheric water vapor content and cloud amount also fit well.

Significant differences are observed in comparing mean annual fields of the zonal wind, but results are better with a seasonal change of insolation taken into account. Calculated meridional circulation reveals only one Hadley cell instead of the observed three cells of circulation, but the calculated meridional energy transfer agrees well with that observed, with advection prevailing in low latitudes and macroturbulent transfer prevailing in middle and high latitudes. ZMA-2 satisfactorily reproduces the values of heat budget components for the surface and the atmosphere.

On the whole, an examination of the adequacy of this model (at a fixed mean annual insolation) testifies to its usefulness for the estimation of climate sensitivity to various factors. Calculations of the effect of aerosol content variations in the layer from 75 to 150 mb were made for aerosol contents 0 and 0.43 μg/cm^2 (normal content) and the values exceeding this normal content by factors of 2 and 4. The complex refraction index of particles is assumed to be 1.45 to 0.005i, which corresponds to weakly absorbing particles. Estimates were also obtained of the climatic impact of an 11.7% decrease in ozone content, which could have been caused by the flights of 500 SSA.

Analysis of the calculated surface temperature showed that the effect of aerosol is stronger and less regular in high latitudes. With a fourfold increase in the aerosol content, which is practically equivalent to a 1% decrease in the solar constant (such calculations were made for comparison), the surface temperature decreases by about 1° in the low latitudes, with a maximum decrease of several degrees at about latitude 70°N. This causes the formation of snow cover on land near the southern boundary of which a warming effect (about 2 K) is observed associated with decreasing cloud amount and planetary albedo. This decrease has been caused, in its turn, by decreasing evaporation in midlatitudes.

Precipitation also decreases here by 3.6%. As is seen, a consideration of varying cloudiness leads to some radical changes, for example, a decrease (but not an increase) of the planetary albedo with the atmospheric dust loading. Calculations made for mean global conditions showed that with a tripled content of stratospheric aerosol as compared to a normal one (which apparently characterizes the maximum possible effect of SSA), the planetary albedo rises by 0.4 to 0.6% due to increasing scattering by aerosols. The actual situation turns out to be more complicated, because of an interaction between dust aerosol and cloudiness.

The effect on the hydrologic cycle of increasing the stratospheric aerosol content turned out to be unexpectedly strong. A decrease in the northward transport of energy is significant in the latitudes where cooling takes place. An increase in a dust loading of the stratosphere lowers evaporation from the surface and turbulent heat exchange, but raises the surface net longwave radiation (the latter due to decreasing atmospheric emissions because of reduced water content of the atmosphere). It is significant that the total shortwave radiation absorbed by the surface remains practically constant (even with a decreasing solar constant), which can be explained by a decrease in humidity and cloudiness compensating for the aerosol attenuation. The heat budget of the atmosphere is characterized by a decrease in latent heat and absorbed solar radiation. As far as the stratosphere is concerned, in case of a fourfold increase of the aerosol content its temperature rises by about 4 K. The albedo of the planet remains practically constant, but these results strongly depend on the treatment of cloud feedback processes.

A comparison of estimates of decreasing mean global surface temperature for the case of a fourfold increase in the aerosol content and a 1% decrease of the solar constant with similar assessments performed earlier by Sellers (using a semiempirical energy-balance climate model without a hydrological cycle) and based on the use of a radiative-convective equilibrium model and a GFDL 3-D model of climate (with a "frozen" cloudiness) led to the conclusion that there would be a considerably lower sensitivity of climate to stratospheric aerosols according to the zonal model. The problem of sensitivity requires, however, further serious study.

On the assumption that the mean global content of ozone will decrease by 11.7%, which would probably accompany regular flights of 500 SSA, climatic implications were assessed. The mean surface temperature increases by 0.24 K in this case (in midlatitudes of the Northern Hemisphere this increase is twice as high). The stratospheric temperature decreases by 0.5 to 1.5 K depending on latitude and altitude. The planetary albedo decreases somewhat because of changes in the hydrologic cycle.

Calculations of the effect of increasing the water vapor content in the stratosphere showed that it is very small: an increase of the mixing ratio above the 150 mb by 1 ppm leads to a rise in the surface mean temperature by 0.02 K and a drop in the stratospheric temperature by 0.2 K. MacCracken (72) emphasized that the results obtained were preliminary and further studies were needed, especially from the viewpoint of considering a number of the processes responsible for the water cycle.

MacCracken and Potter (71) undertook numerical modeling of climate changes based on the assumption that either the aerosol content in the global stratosphere increased by a factor of 10 (up to 4.3 $\mu g/cm^2$), which corresponds approximately to a maximum dust loading of the stratosphere after the 1963 Agung eruption, or a solar constant decrease by 3%. The complex refraction index for aerosol particles is assumed to be $1.45 - 0.005i$. Because a stable regime for a perturbed stratosphere has not been reached, the results must be considered preliminary.

In both cases similar climate changes take place. The mean global surface temperature decreases from 291.5 K (test calculations for present conditions) to 286.9 K. In both cases the temperature fields are similar: the temperature decrease varies from 4 K in the equatorial zone to 15 K in the polar regions. The air temperature in the troposphere also declines, but temperature variations in the stratosphere are more complicated. When the stratosphere becomes loaded with dust, it becomes strongly heated (8 K), whereas when the solar constant decreases, the lower stratosphere is slightly warmed and the upper stratosphere becomes cooler. Changes in the constituents of the heat and the water budgets are given in detail.

The surface net longwave radiation increases (especially in polar regions) due to decreasing atmospheric emission combined with a decrease in global radiation and evaporation (the latter decreases by more than 12%). An increase in the planetary albedo takes place as a result of increasing (by 2%) the cloud amount when the solar constant decreases and as a result of the aerosol effect when the stratosphere becomes dust loaded. In both cases the cloudiness increases at the expense of increasing relative humidity in the middle troposphere. The intensity of the Hadley circulation cell increases to some extent. The boundary of the snow cover shifts southward: by 10° of latitude with stratospheric dust loading (the zone of snowfall shifts southward by 10° of latitude, reaching 60° N). When the solar constant decreases, snow accumulates in mountains at 60° N and snowfall reaches 50° N.

Numerical modeling results show that stratospheric dust loading and a decrease of the solar constant are not equivalent from the viewpoint of their climatic impact, as some authors have suggested. This nonequivalency is determined because feedback relationships due to the hydrologic cycle and variations in the heat-budget components are more important than albedo feedback (an increase in albedo with enlarged ice or snow cover, and vice versa). The most significant interacting factors are as follows:

1. A decrease in incoming solar radiation in the troposphere leads to cooling, which entails an intensification of cloudiness, which in turn leads to an increase in albedo and a decrease in absorbed radiation.

2. A decrease in the solar radiation absorbed by the surface, together with

atmospheric cooling, leads to a substantial decrease of evaporation and water content of the atmosphere.

3. The latter determines a decrease in the atmospheric thermal emission and an increase in net longwave radiation, especially in polar regions, where the water content is small.

4. In connection with decreasing meridional energy transfer in the form of latent heat, the meridional circulation is intensified and the role of diabatic heating increases.

The results obtained by MacCracken and Potter (71) underscore the need to take into account various feedbacks when studying the factors that determine climate changes. MacCracken and Luther (73) performed calculations of the effect of El Chichón volcanic stratospheric aerosols on the atmospheric radiative regime, the results of which were used in the numerical modeling of the climatic impacts of eruptions with a 2-D zonal dynamic statistical climate model in which a latitude-dependent annual change of cloudiness was prescribed. Table 2.10 presents results of calculations of variations in direct (S), scattered (H), and global (G) solar radiation at the surface level, with the prescribed aerosol optical thickness 0.255 at 0.55-μm wavelength (this corresponds to conditions of July 1982 in Hawaii).

These results agree well with observations. In estimating the equilibrium effect on climate it is assumed that the aerosol optical thickness in the infrared constitutes 5.9% with respect to the visible (since the subsequent observational data analysis revealed a substantial underestimation of this parameter, it will be raised to 25%).

A major result of numerical modeling of the equilibrium state consists not only of the conclusion about a gradual global-scale cooling during the year following the eruption but also in revealing a meridional redistribution of rainfall. In the presence of an eruptive aerosol layer at altitudes of 23 to 29 km in the latitude band 5–35°N calculations showed a 1.0 K decrease of surface air temperature at 20°N, increasing to 1.4 K at 70°N. Averaged

TABLE 2.10 Volcanically Induced Changes (%) in Shortwave Radiation Fluxes

	Sun Elevation	
Parameter	60°	30°
S	−19	−30
H		
Without aerosol	0.04	0.07
With aerosol	0.28	0.46
G	−1.2	−4.9

values were 0.9 K (for the Northern Hemisphere) and 0.3 K (for the Southern Hemisphere).

Calculations revealed a southward shift of the ITCZ (Intertropical Convergence Zone) and an intensification of the Hadley cell in the Northern Hemisphere, which favored the "pumping" of heat from the Southern Hemisphere oceans, thereby determining an indirect climatic effect of eruptions in the Southern Hemisphere. The shifting of the ITCZ intensifies rainfall in the subtropical Southern Hemisphere. The intensity of the global moisture cycle weakens and is manifested through decreases in rainfall and evaporation.

A prescribed temporal change in the aerosol optical thickness substantiated by Robock (103) made it possible to calculate the climate dynamics between 1 April 1982 and 1 July 1983. These calculations showed that between late 1982 and July 1983 a uniform 0.3°C cooling took place in the Northern Hemisphere, whereas in the Southern Hemisphere only a slight cooling was observed. A latitudinal intensification of cooling turned out to be much weaker than in the case of a stationary mean annual regime. A year after the eruption the mean global SST rose by only 0.15 K. Calculations suggested the recovery of the mean global temperature by mid-1983.

Since the estimates of the equilibrium climate suggested the possibility of detecting the effects of the El Chichón eruption for several years after the eruption, MacCracken and Luther (74) estimated a time-dependent regime using a seasonal model with a 70-m oceanic layer. In this case a time-dependent meridional profile of the aerosol optical thickness is prescribed, with a maximum of 0.3 in June. Calculations for the period between 1 April 1982 and 1 July 1983 revealed a gradual global-scale temperature decrease of about 0.2 K for the last half year of integration. The results do not permit one to estimate the possible duration of cooling and its subsequent intensification. Numerical modeling showed a decrease of rainfall from 5% to more than 10% north of the ITCZ, during the year following the eruption.

The authors (74) emphasized the necessity to continue numerical experiments to use the "volcanic experiment" in order to check the reliability of climate models applied to estimate the sensitivity of climate to a growing CO_2 concentration. Such experiments will provide a better understanding of the causes of climate changes during the next century.

Charlock (45) calculated the vertical profile of the IR radiative warming of the stratosphere (at altitudes of 12 to 28 km) at the equator in July for conditions simulating a disturbed stratosphere after the El Chichon eruption. A 4-km eruptive aerosol layer (optical thickness 0.25 at 0.69 μm) is assumed to be located at altitudes either 18 or 25 km. In both cases (a cloudless atmosphere is considered) nearly equal relative warming of the aerosol layer takes place, caused by the absorption of the surface thermal ·

emission in the atmospheric transparency window at 760 to 1240 cm^{-1}, which is only partially compensated for by cooling due to the self-emission of the layer.

An absolute value of the radiative heat flux divergence in the lower aerosol layer is about twice as large as in the upper layer. The aerosol-induced warming is determined only by the contribution from the transparency window, since in the remaining spectral intervals the radiative heat flux divergence is negative. The total radiative temperature change in the lower layer is positive, while in the upper layer it is negative. An increase in the total downward thermal emission flux due to the volcanic stratospheric aerosol at the surface level constitutes only 0.1 to 0.3 W/m^2, and at 10-km height it varies within 0.7 to 1.2 W/m^2 (these calculations were made for latitudes 0° and 35°N). No doubt the effect of aerosols on the tropospheric climate is determined by their impact on the shortwave radiation transport. With clouds in the troposphere not only the value but also the sign of the aerosol impact on the stratospheric radiative regime can vary, since here the aerosol-absorbed upward thermal emission flux decreases. The change in the radiative heat flux divergence is in approximate proportion to cloud-top height above 4 km.

Interesting results were obtained by Batten (35) from numerical modeling aimed at the assessment of the climatic impact of a homogeneous dust-cloud layer in the stratosphere uniformly distributed through the 25 – 75°N latitudinal belt, based on the use of an atmospheric general circulation model developed by Mintz and Arakawa. The basic feature of the physical scheme of modeling consists of consideration of not only an attenuation of solar radiation by the dust cloud but also the effect of the dust aerosol on the processes of cloud and rain formation. An extreme case of stratospheric dust loading is considered, equivalent by dust volume (4×10^{-2} km^3) and particle diameter (2 μm) to the 1883 Krakatoa eruption, but with the extent of the stratospheric aerosol layer approximately half as much (the posteruption dust layer had covered the 30°S – 30°N latitudinal belt).

In the Mintz-Arakawa model the solar radiation transfer was considered by dividing the entire shortwave radiation spectrum into two intervals:

1. A wavelength region at $\lambda < 0.9$ μm to which 65.1% of the solar constant S_0 corresponds, where only the Rayleigh scattering is taken into account.

2. $\lambda \leq 0.9$ μm, where only the absorption by a cloudless atmosphere is considered.

It was assumed that in numerical modeling of the effects of the stratospheric dust layer it is sufficient to consider only the region where the scattering effect manifests itself. Mean optical thickness of the layer is assumed to be 0.92, which corresponds to the persistent post-Krakatoa dust loading of the stratosphere. Thus solar radiation is attenuated by the dust layer $e^{-0.92} = 0.4$ times, which is equivalent to a decrease of the subcloud short-wave radiation flux to $0.651 \, (1 - e^{-0.92}) S_0 = 0.39 S_0$. This corresponds to a 5% mean global decrease of the incoming shortwave radiation, provided that the dust cloud is in the Northern Hemisphere , and a 15% increase if it is in the Southern Hemisphere. The resulting decrease in the shortwave radiation absorbed by the surface broadens the zone of the negative surface radiation budget in the given winter period to 30°N (the effect of the dust cloud on the absorbed radiation and radiation budget manifests itself most clearly at low latitudes).

It is assumed that an appearance of the stratospheric dust layer entails the increase in the rain rate by 20% in the 25–75°N zone. This is equivalent to the following changes in the three types of rainfall considered in the Mintz-Arakawa model: (i) the threshold value of the relative humidity 99.67% (instead of 100%) at which rainfall starts, corresponds to large-scale rainfall; and (ii) the intensity of two types of convective precipitation increases by 20%, which is equal to the corresponding intensification of convection.

Batten (35) discussed the results of numerical modeling for a 60-day period (31 December to 28 February) for control and "dust" cases, which may characterize only an initial reaction of the atmosphere to the dust layer. A comparison of data related to these cases was made by using 25-day averages (21 January to 14 February) to exclude the effect of abnormally intensive rainfall observed in different regions in both cases. The under-dust cloud air temperature and land surface temperature lowered by 2 to 4° (the ocean surface temperature in the Mintz-Arakawa model is assumed to be constant). An exception are the latitudes near 70°N, where the temperature remains practically constant, owing to increasing cloud cover. The land-ocean temperature contrasts (typical of winter) intensified.

A decrease in surface temperature should lead to a southward shift in the snow cover boundary, and a resulting increase in albedo should further decrease the solar radiation absorbed by the surface. As a result, the south-ward extending snow cover could have remained longer (including spring and probably summer). Therefore, one can expect that a dust cloud appearing in winter creates strongly changed conditions (as compared to normal conditions) in spring and summer. In the model above the surface albedo was, however, assumed to be constant, which prevented the study of these effects (this limitation will be eliminated in the future).

Despite a 20% increase of the rainfall rate, the total amount of rainfall has generally not changed in the latitudinal belt under consideration. However, its latitudinal distribution has been substantially transformed: precipitation in high latitudes (46 – 74°N) decreased by 19.5%, and in low latitudes (26 – 46°N) increased by 22.8%. Although some peculiar features of the precipitation field can be explained by evaporation anomalies, variations in the circulation and moisture flux are a major factor. So, for example, a decrease in high-latitude rainfall is totally the result of the attenuation of water vapor convergence in this region. As a result of an additional attenuation of solar radiation by the dust cloud and increasing precipitation, an attenuation of the Ferrel circulation cells takes place in midlatitudes, as well as a decrease of the baroclinicity of the atmosphere (and hence the northward macroturbulent transport of water) in high latitudes (in the northern part of the dust cloud). An opposite situation (growing baroclinicity, intensification of the water transport) takes place in low latitudes; that is, the zone of baroclinicity drifts toward the southern boundary of the dust cloud. It was emphasized that in future all of the most significant feedbacks, especially the interaction with the ocean, must be taken into account.

Also of great importance are further studies of physical mechanisms for the shortwave radiation transfer (the aerosol effect) and its impact on precipitation (the role of condensation nuclei). Studies aimed at assessing the meteorological consequences of eruptions were preliminary, since they were based on the assumption of a fixed aerosol content in the stratosphere and on simplified numerical modeling of stratospheric general circulation. In this connection Hunt (65) undertook numerical modeling of the climatic impact of volcanic aerosols, taking into account a realistically reproduced diffusion of eruption products from their source in tropical latitudes (the 1883 Krakatoa eruption was considered, the geographical coordinates of which are 6°S 105°E).

Numerical modeling using a 3-D AGCM permitted an evaluation of both the direct impact of eruption products on the radiative and thermal regimes and the indirect effect on the temperature field through atmospheric dynamics. An 18-level hemispherical GCM (a finite-difference analogue to the system of primitive equations in stereographic projection) used in calculations is a further modification of the GFDL.

An integration was made for mean annual conditions without regard to orography and land-ocean contrasts, but taking the hydrologic cycle into account. For comparison purposes, results of previous numerical modeling for the case of an undisturbed stratosphere were used. Since the model does not take into account the thermal regime of the oceans, it becomes highly sensitive to different influences as compared to real atmospheric conditions.

Radiative heat flux divergences due to shortwave and longwave radiation are considered on the assumption of fixed climatological disturbances of water vapor, carbon dioxide, ozone, clouds, and surface albedo. At the initial stage the source of aerosols is assumed to be located at a height of 23 km in a

3° latitudinal belt encircling the equator. Since the GCM contains a "wall" at the equator, the aerosol diffusion confines itself to the Southern Hemisphere.

It is assumed that having reached the lower level (0.85 km), the aerosol is immediately washed out from the atmosphere (this sink for aerosols is assumed to be the only one). In accordance with the aerosol model proposed by Deirmendjian (171), it was assumed for the first post-Krakatoa eruption phase that the aerosol size distribution is characterized by prevailing large (2-μm) particles. The initial total number of particles determining their mixing ratio constitutes 8×10^{23} particles.

Particles are assumed to be silicate (purely scattering) with the refraction index 1.56 independent of wavelength within the shortwave radiation spectrum. This corresponds to the attenuation coefficient 2.9 km^{-1} at $\lambda = 0.45$ μm for the particle concentration 10^8 m^{-3}. The effect of stratospheric aerosol on the longwave radiation transfer is not considered, and in case of the shortwave radiation it is determined by scattering, which for mean annual conditions, constitutes about 15% of total scattered radiation.

Specific character of the model radiation parameterization is that it does not permit an adequate consideration of the backscattering effect, since part of the backscattered radiation manifests itself as local warming (up to 10°C at the upper level). This disadvantage of the model can, however, be considered as a certain merit, since observations reveal a strong heating of the stratosphere after large-scale eruptions.

Calculations of the atmospheric general circulation were made with a time step of 10 minutes, but the effect of change in the radiative heat flux divergence is taken into account every 24 hours. A "volcanic" experiment began on day 254 of the control numerical modeling and continued for 150 days, when the volcanic aerosol spread all over the hemisphere. For 130 days the control and volcanic experiments were carried out simultaneously. The eruption effects were estimated by comparing 10-day running means with the first day of the volcanic experiment. Additional calculations were also made during 14 days by Batten (35), with the aerosol diffusion assumed to be a passive tracer (taking into account an interaction with radiative factors).

The comparison of variability of calculated values of the mean hemispheric kinetic energy of the atmosphere showed that significant differences (as compared to control calculations) occur only during the second and third weeks, when a higher concentration of aerosols was observed in the tropical latitudes. With the effect of radiative factors on the aerosol distribution not taken into account, the values of kinetic energy vary around the control ones.

The effect of volcanic aerosols on the wind field is weak and consists, as a rule, of a slight attenuation of the mean zonal wind and some changes in synoptic wind fields. These would be difficult to trace in the real atmosphere. The temperature field changes substantially (although these changes are overestimated because of neglecting the thermal inertia of the ocean). Mean surface temperature of the hemisphere decreases by 0.7 K. In the middle and high latitudes the effect of an eruption is masked by variations in local weather conditions (Fig. 2.16).

Such a limited reaction of the atmosphere in the middle and high latitudes is determined by the "hemisphericity" of the model and by the comparatively short duration of the experiment (consideration of global disturbances made in other studies revealed a maximum response to eruptions in the high latitudes). Temperature variations outside the tropical belt are mainly caused by variations in the meridional energy transfer (especially the latent heat of condensation) and not by the direct effect of stratospheric dust loading on the attenuation of solar radiation. An eruption generally does not affect the hydrologic cycle. This conclusion is apparently incorrect if one monitors the evolutionary path of the processes for a longer period.

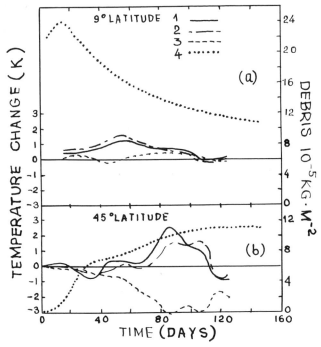

FIGURE 2.16. Zonally averaged temperature differences for the control and "volcanic" cases: (a) at 9°N; (b) at 45°N. 1, 5.35 km (515.8 mb); 2, 9.55 km (283.6 mb); 3, 12.2 km (217.1 mb); 4, dynamics of ejected material.

The main conclusion drawn by Hunt (65) from numerical modeling is that the climatic impact of individual large-scale eruptions cannot be of long duration. More serious effects must occur in case of multiple eruptions. Numerical experiments should continue to assess the effect of volcanic eruptions on climate using a more realistic AGCM which takes into account ocean circulation and annual variability. The radiation parameterization of the model requires specification, especially from the viewpoint of taking into account the aerosol effect on the longwave radiation transfer. Consideration of the hydrologic cycle and surface characteristics should be more adequate. Of great importance is consideration of the effects of different types of volcanoes located at different latitudes. Extremely limited are observational data that characterize the products of volcanic eruptions, their distribution, and transformation.

Using an energy-balance climate model developed by Sellers and Robock (reported in Refs. 101 and 166) estimates of the climatic impact of eruptions were obtained, based on the following assumptions (Robock [166]):

1. The characteristic time of gas-phase reactions of the formation of the stratospheric sulfate aerosol is 60 days.

2. An exponential decay of the eruptive cloud begins in 500 days following the eruption (Fig. 2.17).

3. The spatial and temporal distribution of the cloud optical thickness is retrieved, taking into account maximum optical thicknesses 0.30 to 0.35 (this corresponds to observations in Hawaii).

4. The disturbance of the global air temperature field near the surface is determined only by variations in the radiative regime of the surface-atmosphere system under the influence of the eruptive cloud.

Numerical modeling performed separately for land and the world ocean showed that the initial reaction of climate (cooling) to the eruption is faster

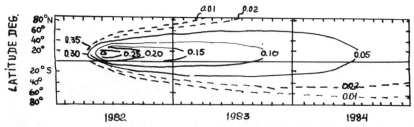

FIGURE 2.17. Model of the spatial and temporal variability of the optical thickness of the eruptive cloud.

over land (which is explained, of course, by large thermal inertia of the ocean). In 2 years, however, this difference is smoothed (Figs. 2.18 through 2.20). Naturally, mean zonal values of cooling ΔT are determined mainly by the contribution from the ocean due to its larger surface area (Fig. 2.21). A most substantial climate cooling must take place not in the region of the eruption but in the North Pole region in spring and fall of 1984 and in spring of 1985. This can be explained by the effect of nonlinear feedbacks determined by the high albedo of the snow and ice cover (the albedo feedback), as well as by the thermal inertia of the ocean. The formation of maximum ΔT

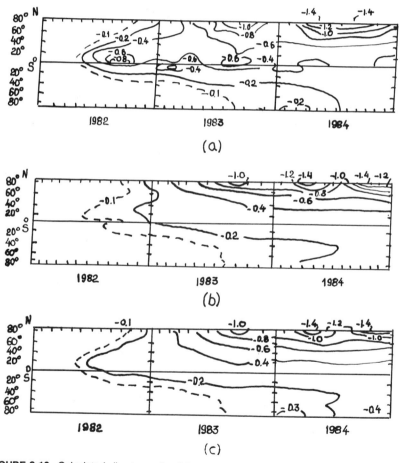

FIGURE 2.18. Calculated climate cooling (air-temperature decrease near the surface) during the first 2.5 years following the volcano eruption of El Chichón: (a) land; (b) ocean; (c) mean zonal values.

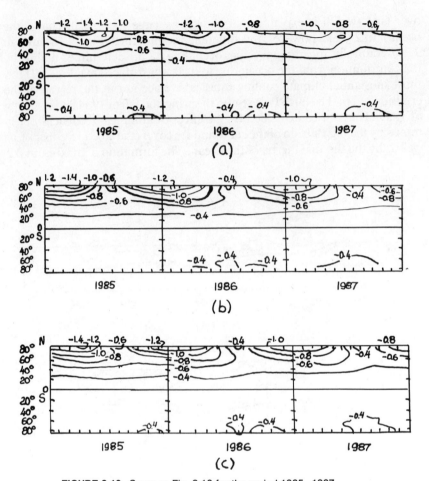

FIGURE 2.19. Same as Fig. 2.18 for the period 1985–1987.

in the spring and fall is determined by the superposition of an increased sensitivity of polar regions to effects on climate (as a result of the feedback noted above) and by the maximum change in the radiative regime at a prescribed optical thickness of the atmosphere. Numerical experiments carried out by Robock (166), without account of feedback, revealed a linear reaction of climate to radiative disturbances and, respectively, a maximum variability in the region of maximum density of the eruptive cloud, without an intensification or phase shift of ΔT at the pole.

The fact that the climate of polar regions with its maximum observed variability is most sensitive to eruptions might be of great importance from the point of view of detecting climate changes due to an increasing CO_2 concentration:

1. The effect of eruptions is characterized by similar spatial distribution but is of the opposite sign, thereby masking the CO_2 effect.

2. The maximum interannual climate variability in high latitudes can result in part from previous eruptions, meaning that by comparing the estimates of the CO_2-induced climate variability (or those due to eruptions)

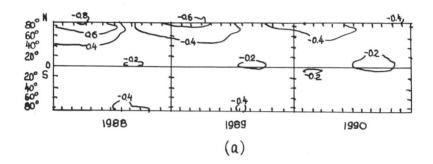

(a)

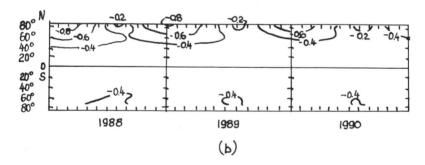

(b)

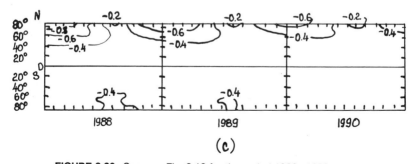

(c)

FIGURE 2.20. Same as Fig. 2.19 for the period 1988–1990.

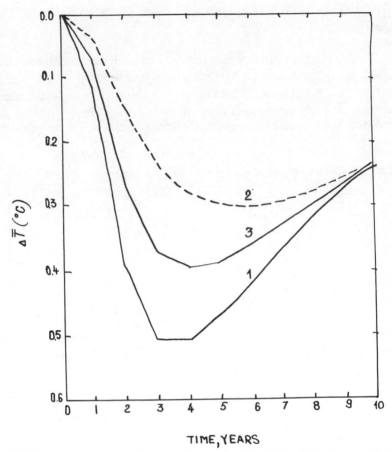

FIGURE 2.21. Calculated climate cooling. 1, Northern Hemisphere; 2, Southern Hemisphere; 3, globe.

and considering the S/N ratio as a criterion of this variability, we are in fact dealing with the $S/S + N$ ratio. Hence detection of the CO_2 signal is possible only upon filtering out the "volcanic" signal.

Calculations of mean hemispheric and mean global climate cooling showed (Fig. 2.21) that the reaction of the Northern Hemisphere (NH) is faster and stronger than that of the Southern Hemisphere (SH), since the NH land area is much greater than in the SH. Upon returning to the undisturbed state, the prevailing influence of the thermal inertia is pronounced: the NH is warmed faster than the SH and therefore by the year 1991 the climate anomalies in both hemispheres will have been similar. Numerical modeling suggested the conclusion that the process of undisturbed state recovery will start in 1985-1987, with numerous anomalies still taking place in the NH polar regions, and disappearing only by 1990.

2.5.3 Conclusions

Principal aspects of studies on volcanically caused climate changes were formulated in the Global Aerosol-Radiation Experiment (GAREX) program (167) and discussed in various papers (12-14,17). A important step in this direction has been publication of the collective monograph edited by Newell and Deepak (83). Without repeating ourselves, we shall enumerate only the key aspects of further studies:

1. Thorough global-scale monitoring of the erupted gaseous and aerosol products with the use of available conventional and satellite techniques, bearing in mind detailed studies of the chemical composition and physical properties of erupted material both close to the place of eruption and in the global atmosphere (with emphasis on sulfur and nitrogen compounds).

2. Further studies of gas-to-particle conversions responsible for the formation of stratospheric aerosols.

3. Numerical modeling of the regional and global impacts on climate (primarily air temperature and precipitation).

4. A search for the most sensible indicators of the climatic impact of volcanic eruptions.

5. Paleoclimatic reconstructions that characterize the effect of volcanic eruptions on climate.

A reliable assessment of the effect of volcanic eruptions on climate is possible by numerical modeling with the use of complete climate models based on adequate empirical information. So far, however, none of the large volcanic eruptions have been adequately documented. Therefore, there is an urgent need for a problem-oriented international program as well as observational subprograms aimed at:

1. Regular observations of active volcanoes and their impact on climate.

2. Observations with the use of a regularly operating network and specific observational means after large-scale volcanic eruptions.

The first steps in these directions have been taken by the USSR within the GAREX program (167).

REFERENCES

1. I. M. Alekseev, S. A. Volovikov, Yu. G. Kaufman, and S. S. Khmelevtsov, 1982. A two-level latitude-averaged model of the impact of stratospheric aerosol on the surface temperature, *Proc. IBM* **28**(101), 65–80.

2. A. S. Grigoryeva and O. A. Drozdov, 1975. On the impact of volcanic eruptions on the Northern Hemisphere precipitation, *Tr. Gl. Geofiz. Obs., Meteorol. gidrol* **354**, 102–108.

3. N. I. Kalitin, 1935. Atmospheric transparency and volcanic eruptions, *Klimat Pogoda*, N 4(61), 8–11.

4. I. L. Karol and E. V. Rozanov, 1984. Impact of volcanic eruptions on the thermal structure of the atmosphere, *Meterol. gidrol.* **6**, 105–107.

5. Yu. G. Kaufman, M. P. Kolomeev, and S. S. Khmelevtsov, 1983. Modeling the impact of stratospheric aerosol on climate, *Meteorol. gidrol.* **6**, 5–12.

6. K. Ya. Kondratyev and L. R. Rakipova, 1974. Radiation and dynamics of the atmosphere: radiative effects of aerosols, *Tr. Gl. Geofiz. Obs.* **344**, 64–82.

7. K. Ya. Kondratyev, 1976. Aerosol and climate, *Tr. Gl. Geofiz. Obs.* **381**, 3–66.

8. K. Ya. Kondratyev and N. I. Moskalenko, 1979. Greenhouse effect of the planetary atmosphere, *Astron. Vestn.* **13**(3), 129–143.

9. K. Ya. Kondratyev, N. I. Moskalenko, and S. N. Parzhin, 1980. Greenhouse effect of Mars' atmosphere during an intensified volcanic activity, *Dokl. Akad. Nauk SSSR* **266**(1), 55–58.

10. K. Ya. Kondratyev, 1980. *Radiative Factors of the Present-Day Global Climate Changes.* Leningrad: Gidrometeorizdat.

11. K. Ya. Kondratyev and D. V. Pozdnyakov, 1980. *Stratospheric Aerosol,* Hydrometeorology, Series Meteorology, Overview, Issue 7. Obninsk: VNIIGMI-MCD.

12. K. Ya. Kondratyev, 1981. *Volcanoes and Climate,* Hydrometeorology, Series Meteorology, Overview, Issue 5. Obninsk: VNIIGMI-MCD.

13. K. Ya. Kondratyev, 1981. *Stratosphere and Climate,* Uspekhi nauki i tekhniki, Meteorologia i klimatologia, Vol. 6. Moscow: VINITI.

14. K. Ya. Kondratyev, N. I. Moskalenko, and V. F. Terzi, 1982. Modeling the optical characteristics of stratospheric aerosol, *Dokl. Akad. Nauk SSSR* **262**(5), 1092–1095.

15. K. Ya. Kondratyev, A. A. Grigoryev, O. M. Pokrovsky, and E. V. Shalina, 1983. *Remote Sounding of Atmospheric Aerosol from Space.* Leningrad: Gidrometeorizdat.

16. K. Ya. Kondratyev, N. I. Moskalenko, and D. V. Pozdnyakov, 1983. *Atmospheric Aerosol.* Leningrad: Gidrometeorizdat.

17. K. Ya. Kondratyev, 1983. *Earth's Radiation Budget, Aerosol, and Clouds,* Uspekhi nauki i tekhniki, Meteorologia i klimatologia, Vol. 10. Moscow: VINITI.

18. K. Ya. Kondratyev and N. I. Moskalenko, 1984. A comparative analysis of the climatic implications of volcanic activity on the Earth and Mars, *Dokl. Akad. Nauk SSSR.* **274**.

19. K. Ya. Kondratyev and N. I. Moskalenko, 1984. Radiative heat exchange in the atmosphere disturbed by volcanic eruption, *Dokl. Akad. Nauk SSSR* **274**(4), 799–801.

20. V. F. Loginov, Z. I. Pivovarova, and E. G. Kravchuk, 1983. An assessment of the contribution of natural and anthropogenic factors to solar radiation variability on the Earth's surface, *Meteorol. Gidrol.* **8**, 55–60.

21. V. F. Loginov, Z. I. Pivovarova, and E. G. Kravchuk, 1983. Variability of direct solar radiation and temperature in the Northern Hemisphere due to volcanic eruptions, *Izv. Vses. Geogr. Obschestva* **115**(5), 401–411.

22. V. F. Loginov, 1984. Possible causes and effects of volcanic eruptions, *Tr. Gl. Geofiz. Obs.* **471**, 103–107.

23. V. F. Loginov, 1984. *Volcanic Eruptions and Climate.* Leningrad: Gidrometeorizdat.

24. S. Ya. Sergin, 1974. Impact of the erupted dust on the global climate, in *Geophysical Studies of the Planetary System Glaciers-Ocean-Atmosphere,* Vladivostok: Dal'navostocnoe Kniznoe Izdaltel'stvo pp. 114–146.

25. L. F. Spirina, 1971. Impact of the volcanic dust on the temperature regime of the Northern Hemisphere, *Meteorol. Gidrol.* **10**, 38–45.

26. E. M. Feigelson, 1984. Impact of the eruption products on the radiative regime of the climate system, *Meteorol. Gidrol.* **5**, 5–11.

27. S. S. Khmelevtsov, A. S. Kabanov, and M. P. Kolomeev, 1981. *Impact of Stratospheric Aerosol on Climate,* Hydrometeorology, Series Meteorology, Overview, Issue 6. Obninsk: VNIIGMI-MCD.

28. J. K. Angell and J. Korshover, 1977. Estimate of global change in temperature, surface to 100 mb, between 1958 and 1975, *Mon. Weather Rev.* **105**, 375–385.

29. J. K. Angell and J. Korshover, 1983. Global temperature variations in the temperature and stratosphere, 1958–1982, *Mon. Weather Rev.* **111**(5), 901–921.

30. J. K. Angell and J. Korshover, 1983. Comparison of stratospheric warmings following Agung and Chichón, *Mon. Weather Rev.* **111**(10), 2129–2135.

31. A. Arakawa, T. Fujita, H. Itoo, Y. Masuda, S. Matsumoto, T. Murakami, T. Ozawa, E. Suzuki, M. Takeuchi, and K. Tomatsu, 1955. Climatic abnormalities as related to the explosions of volcano and hydrogen-bomb, *Geophys. Mag.* **26**(4), 231–255.

32. C. B. Baker, W. R. Kuhn, and E. Ryznar, 1984. Effects of the El Chichón volcanic cloud on direct and diffuse solar irradiances, *J. Climate Appl. Meteorol.* **23**(3), 449–452.

33. J. R. Banister, 1984. Pressure wave generated by the Mount St. Helens eruption, *J. Geophys. Res.* **D89**(3), 4895–4904.

34. C. A. Barth, R. W. Sanders, R. J. Thomas, C. E. Thomas, B. M. Jakosky, and R. A. West, 1983. Formation of the El Chichón aerosol cloud, *Geophys. Res. Lett.* **10**(11), 993–996.

35. E. S. Batten, 1974. *The Atmospheric Response to a Stratospheric Dust Cloud as Simulated by a General Circulation Model,* Report R-1324-ARPA. Santa Monica, CA: 1982. The Rand Corp.

36. H.-J. Bolle, 1982. Aerosol research within the World Climate Research Program, in *Light Absorption by Aerosol Particles.* Culver City, CA: Spectrum Press, pp. 1–51.

37. J. L. Bravo and A. Muhlia, 1984. Heating rates due to direct solar radiation in an atmospheric model, *Z. Meteorol.* **34**(2), 86–99.

38. J. R. Bray, 1974. Glacial advance relative to volcanic activity since 1500 AD, *Nature* **248**, 42–43.

39. J. R. Bray, 1978. Volcanic eruptions and climate during the past 500 years, in *Climate Change and Variability Southern Perspective,* Cambridge: Cambridge University Press pp. 256–262.

40. R. A. Bryson and B. M. Goodman, 1982. The climatic effect of explosive volcanic activity: analysis of the historical data, *Proc. Symp. on Atmospheric Effects and Potential Climate Impact of the 1980 Eruptions of Mount St. Helens,* NASA, NASA Conference Publication 2240, pp. 191–202.

41. R. D. Cadle, 1980. Some effects of the emissions of explosive volcanoes on the stratosphere, *J. Geophys. Res.* **C85**(8), 4495–4498.

42. L. A. Capone, O. B. Toon, R. C. Whitten, R. P. Turco, Ch. A. Riegel, and K. Santhanam, 1983. A two-dimensional model simulation of the El Chichón volcanic eruption cloud, *Geophys. Res. Lett.* **10**(11), 1053–1056.

43. R. D. Cess, J. A. Coakley, Jr., and P. M. Kolesnikov, 1981. Stratospheric volcanic aerosols: a model study of interactive influences upon solar radiation, *Tellus* **33**(5), 444–452.

44. T. P. Charlock and W. D. Sellers, 1980. Aerosol effects on climate: calculations with time-dependent and steady-state radiative-convective models, *J. Atmos. Sci.* **37**(6), 1327–1341.

45. T. P. Charlock, 1983. The effect of volcanic aerosols on the thermal infrared budget of the lower stratosphere, *Fifth Conf. on Atmospheric Radiation, Oct. 31–Nov. 4, Baltimore, Md.* Boston: American Meteorological Society, pp. 350–353.

46. J. A. Coakley, Jr., and G. W. Grams, 1976. Relative influence on visible and infrared optical properties of a stratospheric aerosol layer on the global climate, *J. Appl. Meteorol.* **15**(7), 679–691.

47. J. A. Coakley, Jr., 1981. Stratospheric aerosols and the tropospheric energy budget: theory versus observations, *J. Geophys. Res.* **C86**(10), 9761–9766.

48. J. A. Coakley, Jr., R. D. Cess, and F. B. Yurevich, 1983. The effect of tropospheric aerosols on the Earth's radiation budget: a parameterization for climate models, *J. Atmos. Sci.* **40**(1), 116–138.

49. P. J. Crutzen and U. Schmailzl, 1983. Chemical budgets of the stratosphere, *Planet. Space Sci.* **31**(9), 1009–1032.

50. R. Decker and B. Decker, 1981. *Volcanoes.* New York: W. H. Freeman.

51. J. J. DeLuisi and B. M. Herman, 1977. Estimation of solar radiation absorption by volcanic stratospheric aerosols from Agung using surface-based observations, *J. Geophys. Res.* **82**(24), 3477–3480.

52. J. J. DeLuisi, E. G. Dutton, K. L. Coulson, T. E. DeFoor, and B. G. Mendonca, 1983. On some radiative features of the El Chichón volcanic stratospheric dust cloud and a cloud of unknown origin observed at Mauna Loa, *J. Geophys. Res.* **C88**(11), 6769–6772.

53. C. J. Dittberner, 1978. Climatic change: volcanoes, man-made pollution, and carbon dioxide, *IEEE Trans. Geosci. Electron.* **16**(1), 50–61.

54. H. W. Ellsaesser, 1977. Effect of Mt. Agung on global temperature, *Mon. Weather Rev.* **105**, 1200.

55. H. W. Elsaesser, 1983. Isolating the climatogenic effects of volcanoes, *Preprint UCRL-89161*, Lawrence Livermore National Laboratories, Livermore, Calif.

56. G. Fiocco, G. Grams, and A. Mugnai, 1976. Energy exchange and temperature of aerosols in the Earth's atmosphere (0.60 km), *J. Atmos. Sci.* **33**(12), 2415–2425.

57. C. U. Hammer, H. B. Clausen, and W. Dansguard, 1980. Greenland ice sheet evidence of post-glacial volcanism and its climatic impact, *Nature* **288**, 230–235.

58. P. Handler, 1984. Possible association between volcanic aerosols and El Niño events, *Preprint*, University of Illinois, Urbana, Ill.

59. J. E. Hansen, W. C. Wang, and A. A. Lacis, 1978. Mount Agung eruption provides test of a global climatic perturbation, *Science* **199**(4333), 1065–1068.

60. Harshvardhan and R. D. Cess, 1976. Stratospheric aerosols: effect upon atmospheric temperature and global climate. *Tellus* **28**(1), 1–10.

61. Harshvardhan, 1979. Perturbation of the zonal radiation balance by a stratospheric aerosol layer, *J. Atmos. Sci.* **36**(7), 1274–1285.

62. A. Henderson-Sellers, 1982. The climatic effect of a persistent volcanic plume: a possible model for anthropogenic perturbations, *Atmos. Environ.* **16**(2), 367–370.

63. D. J. Hofmann and J. M. Rosen, 1984, On the temporal variation of stratospheric aerosol size and mass during the first 18 months following the 1982 eruptions of El Chichón, *J. Geophys. Res.* **89**(3), 4883–4890.

64. D. V. Hoyt, 1979. Atmospheric transmission from the Smithsonian Astrophysical Observatory pyrheliometric measurements from 1923 to 1957, *J. Geophys. Res.* **C84**(8), 5015–5028.

65. B. G. Hunt, 1977. A simulation of the possible consequences of a volcanic eruption on the general circulation of the atmosphere, *Mon. Weather Rev.* **105**(3), 247–260.

66. K. Labitzke, B. Naujokat, and M. P. McCormick, 1983. Temperature effects

on the stratosphere of the April 4, 1982 eruption of El Chichón, Mexico, *Geophys. Res. Lett.* **10**(1), 24–26.

67. H. H. Lamb, 1972. *Climate: Present, Past, and Future,* Vol. I: *Fundamentals and Climate Now.* London: Methuen.

68. H. H. Lamb, 1977. *Climate: Present, Past, and Future,* Vol. II: *Climatic History and Future.* London: Methuen.

69. J. Lenoble, D. Tanre, P. Y. Deschamps, and M. Herman, 1982. A simple method to compute the change in Earth-atmosphere radiative balance due to a stratospheric aerosol layer, *J. Atmos. Sci.* **39**(11), 2565–2576.

70. J. Lenoble and C. Brogniez, 1984. A comparative review of radiation aerosol models, *Contrib. Atmos. Phys.* **57**(1), 1–20.

71. M. C. MacCracken and C. L. Potter, 1975. Comparative climatic impact of increased stratospheric aerosol loading and decreased solar constant in a zonal climate model, *Preprint UCRL-76132,* WMO/IAMAP Symp. on Long-Term Climate Fluctuations, Aug. 17–23, University of East Anglia, England.

72. M. C. MacCracken, 1976. Climate model results of stratospheric perturbations, *Proc. 4th Conf. on CIAP, Feb 1975.* Washington, D.C.: U.S. Department of Transportation, pp. 183–194.

73. M. C. MacCracken and F. M. Luther, 1983. Radiative and climatic effects of the El Chichón eruption, *Proc. 5th Conf. Atmospheric Radiation, Oct. 31–Nov. 4, 1983, Baltimore, Md.* Boston, Mass.: American Meteorological Society, pp. 346–349.

74. M. C. MacCracken and F. M. Luther, 1983. Radiative and climatic effects of the El Chichón eruption, *Res. Activ. Atmos. Ocean Model.* **5**, 6.28.

75. C. Mass and S. H. Schneider, 1977. Statistical evidence on the influence of sunspots and volcanic dust on long-term temperature records, *J. Atmos. Sci.* **34**(12), 1995–2004.

76. C. Mass and A. Robock, 1982. The short-term influence of the Mount St. Helens volcanic eruption on surface temperature in the northwest United States, *Mon. Weather Rev.* **110**(6), 614–622.

77. M. P. McCormick, 1983. Global distribution of stratospheric aerosols by satellite measurements, *AIAA J.* **21**(4), 633–635.

78. M. P. McCormick, 1983. Aerosol measurements from earth orbiting spacecraft, *Adv. Space Res.* **2**(5), 73–86.

79. B. G. Mendonca, K. J. Hanson, and J. J. DeLuisi, 1978. Volcanically related secular trends in atmospheric transmission at Mauna Loa Observatory, Hawaii, *Science* **202**(4367), 513–514.

80. M. K. Miles and P. B. Gildersleeves, 1978. Volcanic dust and changes in Northern Hemisphere temperature, *Nature* **271**(5647), 735–736.

81. J. M. Mitchell, Jr., 1975. Notes on solar variability and volcanic activity as potential sources of climatic variability, in *The Physical Basis of Climate and Climate Modeling,* GARP Publications Series 16. Geneva, WMO, pp. 127–131.

82. J. M. Mitchell, 1982. El Chichón: weather-maker of the century? *Weatherwise* **35**(6), 252–262.

83. R. E. Newell and A. Deepak (Eds.), 1982. *Mount St. Helens Eruptions of 1980: Atmospheric Effects and Potential Climatic Impact*, NASA SP-458, NASA, Washington, D.C.

84. R. E. Newell, 1970. Stratospheric temperature change from the Mt. Agung volcanic eruption of 1963, *J. Atmos. Sci.* **27**(4), 977–978.

85. R. E. Newell, 1981. Further studies of the atmospheric temperature change produced by the Mt. Agung volcanic eruption in 1963, *J. Volcanol. Geotherm. Res.* **11**(1), 61–66.

86. R. E. Newell, 1982. Workshop on Mount St. Helens eruptions of 1980: atmospheric effects and potential climatic impact, *Bull. Am. Meteorol. Soc.* **64**(23), 154–156.

87. R. E. Newell, 1984. Volcanism and climate, *1985 Yearbook of Science and Technology*, pp. 206–225. New York: McGraw-Hill.

88. G. Ohring, 1979. The effect of aerosols on the temperatures of a zonal average climate model, *Pure Appl. Geophys.* **117**(5), 851–864.

89. R. C. Oliver, 1976. On the response of hemispheric mean temperature to stratospheric dust: an empirical approach, *Proc. 4th Conf. on CIAP, Feb. 1975*. Washington, D.C.: U.S. Department of Transportation, pp. 335–353.

90. D. E. Parker and J. L. Brownscombe, 1983. Stratospheric warming following the El Chichón volcanic eruption, *Nature* **301**(5899), 406–408.

91. D. E. Parker, 1984. The influence of the southern oscillation and volcanic eruption on temperature in the tropical troposphere, *Trop. Ocean-Atmos. Newslett.* **26**, 4–51.

92. E. M. Patterson, C. O. Pollard, and I. Galindo, 1983. Optical properties of the ash from El Chichón volcano, *Geophys. Res. Lett.* **10**(4), 317–320.

93. J. B. Pollack and Th. P. Ackerman, 1983. Possible effects of the El Chichón volcanic cloud on the radiation budget of the northern tropics, *Geophys. Res. Lett.* **10**(11), 1057–1060.

94. J. B. Pollack, O. B. Toon, E. F. Danielsen, D. J. Hofmann, and J. M. Rosen, 1983. The El Chichón volcanic cloud: an introduction, *Geophys. Res. Lett.* **10**(11), 989–992.

95. R. S. Quiroz, 1983. The isolation of stratospheric temperature change due to the El Chichón volcanic eruption from nonvolcanic signals, *J. Geophys. Res.* **C88**(11), 6773–6780.

96. W. R. Bandeen and R. S. Fraser (Eds.), 1982. *Radiative Effects of the El Chichon Volcanic Eruption: Preliminary Results concerning Remote Sensing*, NASA Technical Memo 84959. Greenbelt, Md.: GSFC.

97. M. R. Rampino and S. Self, 1982. Historic eruptions of Tambora (1815), Krakatao (1883), and Agung (1963)—their stratospheric aerosols and climatic impact, *Quart. Res.* **18**(2), 127–143.

98. C. R. N. Rao and W. A. Bradley, 1983. Effects of the El Chichón volcanic dust

cloud on insolation measurements at Corvallis, Oregon (USA), *Geophys. Res. Lett.* **10**(5), 389–391.

99. R. A. Reck, 1979. Influence of airborne particles on the Earth's radiation balance, in *Report of the JOC Study Conf. on Climate Models: Performance, Intercomparison and Sensitivity Studies,* GARP Publications Series, Vol. 1, No. 22, part 2. Geneva: WMO, pp. 947–973.

100. A. Robock, 1978. Internally and externally caused climate change, *J. Atmos. Sci.* **35**(6), 1111–1122.

101. A. Robock and C. Mass, 1982. The Mount St. Helens volcanic eruption of 18 May 1980: large short-term surface temperature effect, *Science* **216**(4546), 628–629.

102. A. Robock, 1983. Ice and snow feedbacks and the latitudinal and seasonal distribution of climate sensitivity, *J. Atmos. Sci.* **40**(4), 986–997.

103. A. Robock, 1983. The dust cloud of the century, *Nature* **301**(5899), 373–374.

104. S. H. Schneider and C. Mass, 1975. Volcanic dust, sunspot, and temperature trends, *Science* **190**(4216), 741–746.

105. S. H. Schneider, 1983. Volcanic dust veils and climate: how clear is the connection?—an editorial, *Clim. Change* **5**(2), 111–113.

105. S. H. Schneider, 1984. Atmospheric double exposure, *Nat. Hist.* **93**(4), 100–101.

107. A. Schwedfeger, L. L. Stowe, and A. Gruber, 1983. Sensitivity of earth radiation budget parameters for El Chichón volcanic aerosol as estimated from NOAA-7 AVHRR data, *5th Conf. on Atmospheric Radiation, Oct. 31–Nov. 4, Baltimore, Md.* Boston: American Meteorological Society, 354–356.

108. C. B. Sear and P. M. Kelly, 1982. The climatic significance of El Chichon, *Clim. Monit.* **11**(5), 134–139.

109. W. A. Sedlacek, E. J. Mroz, A. L. Lazrus, and B. W. Gandrud, 1983. A decade of stratospheric sulfate measurements compared with observations of volcanic eruptions, *J. Geophys. Res.* **C88**(6), 3741–3771.

110. S. Self, M. R. Rampino, and J. J. Barbera, 1981. The possible effects of large 19th and 20th century volcanic eruptions on zonal and hemispheric surface temperatures, *J. Volcanol. Geotherm. Res.* **11**, 41–60.

111. G. E. Shaw, 1982. Solar spectral irradiance and atmospheric transmission at Mauna Loa observatory, *Appl. Opt.* **21**(11), 2006–2011.

112. R. S. Stolarski and D. M. Butler, Possible effects of volcanic eruptions on stratospheric minor constituents chemistry, *Pure Appl. Geophys.* **117**(3), 486–497.

113. R. B. Stothers and M. R. Rampino, 1983. Historic volcanism, European dry fogs, and Greenland acid precipitation, 1500 BC to 1500 AD, *Science* **222**(4622), 411–412.

114. R. B. Stothers and M. R. Rampino, 1983. Volcanic eruptions in the Mediterranean before 630 AD from written and archaeological sources, *J. Geophys. Res.* **B88**(8), 6357–6372.

115. A. E. Strong, L. L. Stowe, and C. C. Walton, 1983. Using the NOAA-7 AVHRR data to monitor El Chichón aerosol evolution and subsequent sea surface temperature anomalies, *Proc. 17th Int. Symp. on Remote Sensing Environment, May 9–13, Ann Arbor, Mich.* Ann Arbor, Mich.: Environmental Research Institute of Michigan, pp. 107–122.

116. B. L. Taylor, T. Gal-Chen, and S. H. Schneider, 1980. Volcanic eruptions and long-term temperature records: an empirical search for cause and effect, *Quart. J. R. Meteorol. Soc.* **106**(447), 175–200.

117. R. D. Hudson and E. I. Reed (Eds.), 1979. *The Stratosphere: Present and Future,* NASA Research Publication 1049. Greenbelt, Md.: GSFC.

118. R. C. Whitten (Ed.), 1982. *The Stratospheric Aerosol Layer (Topics in Current Physics, No. 28).* Berlin: Springer-Verlag.

119. O. B. Toon, R. P. Turco, P. Hamill, C. S. Kiang, and R. C. Whitten, 1979. A one-dimensional model describing aerosol formation and evolution in the stratosphere, II: Sensitivity studies and comparison with observations, *J. Atmos. Sci.* **36**(4), 718–736.

120. O. B. Toon, 1982. Volcanoes and climate, *Proc. Symp. on Atmospheric Effects and Potential Climatic Impact of the 1980 Eruptions of Mount St. Helens, NASA,* NASA Conference Publication 2240, NASA, pp. 15–36.

121. R. P. Turco, P. Hamill, O. B. Toon, R. C. Whitten, and C. S. Kiang, 1979. A one-dimensional model describing aerosol formation and evolution in the stratosphere, I: Physical processes and mathematical analogs, *J. Atmos. Sci.* **36**(4), 699–717.

122. R. P. Turco, O. B. Toon, R. C. Whitten, R. C. Keese, and P. Hamill, 1982. Simulation studies of the physical and chemical processes occurring in the stratospheric clouds of the Mount St. Helens eruptions of May and June 1980, *Proc. Symp. on Atmospheric Effects and Potential Climatic Impact of the 1980 Eruptions of Mount St. Helens,* NASA Conf. Publication 2240, NASA, pp. 161–190

123. R. P. Turco, R. C. Whitten, and O. B. Toon, 1982. Stratospheric aerosols: observations and theory, *Rev. Geophys. Space Phys.* **20**(2), 233–279.

124. R. P. Turco, O. B. Toon, R. C. Whitten, P. Hamill, and R. C. Keese, 1983. The 1980 eruptions of Mount St. Helens: physical and chemical processes in the stratospheric clouds, *J. Geophys. Res.* **C88**(9), 5299–5320.

125. S. A. Weisrose and G. Shadmon, 1984. Radiative transfer calculations through an aerosol cloud, *J. Quant. Spectrosc. Radiat. Transfer* **31**(1), 63–70.

126. R. J. Whittaker, K. Richards, et al., 1984. Krakatao 1883–1983: a biogeographical assessment, *Prog. Phys. Geogr.* **8**(1), 61–82.

127. R. Yamamoto and T. Iwashima, 1975. Change of the surface air temperature averaged over the Northern Hemisphere and large volcanic eruptions during the years 1951–1972, *J. Meteorol. Soc. Jpn.* **53**(6), 482–486.

128. G. I. Marchuk, 1982. *Mathematical Modeling in the Environmental Problem.* Moscow: Nauka.

129. P. M. Kelly and C. B. Sear, 1982. The formulation of Lamb's dust veil index, *Proc. Symp. on Atmospheric Effects and Potential Climatic Impact on the 1980 Eruptions of Mount St. Helens,* NASA Conference Publication 2240, NASA, pp. 293–298.

130. G. I. Marchuk, K. Ya. Kondratyev, and V. P. Dymniklov, 1981. *Some Problems in Climate Theory,* Uspekhi nauki i tekhniki, Meteorologia i klimatologia, Vol. 7. Moscow: VINITI.

131. G. I. Marchuk, V. P. Dymnikov, V. B. Zalesny, V. N. Lykosov, and V. Ya. Galin, 1984. *Mathematical Modeling of the General Circulation of the Atmosphere and Ocean.* Leningrad: Gidrometeorizdat.

132. P. J. Murrow, W. I. Rose, Jr., and S. Self, 1980. Determination of the total grain size distribution in a volcanian eruption column, and its implications to stratospheric aerosol perturbation, *Geophys. Res. Lett.* 7(11) 893–896.

133. R. W. Stewart, 1982. *Summary of Sulfur Oxidation,* NASA Technical Memo 84959. Greenbelt, Md.: GSFC, pp. 2.18–2.23.

134. C. Boutron, 1980. Respective influence of global pollution and volcanic eruptions on the trace metals content of Antarctic snows since 1880s, *J. Geophys. Res.* **C85**(12), 7426–7432.

135. F. Arnold, 1980. Multi-ion complexes in the stratosphere—implications for trace gases and aerosol, *Nature* **284**(5757), 610–611.

136. N. H. Farlow, K. G. Snetsinger, D. M. Hayes, H. Y. Lem, and B. M. Tooper, 1978. Nitrogen-sulfur compounds in stratospheric aerosols, *J. Geophys. Res.* **C83**(12), 6207–6211.

137. R. A. Cox and D. Sheppard, 1980. Reactions of OH radicals with gaseous sulphur compounds, *Nature* **284**(5754), 330–331.

138. K. Ya. Kondratyev and G. E. Hunt, 1982. *Weather and Climate on Planets.* Oxford: Pergamon Press.

139. M. Settle, 1979. Formation and deposition of volcanic sulfate aerosols on Mars, *J. Geophys. Res.* **B84**(14), 8343–8354.

140. J. M. Hoffer, F. Gomez, and P. Muela, 1982. Eruption of El Chichón volcano, Chiapas, Mexico, 28 March to 7 April 1982, *Science* **218**(4579), 1307–1308.

141. M. Matson, 1982. *GOES and NOAA-6 Observations of El Chichón eruptions,* NASA Technical Memo 84959. Greenbelt, Md.: GSFC, pp. 2.1–2.7.

142. A. Robock and M. Matson, 1983. Circumglobal transport of the El Chichón volcanic dust cloud, *Science* **221**, 195–197.

143. A. J. Krueger, 1982. Observations of El Chichón cloud characteristics, in W. R. Bandeen and F. S. Fraser (Eds.), *Radiative Effects of the El Chichón Volcanic Eruption: Preliminary Results concerning Remote Sensing,* NASA Technical Memo 84959. Greenbelt, Md.: GSFC, pp. 2.8–2.17.

144. J. P. Kotra, D. L. Finnegan, W. H. Zoller, M. A. Hart, and J. L. Moyers, 1983. El Chichón composition and plume gases and particles, *Science* **222**(4627), 1018–1021.

145. J. C. Wilson, E. D. Blackshear, and J. H. Hyan, 1983. Changes in the sub-2.5 micron diameter aerosol observed at 20 km altitude after the eruption of El Chichón, *Geophys. Res. Lett.* **10**(11), 1029 – 1032.

146. B. W. Gandrud, M. A. Kritz, and A. L. Lazrus, 1983. Balloon and aircraft measurements of stratospheric sulfate mixing ratio following the El Chichón eruption, *Geophys. Res. Lett.* **10**(11), 1037 – 1040.

147. J. F. Vedder, E. P. Condon, E. C. Y. Inn, K. D. Tabor, and M. A. Kritz, 1983. Measurements of stratospheric SO_2 after the El Chichón eruptions, *Geophys. Res. Lett.* **10**(11), 1045 – 1048.

148. E. Dutton and J. DeLuisi, 1983. Spectral extinction of direct solar radiation by the El Chichón cloud during December 1982, *Geophys. Res. Lett.* **10**(11), 1013 – 1016.

149. F. C. Wittenborn, K. O'Brien, H. W. Green, J. B. Pollack, and K. H. Bilski, 1983. Spectroscopic measurements of the 8- to 13-micrometer transmission of the upper atmosphere following the El Chichón eruptions, *Geophys. Res. Lett.* **10**(11), 1009 – 1012.

150. E. Dutton and J. DeLuisi, 1983. Optical thickness features of the El Chichón stratospheric debris cloud, *5th Conf. on Atmospheric Radiation, Oct. 31 – Nov. 4, 1983, Baltimore, Md.* Boston: American Meteorological Society, pp. 361 – 363.

151. D. J. Hofmann and J. M. Rosen, 1983. Measurement of the sulfuric acid weight percent in the stratospheric aerosol from the El Chichón eruption, *Papers 18th Gen. Assemb. IUGG, Aug, Hamburg,* Laramie, Wyo.: University of Wyoming, pp. 1 – 12.

152. D. J. Hofmann and J. M. Rosen, 1983. *Balloon-Borne Particle Counter Observations of the El Chichón Aerosol Layers in the 0.01 – 1.8 μm Radius Range,* Report AP-77, Department of Physics and Astronomy. Laramie, Wyo.: University of Wyoming.

153. D. J. Hofmann and J. M. Rosen, 1983. Stratospheric sulfuric acid fraction and mass estimate for the 1982 volcanic eruption of El Chichón, *Geophys. Res. Lett.* **20**(4), 313 – 316.

154. D. J. Hofmann and J. M. Rosen, 1983. Sulfuric acid droplet formation and growth in the stratosphere after the 1982 eruption of El Chichón *Science* **222**(4621), 325 – 326.

155. G. W. Lockwood, 1982. Spectrally resolved measurements of the El Chichón cloud from Flagstaff, Arizona, *EOS* **63**, 897.

156. D. J. Hofmann and J. M. Rosen, 1981. Stratospheric aerosol and condensation nuclei enhancements following the eruption of Alaid in April 1981. *Geophys. Res. Lett.* **8**(22), 1231 – 1234.

157. G. W. Grams and J. M. Rosen, 1978. Instrumentation for in situ measurements of the optical properties of stratospheric aerosol particles, *Atmos. Technol.* **9**, 35 – 54.

158. S. D. Clarke, R. J. Charlson, and J. A. Ogren, 1983. Stratospheric aerosol light

absorption before and after El Chichón, *Geophys. Res. Lett.* **10**(11), 1017–1020.

159. M. P. McCormick, 1982. Ground-based and airborne measurements of Mount St. Helens stratospheric effluents, *Proc. Symp. on Atmospheric Effects and Potential Climatic Impact of the 1980 Eruptions of Mount St. Helens,* NASA Conference Publication 2240, NASA, pp. 125–130.

160. M. D. King, 1982. *Radiative Characteristics of the Aerosols,* NASA Technical Memo 94859. Greenbelt, Md.: GSFC, pp. 3.3–3.13.

161. J. J. DeLuisi, B. G. Mendonca, and K. J. Hanson, 1982. Measurements of stratospheric aerosol over Mauna Loa, Hawaii, and Boulder, Colorado, *Proc. Symp. on Atmospheric Effects and Potential Climatic Impact of the 1980 Eruptions of Mount St. Helens,* NASA Conference Publication 2240, NASA, pp. 117–124.

162. J. K. Angell and J. Korshover, 1978. Comparison of stratospheric trends in temperature, ozone and water vapor in northern temperature latitudes, *J. Appl. Meteorol.* **17**(9), 1397–1401.

163. J. B. Pollack, O. B. Toon, and D. Wiedman, 1981. Radiative properties of the background stratospheric aerosols and implications for perturbed conditions, *Geophys. Res. Lett.* **8**(1), 26–28.

164. R. C. Whitten, O. B. Toon, and R. P. Turco, 1980. The stratospheric sulfate aerosol layer: processes, models, observations, and simulations, *Pure Appl. Geophys.* **118**(1–2), 86–127.

165. R. D. Cadle and G. W. Grams, 1975. Stratospheric aerosol particles and their optical properties, *Rev. Geophys. Space Phys.* **13**, 475–501.

166. A. Robock, 1982. El Chichón could cause considerable cooling, *Preprint,* University of Maryland, College Park, Md.

167. K. Ya. Kondratyev, O. B. Vasilyev, and L. S. Ivlev, 1976. *The Global Aerosol-Radiation Experiment (GAREX).* Overview, Series Meterology. Obninsk: VNIIGMI-MCD.

168. G. S. Golitsyn and A. S. Ginsburg, 1985. Comparative estimates of climate consequences of Martian dust storms and of possible nuclear war, *Tellus* **37B,** 173–181.

169. N. Braslau and J. A. M. Dave, 1973. Effect of aerosols on the transfer of solar energy through realistic atmospheres, I: Non-absorbing aerosols, *JAM* **12,** 601–615.

170. T. H. Vonder Haar and V. E. Suomi, 1971. Measurements of the earth's radiation budget from satellites during a five-year period, I: Extended time and space means, *J. Atm. Sci.* **28,** 305–313.

171. D. Deirmendjian, 1969. *Electromagnetic Scattering on Spherical Polydispersions.* New York: Elsevier.

172. J. I. Borzenkova, et al., 1976. Change in the air temperature of the northern hemisphere for the period 1881–1975. *Meteorologiya y Gidrologiya,* **7,** 27–35.

Chapter 3

Possible Climatic Impacts of a Nuclear War

This chapter was written by K. Ya. Kondratyev and G. A. Nikolsky.

Many catastrophic consequences of a global-scale nuclear war threaten the civilization of our planet, but the possible impacts on the environment and climate are, perhaps, a most threatening danger fraught with an ecological collapse. Confining ourselves to this aspect of this problem, we draw the reader's attention to the most serious consequences for the global environment of a nuclear war (1,2).

Wars in the past had only marked (so to speak) the locations where they had been waged, generally without any long-term impacts on the environment and World Ocean. The geographic scope of multiple nuclear impacts on the environment, however, will be so large that it will affect the entire globe (land, ocean, and atmosphere). Changes in the chemical composition of the upper atmospheric layers after nuclear explosions in the atmosphere will be of particular importance. The atmosphere of our planet has been formed in the long process of geological evolution. Its present chemical composition is largely the result of its interaction with the biosphere. Although the atmosphere consists mainly of nitrogen and oxygen, its optically active components are certain trace gases (water vapor, carbon dioxide, ozone, nitrogen oxides, hydrocarbons, and so forth). This results in a very high sensitivity of the climate to external forcing (e.g., pollution by industrially produced carbon dioxide).

The presence of ozone in the atmosphere is vitally important for life on Earth. This gas, concentrated in a stratospheric layer (at a height of 10 to 50 km), creates a shield that protects every living being on the planet from the

173

destructive impact of extreme UV solar radiation, because ozone absorbs this radiation.

Studies on the effect of the oxides of nitrogen on the ozone layer are of great importance: they have created the basis for the estimating the possible impact of atmospheric nuclear explosions on ozone. Several studies have shown that nitrogen dioxide formed in the nuclear fireball can play an independent role in an antigreenhouse effect by absorbing shortwave radiation and promoting the creation of a solar energy deficit in the lower atmosphere and on the surface. Estimates have shown that the presence of 1.5×10^{17} molecules of NO_2/cm^2 in an air column will lead to an additional absorption of solar radiation in the visible, reaching 2% (3). The correlation between the NO_2 content in an air column and the solar radiation extinction ΔS is shown schematically in Fig. 3.1. The second curve drawn from Reid's

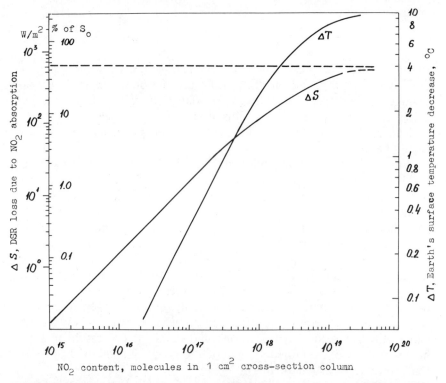

FIGURE 3.1. Variations in the extinction of the total direct solar radiation flux (related to solar constant value) ΔS and air surface temperature for various contents of NO_2 in an air column, ΔT. Dependencies are obtained from data of Refs. 3 and 4. The dependence $\Delta_T = f(NO_2)$ is determined from measurements at the NH meteorological network (ref. 5).

data (4) and specified from measurements of air surface temperature at the Northern Hemisphere meteorological network (5) relates variations in the NO_2 content to possible temperature variations ΔT, provided that new values of the NO_2 content are preserved for at least 2 to 3 years. Two to three years is characteristic of the residence time for NO_x molecules in the lower stratosphere and of the time for the temperature signal to lag behind the varying input of solar radiation to the lower troposphere. A very important perspective on this problem is an estimation of the impact of nuclear tests in 1958, 1961, and 1962.

3.1   THE IMPACT OF NUCLEAR TESTS ON THE ATMOSPHERE AND CLIMATE

Observational evidence for abnormal phenomena in the atmosphere observed between 1958 and 1965 is most convincing that a nuclear conflict can result in an ecological catastrophe and eventually in the annihilation of civilization on Earth. The last act of human life can start with the devastation of the Northern Hemisphere and end with a global ecological collapse. This is our opinion. It is based on analysis of observational evidence and, in principle, runs counter to the idea of the acceptability of nuclear weapons as expressed in a recent paper by Edward Teller (90) entitled, "Deadly Myths About Nuclear Arms" (see also Ref. 91). The argument of Teller's paper proceeds from an erroneous view about the physicochemical processes that might accompany nuclear explosions in the stratosphere. A continued disregard of scientific facts that indicate the enormous danger of using nuclear weapons can jeopardize the very existence of humankind.

Particularly important are thorough and exhaustive studies (on an international cooperation basis) of the possible impacts of nuclear explosions in the stratosphere on its chemical composition, temperature, regime, and dynamics aimed at finding indisputable evidence and averting disastrous consequences of irreversible changes in the atmosphere and climate.

From periodicals and scientific publications [e.g., by Carter and Moghissi (6)], the total cumulative number of announced nuclear bomb testings by the end of 1975 had reached 801 [with 383 explosions in the atmosphere, 579 (with 193 explosions in the atmosphere) in the United States], with a total energy yield of about 340 megatons. The largest annual yield of nuclear tests occurred in 1962 (108 megatons), including two explosions of 30 megatons. The largest yield was an explosion in 1961 — 58 megatons.

Forty years have passed since the beginning of nuclear tests. The tests have grown in number, but their cumulative yield has not been as large, because there is an agreement between the United States and the USSR as to a limitation on the nuclear arsenal and on the types of nuclear tests. According to this agreement, only underground explosions are allowed, with an explosive yield of not more than 150 kilotons and with the yield of individual explosions for industrial purposes not to exceed 1.5 megatons.

The largest annual number of explosions was reached in 1962 with 143, followed by the year 1958 with 96. During the period including the years 1958, 1961, and 1962 the total yield, according to Carter and Moghissi (6), was about 200 megatons, although estimates by Crutzen and Birks (7) point to a larger explosive yield of 300 megatons, even in the period from 1961 to 1962. This disparity highlights the still existing substantial uncertainty about indirect estimates of the yields and heights of explosions.

According to Carter and Moghissi (6), about 75% of the total explosive yield during these years were explosions in the atmosphere, with 50% being of the megaton class. The bulk of these detonations was made in the upper troposphere and the stratosphere, but the test of Starfish with a yield of 1.4 megatons occurred in the thermosphere at a height of 400 km.

Information on the height of the explosion and on the construction of the warhead shell is important for the estimation of the amount of nitrogen oxides NO_x (here $NO_x = NO + NO_2$), a major direct product of a nuclear explosion in the atmosphere. The oxides of nitrogen are mainly produced in the fireball, with heating and cooling of the captured air. Shockwaves and strong UV radiation penetrating the environment create a dissociating effect on air molecules. As a result, apart from NO_x, a significant amount of ozone can also be produced.

Note should be taken here about radioactive nuclei debris of fissionable material and ionized atoms of the warhead shell material which create a very high level of air ionization (two or three orders of magnitude greater in contrast to the environment) in the explosion cloud spreading for a distance of 250 to 300 km for 10 to 15 hours. The high ionization promotes production of additional amounts of NO_x and ozone, the latter being produced in amounts that not only compensate for the depletion of O_3 resulting from reactions with NO, but also produce (in addition to the UV radiation of the explosion) an increase reaching 10 to 30% of the O_3 concentration in the cloud. Processes of conversion of gaseous components in the explosion cloud are illustrated by data on variations in the total ozone content (TOC), during the passage of the Chinese warhead explosion cloud over Sapporo (Japan) (8).

Results of direct and remote soundings of the explosion cloud after the 1970 tests of the French 1- and 2-megaton warheads (9,10) point to the absence of an ozone concentration decrease in that cloud.

Analysis of an average ozone content over the Northern Hemisphere showed, according to (11–13) (from observations at the ozonometric network), that there was no ozone depletion following the 1961–1962 tests. As Johnston et al. (13) showed, the reason for false trends of ozone depletion discovered in many model estimates of the effect of nuclear tests on ozone is an inadequacy of the photochemical block of the models, namely, the omission of reactions of HO with NO_x which block the reactions of ozone depletion by active components of the hydrogen-hydroxyl cycle.

Apparently, in some current models the photochemical block contains incomplete or inadequate schemes of reactions, because possible trends of postnuclear ozone depletion in the stratosphere (7,14), along with other major impacts on the atmosphere [such as the Tunguska meteor (TM) fall (15) are still being determined. Probably, the needed revision of the photochemical "thinking" was delayed in part because of work by London et al. (16), who, in analyzing the errors of ozone network observations, suggested that all Soviet ozonometric network data between 1961 and 1964 be rejected as unreliable. These data, however, have given an objective pattern of anomalous changes in the ozone content.

This practice of careless "filtering" of experimental data has often led to distortion and even to casting doubt on the truth of reports. For example, the sampling of turbidity data obtained at Mauna Loa from 1958 to 1962 was too subjective (17).

Turning back to estimates of the heights of nuclear explosions, which are very important for the calculation of NO_x production, note that fireballs penetrate the stratosphere only with nuclear ground bursts of more than 1 megaton yield. According to Peterson (18), the dependence of the cloud top height H_T on the explosive yield Y is determined by

$$H_T = 22Y^{0.2}$$

and for the bottom of the cloud there is a relationship

$$H_B = 13Y^{0.2}$$

(where Y is in megatons and $H_{T,B}$ in kilometers.)

In the lower atmosphere, about 10^{32} molecules of NO is produced per megaton of explosive energy, depending on the type (thickness) of the warhead shell (19,20). But in the upper stratosphere, due to the contribution by γ and the X-ray explosive energy to NO production, the latter is increased up to 4.8×10^{32} molecules of NO per megaton (7).

Thus far, the conclusions by Hampson, who as early as 1974 pointed out the possibility of heavy destruction of the ozone layer during a nuclear war (21), have not been properly verified. According to Hampson, one of the reasons for the variance of opinions about the influence of nuclear explosions on ozone is a possible underestimation of the NO_x per megaton yield. Based on the observational data of balloon soundings carried out from 1962 to 1967 at the University of Leningrad (22,23), Hampson believes that the previously adopted output of the oxides of nitrogen per unit yield of a nuclear warhead must have been underestimated by approximately a factor of 4 and that in fact it constitutes about 2.5×10^{32} molecules of NO_x per megaton. Another important assumption by Hampson consists of the view that nuclear detonations in the upper stratosphere are much more efficient in producing NO_x due to the gamma radiation involved in the process of NO_x formation. A considerable amount of nuclear explosive energy is spent on the generation of this gamma radiation. As a result of this process, the high-altitude detonation NO_x per megaton yield can increase 20-fold; the NO release can grow by a factor of 5 in the thermosphere.

These assumptions by Hampson are extremely important for estimating the impact of nuclear explosions on the atmospheric composition and dynamics and, consequently, on climate. In addition, they may even influence the conceptions of war strategy and defense systems for industrial centers and military targets. No doubt, these problems require careful examination. The assumption of correlations between the NO_x per megaton yield and atmospheric pressure at the level of detonation should be verified.

Hampson (24) considers a mechanism of the effect on the ozone layer of large quantities of NO_x formed in the upper stratosphere with serial nuclear explosions of the intercontinental rocket interceptors. The process develops in the following way. The destruction of ozone in the upper stratosphere may lead to a rapid cooling of these layers. Rapid cooling of an air layer is known to stimulate atmospheric instability. Downward motions appearing in this process force the cloud of the oxides of nitrogen to proceed downward at a speed of about 1 km per day, and in 10 to 15 days it can reach the layer of maximum ozone concentration. The resulting catalytic destruction of ozone may be very substantial. If one takes into account the prevailing winds at heights between 30 and 50 km, the destruction of ozone over a given territory may take place in 2 weeks following a series of detonations in the stratosphere. Thus a "window" or "hole" may appear in the ozone layer under the influence of detonations, at a distance of 1500 to 2000 km from the point of explosion. High levels of UV radiation are dangerous not so for people, who can protect themselves, but for vegetation (e.g., crop and garden cultures) and animals.

Hampson believes that the combined efforts of the experts are urgently needed in the kinetics of nonequilibrium reactions, the chemistry and dynamics of the stratosphere, atmospheric evolution, and even in nuclear strategy to obtain and substantiate reliable estimates of the potential impact of nuclear explosions on the environment.

3.1.1 Studies in the USSR of the Impact of Nuclear Tests on the Atmosphere and Climate

Here we discuss briefly results of some observational studies carried out in the USSR. The high-altitude balloon, rocket, and satellite observations undertaken in 1958–1959 (15) and 1961–1967 (25,27,28) made it possible to monitor the variability of the abnormally high infrared emission of the thermosphere and the anomalous extinction of direct solar radiation (DSR) fluxes in the stratosphere above 30 km (29). Markov et al. (25) examined the results of the 27 August 1958 atmospheric rocket sounding up to a height of about 500 km. Infrared atmospheric emission fluxes were measured in directions close to the horizontal one.

In the thermosphere (at heights of 290 and 425 km), layers with a powerful IR emission (over $600 W/m^2$) have been found. Measurements were taken in midlatitudes 2.5 hours after the high-altitude nuclear explosion of Argus-1. During this period the perturbation of the Earth's geomagnetic field was high (the index $K_p \sim 5$). Markov et al. (25) noted that only a combination of high magnetic perturbation and increased NO concentration might have provided the conditions for a powerful IR emission. The low spectral resolution of the instruments has not permitted reliable identification of the emitting components. Further observations (30) revealed a maximum of emission to be near $5.3-\mu m$ wavelength.

Analysis of data of subsequent observations from rockets and satellites in 1962–1965 showed that the thermospheric emission decreased stepwise down to 160 to 230 W/m^2 in 1962–1963, 100 W/m^2 in 1964, and 30 W/m^2 in 1965 (25). Measurements in 1976 (Fig. 3.2) have shown that maximum emission values do not exceed 0.8 W/m^2 (27).

Gordiets et al. (26) and Markov (30) undertook a search for possible mechanisms and physical processes that account for the energetic aspect of this phenomenon. The search revealed that apart from a powerful emission-forcing mechanism, concentrations of the emitting gas molecules which would largely exceed the background level in respective thermospheric layers are necessary.

Nuclear explosions in high atmospheric layers can generate huge amounts of NO molecules, especially if one takes into account the dependence of the NO_x per megaton yield on altitude assumed by Hampson. The explosions of Argus-1,2,3 took place at a height near 480 km (25,31), where the efficiency of the NO_x per megaton yield must be very high compared with the ground-based efficiency (exceeding the latter, perhaps, by a factor of 10^2). Approximate calculations lead to an estimate of the total postexplosion NO_x amount of about 3000 tons. For the thermospheric conditions (300 to 400 km) this is a substantial addition. But even in this case the values of IR emission fluxes presented by Markov et al. (25) are important.

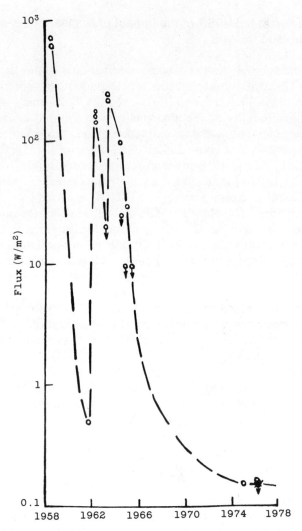

FIGURE 3.2. Results of rocket and satellite measurements of horizontal IR emission fluxes in the upper atmosphere in the period 1958–1977. Data from Ref. 25 are marked by circles; measurements from Ref. 27 are marked by an asterisk.

Gordiets et al. (26) mentioned that the maximum emission flux during a geomagnetic storm increased to 10^2 to 10^3 ergs/cm^2, that is, it constituted 0.1 to 1.0 W/m^2. Although it is also a large value for the thermosphere, it is still within the limits of real physical processes. The transition from measured radiances to radiance fluxes was performed correctly by multiplication by π. The values of fluxes obtained from this transition are conditional and should not be considered as hemispheric fluxes to the horizontal surface in the lower thermosphere.

The transition to such fluxes from the values presented by Gordiets et al. is realized through an account of the ratio between the horizontal and vertical thicknesses of the emitting layer (about 100) and the cosine law. This will result in a decrease of the fluxes found by Gordiets et al. by a factor of about 200 more. However, these remarks by no means reject the role of IR emissions as an indicator of variations and a pacticipator emission-forcing processes in the upper atmosphere.

Temporal variations in the IR emission and the amplitudes, defined more accurately in absolute units, indicate essential disturbances of the radiative regime of the upper atmospheric layers by nuclear explosions and thus reveal the critical danger of a series of powerful explosions in the stratosphere in the case of a nuclear conflict.

According to data of Hampson (21), about 20 megatons was exploded in the upper stratosphere in 1958 (USA and USSR 10 megatons each). Proceeding from an assumed lifetime for NO_x a negligible amount of NO_x must remain in the atmosphere by the year 1964. Nevertheless, data presented by Nikolsky (32) (Fig. 3.3) point to a much more efficient formation of NO_x per megaton of explosive energy than had been assumed, and point to a much longer lifetime of NO_x (about 4 years).

The lifetime of a component can be estimated from the power of global sources and sinks, as well as from the rates of reactions of relevant photochemical interactions in the system. However, many uncertainties appear with regard to input parameters. Therefore, it is worthwhile to consider experimental data that characterize temporal variations in the content of the components of interest. The rate of reducing the concentration of NO_2 molecules in the stratosphere of the Northern Hemisphere depends not only on the lifetime of NO_2 molecules but also on the intensity of the diffusion of these molecules to the Southern Hemisphere.

Postponing for the moment an explanation of the origin of the graph in Fig. 3.3, examine the curve for the NO_2 content over the Northern Hemisphere. Figure 3.3 shows that because of the moratorium on atmospheric nuclear tests, the NO_2 content in the Northern Hemisphere has halved (from 1.5×10^{17} to 0.75×10^{17}) between mid-1964 and early 1967, that is, during 2.5 years. Hence the lifetime of the NO_x group is, at least, about 3.3 years. Considering the transport of NO_x molecules to the Southern Hemisphere, one may assume that the photochemical lifetime of NO_x is about 4 years.

Bearing in mind a possible increase of the efficient NO_x production, we find that in 1958 the amount of injected oxides of nitrogen could have exceeded the background NO_x content in the stratosphere of the Northern Hemisphere.

Thus it follows that the upper stratosphere, lower mesosphere, and the

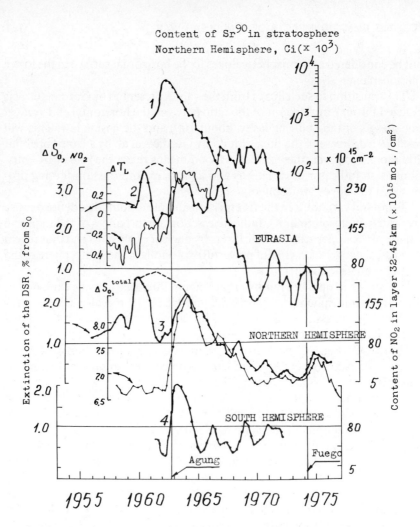

Content of Sr^{90} in stratosphere Northern Hemisphere, Ci($\times 10^3$)

FIGURE 3.3. Variations in the NO_2 content in the stratosphere of the Northern and Southern Hemispheres. 1, ^{90}Sr content in the NH stratosphere from global network data (aircraft and balloons). In 1963 the content of ^{90}Sr was 6×10^6 Ci; 2, NO_2 content in an air column of unit cross section in the layer from 32 to 45 km over the Eurasian territory (in the winter of 1961–1962 it reached 2.7×10^{17} molecules/cm^2) can be counted from the right-hand scale. Excess DSR extinction due to the absorption by respective amount of NO_2 is given in percent of solar constant (S_0) on the left-hand scale. The left-hand scale inside the graph is for variations in the Linke coefficient of turbidity T_L measured at the Mauna Loa background station; 3, NO_2 content in the layer from 32 to 55 km in an air column of unit cross section over the Northern Hemisphere. The left-hand scale inside the graph is for variations in the extinction of the total DSR flux $S_{0, total}$ for Mauna Loa; 4, NO_2 content in the layer from 32 to 45 km over the Southern Hemisphere. The vertical lines show the moment of the Agung (8°S) and Fuego volcano eruptions (After Ref. 32.) Dashed lines overlapping the minima 1961–1962 (curves 2 and 3) indicate additional content of NO_2 in the layer 46–55 km.

thermosphere were strongly disturbed in the year 1958. Data in Fig. 3.3 for the Northern Hemisphere show that between 1958 and 1960 a rapid stepwise increase of the NO_2 content was observed (discussed below). In mid-1960 an absolute maximum of the NO_2 content was reached: 2×10^{17} molecules of NO_2 per square centimeter.

The nuclear test series in 1961–1962 resulted in an increased IR emission of the thermosphere, with a maximum in 1962–1964 (Fig. 3.2). The NO_2 content in the layer 32–55 km also reached a maximum in late 1963–early 1964 (Fig. 3.3). In the subsequent years a gradual removal of NO_x from the NH took place. From data in Figs. 3.2 and 3.3 the stratospheric cleaning continued at least until 1974.

Let us focus on the results of balloon soundings carried out from 1962 to 1967, when the stratosphere was most disturbed (1962), and on the period of its gradual recovery. In one of our early papers (22) temporal variations of direct solar radiation (DSR) fluxes measured at balloon ascent below 34 km height were interpreted as a manifestation of variability of the astronomical solar constant $S_{0,astr}$ due to variations in solar activity. Despite some anomalies observed in the stratospheric transparency, variations in the DSR flux extrapolated beyond the atmosphere were attributed to variations in the intensity of the radiation source, that is, the sun. For lack of information at the time about the amount of material produced by nuclear explosions (e.g., NO_x) it was impossible to identify the reasons for anomalous DSR extinction. Nevertheless, each time we mentioned the observed correlations of an additional DSR extinction with nuclear explosions and with the post-eruption aerosol component, we pointed out that the total DSR flux after some nuclear tests in 1962 decreased by 6 to 8% against an extra-atmospheric value of S_0.

Subsequent detailed analysis of balloon sounding data by Kondratyev and Nikolsky (29) confirmed the supposition that an anomalous extinction of DSR at altitudes above 26 km had been caused by nuclear explosive material residing in the upper stratosphere.

Let us estimate the cumulative amount of NO_2 residing in the NH stratosphere by the end of 1962. Note that the estimates are based on yet-to-be documented suppositions about strongly increased NO releases at high-altitude nuclear explosions, as well as on the remaining uncertainties in the estimation of the yields and heights of nuclear explosions during the periods 1958 and 1961–1962.

During high-altitude tests in 1958 the total explosive energy constituted 20 megatons trotyl equivalent (t.e.). With increased NO production at high altitudes, the equivalent explosive energy should be increased 200-fold. From the first tests in 1958 until the end of 1962, 4 years passed. During this period the initial amount of NO_x ought to have decreased by a factor of 2.7.

Thus, by the end of 1962, $2 \times 10^{35} : 2.7 = 7.4 \times 10^{34}$ molecules of NO_x could have remained.

Data given by Nikolsky (32) (Fig. 3.3) illustrate still higher NO_x production per megaton. For mid-1960 the NO_x content at NH midlatitudes, determined from the curve "North Hemisphere" (Fig. 3.3), constituted about 2×10^{17} molecules of NO_2 per square centimeter. To explain this, the efficiency of NO_x production must be increased not by two orders of magnitude but by a factor of 170. By the end of 1962 the NO_x content had to decrease (at $\tau \sim 4$ years) $2 \times 10^{17} \times 0.61 = 1.22 \times 10^{17}$, but with an efficiency increased by two orders of magnitude the content will range from 4.35×10^{16} to 5.1×10^{16} molecules of NO_2 per square centimeter, depending on the area of the latitudinal belt in which the nuclear products reside ($1.7-1.46 \times 10^{18}$ cm^2). For our calculations we shall take an average of 0.8×10^{17} NO_2 per square centimeter.

The explosive energy of the 1961–1962 nuclear tests was estimated by Crutzen and Birks (7) at 300 megatons, 200 megatons of which were detonated in the atmosphere. Since the atmospheric tests were made partly at low altitudes, only 185 megatons was considered in the calculations. The NO_x production will be higher in the polar stratosphere. Most of the Soviet tests in 1961–1962 are known (31) to have been made in polar regions in late fall, which led to an intensified accumulation of NO_x (the low tropopause, the circumpolar motion of air masses, the seasonal minimum of ozone, increased atmospheric ionization, etc.). We shall take the lower limit of the efficiency proposed by Hampson: $185 \times 20 \times 10^{32} = 3.7 \times 10^{35}$ molecules of NO.

According to Crutzen, in a small period most of the NO in the stratosphere is transformed into NO_2. We take three-fourths of NO molecules to be transformed into NO_2. Then the 1961–1962 nuclear tests produced in the stratosphere about 2.5×10^{35} molecules of NO_2. Note that by mid-1964 about 1.52×10^{35} molecules of NO_2 had remained.

Taking the explosive material to be concentrated in the ring layer of the stratosphere, covering two-thirds of the NH area, we obtain the NO_2 content per unit cross section:

$$2.5 \times 10^{35} : 1.7 \times 10^{18} = 1.47 \times 10^{17} \ NO_2/cm^2.$$

Our calculations show that NO_2 in an air column of unit cross section within the postnuclear semiglobal ring cloud in the NH middle and high latitudes should reach an amount of 2×10^{17} molecules to cause an additional 2.6% extinction of total DSR flux. This value agrees well with the calculated value 1.47×10^{17}.

The reality of these amounts of NO_2 in an air column at 50°N in the second half of 1962 is confirmed by simultaneous balloon filter measurements of DSR spectral fluxes (in the wavelength range between 305 and 370 nm). So, for example, for the 7 July 1962 flight (flight 9, Fig. 3.4), the NO_2 amount above 26 km constituted 3.4×10^{17} molecules/cm^2. Thus the conclusion can be drawn about the global scale of the impact of the 1961–1962 nuclear tests on the atmosphere manifested through considerable DSR extinction during a long period (1962–1968). Figure 3.4 illustrates the rate of decrease of an excess reduction of DSR in the midlatitude upper stratosphere, constituting 0.4% per year (before the year 1968).

A persistent deficit of solar radiation in the troposphere and on the Earth's surface in the NH midlatitudes in 1962–1963 has influenced synoptic processes and weather conditions (33). Such anthropogenic effects break the normal circulation in the stratosphere and troposphere; that is, they lead to disturbances in climate.

The temporal variation in DSR reduction should be attributed not only to the natural NO_2 sink in the NH troposphere but also to the NO_2 transport in the upper troposphere to the Southern Hemisphere, where an intensified anomalous extinction of DSR took place between 1963 and 1966. Comparison of observed and calculated data suggests the conclusion that in the Southern Hemisphere the nuclear and volcanic effects combine. Results of

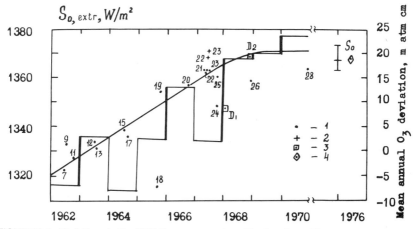

FIGURE 3.4. Variations in the DSR flux measured at altitudes about 30 km and extrapolated beyond the atmosphere, taking into account the attenuating properties of the layer above, typical of the periods of stratospheric self-cleaning (1967–1968). Figures at points correspond to the number of the balloon launch. 1, Model extrapolated data; 2, extrapolation based on actual state of the stratosphere; 3, D_1 and D_2 are the Denver University flights (Ref. 23); 4, S_0 from Aerobee rocket measurements (Ref. 43). The stepped graph illustrates deviations of mean annual global TOC from averages over a long period; bold vertical lines mark the moments of strongest sudden heatings of the stratosphere.

rocket spectral measurement ($\lambda = 3800A$) at Woomera, Australia (30°S) in 1965 revealed an attenuating component at heights of 20 to 40 km with Rayleigh optical thickness at 28 km, but a 2.5-fold greater optical thickness at 22 km (34). This attenuation cannot be attributed only to the aerosol component (34).

A decrease of an additional attenuation in the stratosphere between 1962 and 1969 was followed, as seen in Fig. 3.4, by considerable growth (by about 8%) of annual mean global values of total ozone content (TOC) (35).

Data of Angell and Korshover (36) also point to an increase (up to 12%) of the ozone content at altitudes above 32 km. Besides, during this period the water vapor content in the NH stratosphere is known to have doubled. Hence neither ozone nor water vapor would have been the components that had created an additional 2.5% DSR reduction (in units of S_0) in 1963.

There is evidence that in this period in the NH and SH upper and middle stratospheric layers the usual course of the seasonal and quasibiennial variations in the temperature and pressure fields was disturbed. As rocket soundings show (37), the temperature at altitudes of 46 to 55 km in the equatorial zone of the Western Hemisphere in 1963–1964 exceeded the average for the subsequent 12 years by 6°C, in the subtropical zone by 2°C, and in the polar zone by 3 to 4°C. In the equatorial zone a temperature increase of 3 to 4°C was also registered in the layer at 24 to 45 km. During the next years the temperature in the layer at 46 to 55 km dropped in all three zones, especially in the equatorial belt, where, between 1964 and 1977, a temperature drop constituted about 10°C. The trend of cooling in the equatorial zone is also characteristic of the layer from 16 to 45 km.

On the assumption that a temperature increase during 1963–1964 in the upper and middle stratosphere is connected with an increased concentration of NO_x at these altitudes, it is likely that in the equatorial zone the oxides of nitrogen have moved downward and partly concentrated in the period 1967–1970 in the layer 25 to 35 km, where positive temperature deviations remained until 1971. These data show that the anthropogenic NO_2 had largely been removed from the upper stratosphere by the year 1966. The equatorial and polar stratosphere (the layer from 26 to 45 km) had been rapidly cleansed. Additional amounts of nitrogen dioxide of "solar" origin were accumulated (due to solar proton events and solar cosmic rays) with a quasibiennial periodicity in the polar zone between 1967 and 1969 (increase in solar activity).

Now consider data on the temperature of lower tropospheric layers. Figure 3.5 shows temporal variations of surface air temperature (against an average) in the NH polar regions ($\varnothing \geq 70°$), air temperature at the 700-mb level (38) for the same latitude, and SST in the Pacific (in the latitude belt 20–55°N).

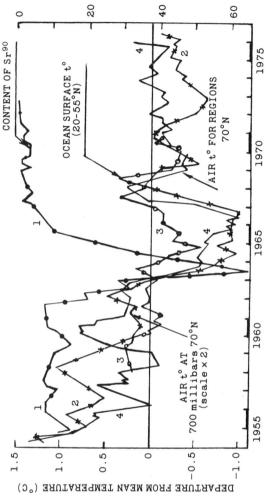

FIGURE 3.5. Temporal variations in air temperatures in polar latitudes and sea surface temperature in the NH midlatitudes. 1, ^{90}Sr content in the NH atmosphere (from Ref. 39); 2, air temperature at the 700-mb level (temperature values should be doubled); 3, mean seasonal Pacific surface temperature (20–55°N); 4, air surface temperature in the NH polar regions (12-month running means).

187

Beginning from 1954, when a rapid increase in the ^{90}Sr content occurred, the air temperature in polar regions decreased. By 1961 the decrease had reached 1.0 to 1.5°C. Beginning in mid-1962, a further drop in the air temperature took place in the lower troposphere. Note that the time dependence of the ^{90}Sr content is considered here only because this long-lived isotope is a tracer for other nuclear products, namely the oxides of nitrogen.

In mid-1963, when the ^{90}Sr content (or NO_2) reached its maximum, a decrease in the sea surface temperature began. It reached its minimum in the autumn of 1964 ($\Delta T \approx -0.6$°C) and the sea surface temperature began to rise. The air temperature had only reached its minimum by mid-1966 (-1°C from the mean). During the following year and a half the air temperature again approached (as in 1962) its mean values. However, in the winter of 1967–1968 again, the temperature dropped because of oscillating weather processes. Another possible reason for this drop was an accumulation of nitrogen oxide due to increased solar activity.

Data on temperature variations in polar regions show that nitrogen dioxide injections to the NH high-latitude stratosphere brought about a 40-month cooling in these regions: the air temperature decrease reached 1°C in the surface layer and 2°C at a level of 700 mb (38).

Data on the spatial distribution of air temperature anomalies in the NH extratropical belt in January 1963 point to the existence of six alternating centers of heat and cold that determined the cold weather in North America, Europe, southeast Asia, and the relatively warm (winter) weather in central Asia, Greenland, the North Atlantic, and the Pacific. On the whole, the mean temperature in the Northern Hemisphere during the winter of 1962–1963 dropped by 0.4°C. The winter of 1963–1964 was also severe, with a surface temperature decrease by 0.6°C. Walsh (38) points out that the 1963–1964 cooling was mostly over Eurasia and that North America did not experience the same cooling. In 1963 a cooling was also observed in the tropics (the temperature dropped by 0.2°) and in the SH extratropical latitudes (by 0.4°). Beginning in 1963, the frequency of negative temperature anomalies increased around the globe. Maximum values of the temperature decrease were observed in 1964–1966.

Thus, as data show, nuclear tests have affected the thermal regime not only of the surface but also of the troposphere and stratosphere, which points to the interactive global nature of the 1958–1962 nuclear tests impacts. So, for example, a temperature drop in the layer from 850 to 300 mb and temperature increase in the layer from 300 to 100 mb started almost simultaneously in the Northern and Southern Hemispheres (33), but temperature variations in the Southern Hemisphere constituted only about two-thirds of those in the Northern Hemisphere, and the cooling period in the Southern Hemisphere was half as long as that in the Northern Hemisphere.

A change of the period of the quasibiennial air temperature variations in the stratosphere is also an important phenomenon. As is well known, during minimum solar activity they weaken and then again, increase their amplitude.

Data on temperature in the surface 100-mb layer for extratropical latitudes given by Angell and Korshover (33) suggest that the 1961–1962 nuclear tests affected thermodynamic processes in the stratosphere, causing a 3-year instead of a 2-year oscillation. However, the quasibiennial oscillation has remained in the layer from 25 to 35 km in the northern polar region. No doubt, the temperature anomalies in the Northern and Southern Hemispheres are a chain of cause-and-effect events, determined to a great extent by the impact of nuclear tests on the atmosphere.

Variations in the DSR extinction in the middle and upper stratosphere between 1962 and 1968, shown in Fig. 3.4 as an increase of the meteorological solar constant, were approximated by a linear dependence because of a limited data base. In a first approximation this dependence can be considered to be an effect of the stratospheric NO_2 decrease with time.

A more detailed temporal course of extinction could be retrieved only after the publication of the work by Angell and Korshover (36) in which data are given on temporal variations in the ozone content in the layer from 32 to 45 km for the Northern and Southern Hemispheres and TOC for the Eurasian territory. Available information on photochemical processes in the middle stratosphere shows that in the layer from 30 to 50 km the concentrations of ozone and nitrogen dioxide are interrelated and that the concentration of O_3 is in inverse proportion to the square root of the NO_2 concentration. Based on these results, it was decided to use the inverted curve of the O_3 temporal change (36) as a relative temporal dependence of the NO_2 content variations in this layer, and then to calibrate this dependence. Results of balloon and rocket soundings of the atmospheric spectral transparency in 1962 (32) and 1964 (34) as well as some values of the NO_2 content obtained in various experiments between 1973 and 1977 (92–98) were used as reference data for purposes of calibration to estimate the NO_2 content. Figure 3.3 shows the temporal variations of the NO_2 content in the atmosphere and in a layer above 32 km. In the period 1960–1964 the curves of NO_2 variations for the Northern Hemisphere and the Eurasian territory show a decrease of NO_2 (or an increase of O_3). These minima are overlapped by a dashed line drawn with account of additional amounts of NO_2 in the layer from 46 to 55 km, whose existence is confirmed by a warming of this layer in 1963–1964. In addition, an increase of the NO_2 content in the upper stratosphere can cause not a decrease of the O_3 content but its growth, since here, as in the lower stratosphere, reactions of NO_2 with OH and HO_2 (which can reduce the ozone depletion due to reactions with free hydrogen) play a substantial role.

(In the authors' opinion, the cometary substance, continuously coming to this layer and to the mesosphere as microbolide intrusion, is a source of HO_x for the upper stratosphere.)

Temporal dependencies of the NO_2 content are supplemented in Fig. 3.3 by the chronology of the ^{90}Sr content in the NH stratosphere (39). A maximum of the NH-averaged ^{90}Sr content was observed in early 1963. For lack of new sources, by spring of 1967 ^{90}Sr had been removed from the stratosphere by gravitational settling, which explains the exponential dynamics of ^{90}Sr in that period. Then, during 1967–1970, several thermonuclear detonations took place, which delayed the stratospheric cleansing.

Of particular importance is the fact that the maximum spreading of the oxides of nitrogen (using ^{90}Sr as an indicator) in the Northern Hemisphere and, respectively, the maximum effect on the thermal regime of the stratosphere and other layers were followed (with a small time shift of about 2 months) by another event—the Agung volcano eruption (30 March 1963). Part of the erupted material penetrated into the stratosphere (mainly, gas and finely dispersed dust components). The location of the volcano (8°S) and seasonal conditions favored the spreading to the Southern Hemisphere of most of the material ejected into the stratosphere. Only 12 to 15% of the products injected penetrated the NH stratosphere. It was in October 1963 that the U.S. Optical Network registered the eruptive aerosols.

Whereas in the Northern Hemisphere the effects of the two events can be separated, it is much more difficult to do so for the Southern Hemisphere. Data on the ^{90}Sr content in the SH stratosphere show that the phase of maximum propagation of ^{90}Sr falls in early 1964. Thus in the Southern Hemisphere the effects of volcanic and nuclear events on the thermal and radiative regimes of the stratosphere also combine.

For more than 10 years a number of scientists have tried (in our view, using an overestimated impact of the Agung eruption—discussed below) to find agreement between observed and calculated data on the thermal and radiative effects of the eruption. They failed to do so, however. A simple comparison of estimates of the injected dust matter and the effects of solar radiation extinction for Krakatoa and Agung volcanoes points to an overestimated impact of the Agung eruption. So, for example, the Krakatoa volcano eruption ejected to the stratosphere (indirect estimates) about $(20-50) \times 10^6$ tons of ash, gases, water, and so on, whereas Agung ejected $(5-15) \times 10^6$ tons. Estimates for Krakatoa and observations during the Agung eruption showed that maximum additional extinction of solar radiation due to these eruptions is practically equal (about 25%). This means that the effects ascribed to Agung are twice overestimated, since in the Southern Hemisphere the volcanic and nuclear effects have almost equally contributed to the DSR extinction and created similar climatic effects.

Calculations by Oliver (40) show good agreement between measured and calculated anomalies of the annual mean surface air temperature in the period before 1961. During nuclear tests, however, this agreement was broken. Based on the optical properties of volcanic aerosols from solar radiation fluxes measured at Aspendale, Australia, DeLuisi (41) concluded that the absorption of radiation by stratospheric aerosols is very small.

Nearly 85% of the volcanically reduced DSR flux must return to the Earth's surface due to increased scattered radiation. Then, for the DSR flux reduced by the Agung eruption, estimated by the authors at 25%:2 = 12.5%, we obtained a climatologically significant decrease of the global radiation of about 1.7%. It must be borne in mind that maximum extinction took place only in some regions of the Southern Hemisphere and only during a short period. Thus the conclusion can be drawn that even in the Southern Hemisphere the effect of the Agung eruption was less important than the effect of nuclear explosions manifested through an increased NO_2 concentration in the middle and upper stratosphere.

Let us return to the analysis of observational data given in Fig. 3.3. To determine the relationships and succession of events, in addition to temporal variations in the NO_2 content in the layer from 32 to 55 km for the Northern and Southern Hemispheres and separately for the Eurasian territory, the figure presents variations in the Linke turbidity coefficient T_L in relative units was well as the change of the total DSR extinction $\Delta S_{0,\text{total}}$ (42). The turbidity coefficient characterizes the atmospheric turbidity as a ratio of logarithms of transparency for actual and ideal (Rayleigh) atmospheres and, therefore, is a more sensitive characteristic of turbidity than is extinction. This can easily be seen when comparing the two dependencies for the period 1958–1962. In these years the DSR extinction had no trend, but the turbidity coefficient changed markedly, increasing by 0.2.

It is of interest to monitor specific features in the course of ΔT_L that illustrate the effects of nuclear tests in 1957–1958 (an appearance in late 1958 of the trend of transparency decrease) and 1961–1962 (this trend intensified in late 1962) on the atmospheric turbidity, with emphasis on a certain similarity of events in the atmosphere in the regions of Hawaii and Eurasia manifested through similar changes of ΔT_L and $\Delta S_{0,NO_2}$ between 1963 and 1967. The similarity of changes in the turbidity coefficient over Mauna Loa and in excess radiation extinction $\Delta S_{0,NO_2}$ in the entire atmosphere over the Eurasian territory can be explained by the post-Agung increase of scattered radiation flux partly due to the increased intensity of the circumsolar radiation, which, apparently, has not been considered in calculations of ΔT_L.

Because of different changes in these parameters over Eurasia and the Northern Hemisphere, on the whole it is important that balloon data in Figs.

3.3 and 3.4 on the change of $\Delta S_{0,NO2}$ show the existence of a temporal change, different from that for Eurasia and almost coinciding with the change in $\Delta S_{0,NO2}$ for the Northern Hemisphere. This suggests the conclusion that variations in the scattered radiation flux in this period caused errors in the total ozone content measured with the ozonometers M-83. One must, however, use the TOC data for the 1966–1967 with caution. However, it does not follow from this that the M-83 data must be totally rejected. From data of the USSR actinometric network, the scattered radiation flux had increased by 16% by the end of the winter of 1964–1965 compared to October 1963.

Now analyze variations in the NO_2 content in the NH midlatitude stratosphere. As follows from Fig. 3.3, the curve $\Delta S_{0,NO_2}$ for the Northern Hemisphere shows that by the year 1960 the NO_2 content had reached 2×10^{17} molecules of NO_2 per square centimeter. The estimates show that these figures are realistic. By 1958 the NO_2 content had also been high enough (about 1×10^{17} molecules of NO_2 per square centimeter). This supports the supposition that the 1956–1957 nuclear tests of a small explosive yield substantially contributed, however, to the stratospheric NO_2 reservoir. An impression is that the stratosphere had reacted to the first nuclear tests with a larger NO_2 release than during the following years. The dotted curve (Fig. 3.3) overlapping the $\Delta S_{0,NO_2}$ minimum in 1961–1963, and the additional dashed curve above, determine maximum values of the NO_2 content.

From 1964, the NO_2 content in the layer from 32 to 46 km dropped practically linearly until 1967 at a rate of about 0.4% per year. An increase of $S_{0,M}$ (balloon data) and a decrease of $\Delta S_{0,NO2}$ from data on ozone variability in the layer from 32 to 46 km obtained from ground-based measurements with Dobson's spectrometers using the inversion technique and transformed into the NO_2 variability for the same layer, are consistent (0.4%/year). The 1967–1969 nuclear tests reduced the rate of atmospheric self-cleansing. Minimum NO_2 contents were observed in early 1974 (about 5×10^{15} molecules of NO_2 per square centimeter) (96,97). A subsequent increase of $\Delta S_{0,NO2}$ corresponds to the change of $\Delta S_{0,total}$ (Mauna Loa). Apart from nuclear tests, the Fuego eruption occurred in this period (the fall of 1974). Therefore, identification of the source of the NO_2 increase in 1975–1976 requires further efforts, even more so because measurements of ozone could be affected by the scattered radiation.

It is difficult to compare $\Delta S_{0,NO2}$ and $S_{0,total}$, since the first dependence characterizes the state of the stratosphere and the second the state of the middle and upper troposphere. Two periods can be selected: in one period (1963–1976) these dependencies had a very similar course, and in the other (1958–1962) there was no correlation whatever. If we take unsmoothed monthly means for $\Delta S_{0,total}$ (42), we shall find peak values of 7.5% for the year 1960 and values of 8% for 1958. It is clear that trying to find a good reference period in 1958–1962, Mendonca et al. (42) applied a very strong smoothing (thereby removing all characteristic features of this period). With these remarks, a very good consistency of the dependencies for $\Delta S_{0,NO2}$ and $S_{0,total}$ can be stated, without any constraints. This confirms the prevailing contribution of nuclear tests to anomalous variations in stratospheric transparency, at least between 1956 and 1974.

Additional amounts of NO_2 in the stratosphere heat the layers containing an excess NO_2 and create a deficit of solar radiation both in the lower troposphere and at the Earth's surface. Balloon data point to an average deficit of 2.6% in the solar radiation during 1962–1963. On the other hand, we have mentioned a 0.4° decrease of the surface temperature in Europe in the winter of 1962–1963.

In the following winter the temperature decrease constituted 0.6°C. The same values of cooling were obtained for the entire globe (36,37) as compared with the year 1958. But as seen from Fig. 3.3, in 1958 an excess NO_2 created a 1.3% deficit in the meteorological solar constant $S_{0,M}$. Beginning in 1958 (probably, some years earlier), the mean global surface air temperature started to decrease. Hence we can estimate the parameter ΔT_I characterizing the surface air temperature drop at a 1% decrease of the solar constant. With a decrease of $S_{0,M}$ in 1958–1960 constituting 1.3%, upon a respective delay (about 3 years) due to the thermal inertia of the troposphere-surface system, the global surface temperature dropped by 0.6°; that is, $\Delta T_I \approx 0.5°$.

The pre-1958 temperature course can be found from work by Budyko (5). Figure 3.6 shows an average trend of temperature variations in the NH midlatitudes.

Let us examine the behavior of the warming and cooling constituents for the trend. They are shown in Fig. 3.6 by thin dashed lines that cross near the year 1940. It is clear that the cooling trend had started not in 1940 but earlier, and hence the warming even approximated by maxima is still underestimated. Without the cooling trend, the warming would have reached at least 1°C by 1970. The cooling trend appearing in the 1930s had probably also been much stronger than that shown in Fig. 3.6, since the cooling trend was still preserved in the midlatitudes but gradually decreased. What is the cause of the cooling trend? There are some circumstances which suggest that both this trend and the warming trend have been caused mainly by inadvertant human impact on the environment.

In the 1930s, due to an acceleration of the scientific-technological revolution, an intensified development of heavy (including military) industry, increasing automotive and aviation transport, and modern mechanized agriculture started in Europe and America. This intensified the input of the oxides of nitrogen to the atmosphere. During World War II the inputs of these oxides of nitrogen and carbon had increased, because of a 100-fold

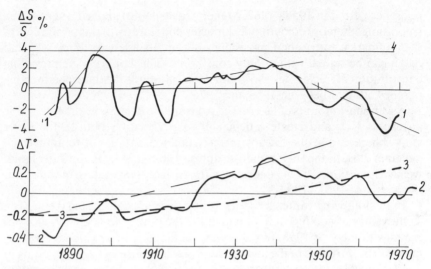

FIGURE 3.6. Relative variations in DSR fluxes and air surface temperatures from data of a NH midlatitude station. 1, DSR variations (%); 2, average Northern Hemisphere temperature variations (°C); 3, averaged trend of temperature variations; 4, solar radiation level in the case of maximum atmospheric self-cleaning. Dashes on curve 1 mark variation trends for individual periods. Curve 2 shows two constituents of an averaged trend—warming and cooling—between 1880 and 1951 in the NH midlatitudes. (Data from Ref. 5.)

increase in the number of engines in military action on the land, in the water, and in the air. Beginning in 1945, a new period of rapid industrial and agricultural development began in Europe and the United States, one that led to a further acceleration of the input of nitrogen to the atmosphere. In the late 1940s and early 1950s the rate of extraction and combustion of fossil fuels, as well as the rate of manufacture of high-temperature engines, was such that it took only 10 years for the rate of NO_x production to double. An acceleration in industrial and agricultural development, as well as an increase in the number of transport units (and their increasing speeds), led to an intensified input of the oxides of nitrogen to the atmosphere. But the rate of CO_2 production also grew by 4 to 5% per year. This could have been enough to compensate for the cooling trend due to increasing NO_2 content.

A correlation between variations in the DSR flux and those of temperature can easily be detected in Fig. 3.6. For this purpose it is enough to take into account the lagging of temperature variations behind those of radiation (about 3 years).

Consider specific features of DSR variations in different time periods. Atmospheric transparency has deteriorated many times. After each case of atmospheric pollution followed a period of self-cleansing. For example, after the eruptions of Krakatoa, Bandaisan, Bogoslov, and so on, the process of cleansing continued for 10 years, and in 1896 the atmosphere was, apparently, close to the undisturbed state. Later, there was no such rapid removal of perturbation. Even during the period of a relatively small dust loading (1920–1940), the atmosphere could not cleanse itself to an undisturbed level. This points to the appearance of an additional constant component in the DSR extinction.

Approximately about 1933 the atmospheric transparency started to decrease, with the rate of decrease becoming very high in 1945–1946. A still faster decrease was observed in 1961–1962. A maximum reduction took place in 1966 (Fig. 3.5). Comparing the data, one can understand the nature of the extinction components and its role in present climate changes.

The Solar Constant Value. Again, let us examine Fig. 3.4, which shows some data of solar constant measurements. Data of rocket measurements of the solar constant in 1976 (43) are given for comparison purposes. Rocket measurements of $S_{0,astr}$ were made with four absolute cavity radiometers of different design. An average over data of three consistent radiometers constitutes $1367 \pm 6 W/m^2$. From data of Mariner-6, Mariner-7, Nimbus-6,7, and SSM satellites, variations in $S_{0,astr}$ do not exceed +0.15%, −0.25%.

Reduction of our balloon data beyond the atmosphere was made according to a model simulation of the upper stratospheric layers transparency for 1968. Therefore, all the extrapolated data reflect the variability of the upper stratosphere transparency against the year 1968, and they should be considered as data on the meteorological solar constant $S_{0,M}$ referenced to an absolute value of the astronomical solar constant $S_{0,astr} = 1371\ W/m^2$, determined from balloon data for 1967–1968 and almost independent of solar activity. However, the meteorological solar constant does exhibit this dependence, which coincides with that for $S_{0,astr}$ (44). The notion of $S_{0,M}$ should be specified; it must, presumably, characterize the subozone value of $S_{0,astr}$ determined by the DSR flux at a level of the tropopause. Here $S_{0,M}$ who is considered as a characteristic of the effect of an excess stratospheric NO_2 (above 16 km) on $S_{0,astr}$. Based on analysis of actinometric measurements of the solar constant, Soviet scientists obtained $S_{0,astr} = 1371 \pm 7\ W/m^2$. An average value is $1369 \pm 6\ W/m^2$ (45).

3.2 NUMERICAL MODELING OF THE IMPACTS

3.2.1 Assumptions about Direct Effects of a Nuclear Attack

It became clear during the last decade that nuclear tests and volcanic eruptions cannot shed light on every possible consequence of a nuclear conflict, not only because of different total explosive yield (eruptive energy) but also because of their temporal concentrations and, most important, because of changes in nuclear strategy. New military doctrines classifying the industrial centers, large cities, gas and oil fields, and oil refining enterprises as first-priority targets for 1 to 3 megaton warheads (atmospheric explosions) and for 1-megaton warheads (ground bursts, rocket-launching shafts) create new conditions for affecting the atmosphere and climate. Powerful warheads (10 to 15 megatons) can also be used to attack intercontinental rockets in the stratosphere, which will cause huge disturbances in the composition and thermal regime of the middle and upper stratosphere.

Bombing cities and other targets with large supplies of combustibles will start large-scale fires in the form of fire tornadoes ejecting to the atmosphere (according to the 5000-megaton scenario)

1. From 200 to 450 megatons of soot, with 10% ejected to the stratosphere,
2. Fifty megatons of ash (5% to the stratosphere),
3. More than 60 megatons of the oxides of nitrogen (50% to the stratosphere),
4. From 2500 to 5000 megatons of CO_2 or CO

Bombing surface and subsurface targets will produce 900 to 2500 megatons of silicate aerosols ejected to the atmosphere.

An appearance of such amounts of aerosol and gaseous components in the troposphere and stratosphere is comparable in order of magnitude with material erupted by the volcanoes Tambora and Krakatoa. However, by its properties, the soot aerosol differs substantially from the volcanic aerosol, and hence the analogies cannot provide correct results. The only possible approach, then, is to assess the scales of the impact of a possible nuclear conflict, for example, the numerical modeling of radiative, dynamic, and physical processes using nonstationary 1-D, 2-D, and 3-D mathematical models.

3.2.2   Theoretical Models of Delayed Atmospheric Effects

In the early 1980s a number of groups of scientists in the Federal Republic of Germany, the United States and the USSR began a model study of the possible environmental consequences of a nuclear conflict, based on various scenarios.

The 1982 paper by Crutzen and Birks (7) in *AMBIO* was one of the first studies in this direction. The paper was presented in June 1983 at the Third

International Congress of "Physicians for the Prevention of Nuclear War" (Amsterdam). An important feature of this excellent study is its consideration of the impact of numerous fires that would take place in cities, forests, agricultural areas, and gas and oil fields at the onset of a nuclear war. The study was made using a 2-D model supplemented with a photochemical block with 100 reactions. Results were obtained for two scenarios of total yields of 5750 megatons (scenario I) and 10,000 megatons (scenario II):

1. Bomb detonations of unit yield less than 1 megaton, with a total yield of 5750 megatons, which produce 5.7×10^{35} molecules of NO in the troposphere.

2. Detonations of five thousand 1-megaton bombs and five hundred 10-megaton bombs, uniformly distributed in the latitudinal belt $20-60°$N; the oxides of nitrogen are ejected at heights below and above 18 km, depending on the yield of the warheads; most of the explosions are in the atmosphere.

Forest fires in the Northern Hemisphere supposedly will break out over a territory of 10^6 km^2. In this case about $(1.3-2.5) \times 10^{15}$ g of carbon will be injected into the atmosphere, $(2-4) \times 10^{14}$ g of this amount being submicron aerosols. These aerosols, with a high content of carbon (40 to 75%) are, to a great extent, a product of gas-to-particle conversion. With forest fires continuing for 2 months and the aerosol lifetime 5 to 10 days, the aerosol content in an air column will constitute 0.1 to 0.5 g/m^2, which will increase the solar radiation extinction on a summer noon by a factor of 2 to 150. Thus most of the Northern Hemisphere will be in darkness for a long time. Considerable additional pollution of the atmosphere will be caused by fires in cities and industrial centers, where a huge amount of fuel is stored (its global stock amounts to 1.5 billion tons of carbon), as well as by fires at damaged gas and oil pipelines and at continental and maritime oil and gas fields.

A catastrophic decrease of illumination will violate (stop) crop vegetation over vast territories in the Northern Hemisphere. The settling of dark aerosol on plants will also damage them. Naturally, a direct consequence of fires would be a considerable destruction of vegetation. Another important consequence of a marked decrease in surface illumination will be the ruin of phyto- and zooplankton over half the ocean area of the Northern Hemisphere.

With the appearance in the atmosphere of a strongly absorbing aerosol, a decrease of the surface albedo, resulting from aerosol settling, will cause marked changes in the circulation and in the thermal regime. A considerable increase in atmospheric stability will make the removal of pollutants difficult. A multiple increase in the concentration of condensation nuclei must essentially change the conditions for the formation of clouds and precipitation.

Although during forest fires most of the carbon is injected into the atmosphere in the form of CO_2 (7×10^{17} g), far more important will be the CO injections of the order of $(2-4) \times 10^{14}$ g, which will double the present CO content in the atmosphere. As a result, a mean CO concentration in the NH midlatitudes will increase by a factor of 4 (still higher on the continents). Simultaneously, tens of teragrams ($1\ Tg = 10^{12}$ g) of chemically active hydrocarbons (mostly ethylene and propylene) will come into the atmosphere, which play an important role in the formation of photochemical smog and large amounts of the oxides of nitrogen.

According to one scenario of a nuclear war, an injection of 5.7×10^{35} molecules of NO (12 Tg of nitrogen) by nuclear explosions and a gradual input of NO_x due to fires will cause radical changes in the tropospheric photochemical processes manifested: through a great increase in ozone concentrations (up to 160 ppb against the normal 30 ppb), as well as through an accumulation of ethane (50 to 100 ppb) and peroxyacethylnitrate — PAN (1 to 10 ppb against the background value less than 0.1 ppb). The capture of NO_x by rain droplets will result in a marked increase in their acidity against normal pH values of less than 4. An increase in the ozone concentration and larger acidity of precipitation will be disastrous for vegetation.

These effects of a nuclear war refer to summer conditions. In winter, forest fires will be of a smaller scale, the photochemical smog reactions will be slower, but the decrease in the illumination will be more harmful (particularly with regard to its impact on human health).

The basic feature of the second scenario is an injection of NO_x into the atmosphere accompanied by an approximately 20-fold increase of the NO_x concentration and a subsequent large decrease of the ozone content, which will take place initially in the Northern Hemisphere but gradually spread to the Southern Hemisphere. If one assumes that the total yield of explosions is $(5-10) \times 10^3$ megatons and that the war breaks out on June 11 (with the explosions concentrated in the belt $30-70°N$), a maximum drop in the total ozone content would take place 2 to 3 months after the NO_x injections and would constitute 35 to 70%. The time for atmospheric recovery would amount to 3 years (7).

An important aspect of the problem concerning the biosphere is an intensification of the biologically active UV radiation in the wavelength interval 290 to 320 nm (UV-B radiation).

Estimates of changes in the atmosphere point to their destructive impact on the ecosystems and human life during a period of several weeks and months. A summary of research (7) suggests that scientific data available definitely confirm the conclusion that a large-scale nuclear war leaves most inhabitants of the Northern Hemisphere no chance to survive.

The pioneer work by Crutzen and Birks was followed by a very thorough study by a group of American experts (Toon, Ackerman, Pollack, Sagan) headed by Turco (TTAPS). They undertook a multivariant modeling (46) of the impact of nuclear explosions on the atmosphere-surface system using a complex of physicomathematical models consisting of:

1. An aerosol microphysical model in which the temporal evolution of the horizontally propagating smoke clouds is determined
2. An advanced radiative-convective model which incorporates calculated (at given size distributions and refraction indices for aerosol particles) optical characteristics of dust and smoke aerosols in the visible and IR spectral regions, calculated attenuation of light and thermal radiation fluxes, and calculated radiative flux divergence, with resulting data on temperature as a function of time and height

Turco et al. (46) faced great difficulties in the selection of nuclear war scenarios because of the unpredictability of all the possible directions that the development of such a war might take. A version with a 5000-megaton yield was selected as a basic scenario. In other scenarios (30 in total) the yield ranged between 100 and 25,000 megatons. Basic differences between the scenarios, apart from total yield of explosions, included the distribution of the yields between the explosions on the surface and in the air, as well as between such targets as cities with industrial and military sites and isolated military targets.

The distribution of nuclear explosions according to some scenarios is given here (Table 3.1) in a shortened version, that is, only for five scenarios. The TTAPS group aimed at modeling the global-scale atmospheric and climatic consequences of a nuclear war, with due regard to new factors of the tropospheric and stratospheric pollution because of numerous large-scale nuclear ground explosions included in the scenarios in accordance with U.S. military strategy.

In addition to smoke and gas ejections, about 9.6×10^8 tons of transformed particles of soil, minerals, and other substances ("nuclear dust") can be injected into the troposphere and mainly (up to 80%) into the stratosphere. A small-sized aerosol fraction with a long residence time in the stratosphere can be increased by 65 megatons as a result of these injections alone (the basic scenario). If the scenario approaches a "counterforce" one, the mass of the stratospheric nuclear dust can increase by 120 megatons or even by 650 megatons (47).

An increase in the stratospheric aerosol mass by 120 megatons, let alone the "addition" of 650 megatons, approximates ejections into the stratosphere of materials by large-scale volcanic eruptions such as that of Tambora (1815). One of the most substantial differences in these events is that practi-

TABLE 3.1 Distribution of Nuclear Attacks by Types of Targets, following Most Known Nuclear War Scenarios, and Estimates of the Masses of Nuclear Smoke and Nuclear Dust

Scenario	Total Yield (Mt)	Ground Bursts (%)	Cities and Industrial Targets (%)	Warhead Yield (Mt)	Total Number of Detonations	Mass of Submicron Smoke Particles (10^6 t)	Mass of Submicron Dust Particles (10^6 t)	Optical Depth of Smoke Cloud	Optical Depth of Dust Cloud
1. Basic	5,000	57	20	0.1–10	10,400	225	65	4.5	1
2. Large-scale conflict	10,000	63	15	0.1–10	16,160	300	130	6	2
3. Attacks of cities (minimum total yield)	100	0	100	0.1	1,000	150	0	3	0
4. Counterforce (mutual attacks of rocket-launching silos)	5,000	100	0	5–10	700	0	650	0	10
5. Multipurpose	5,000	10	33	0.1–1	22,500	300	15	6	0.2

Source: Ref. 46.   Mt = megatons; t = tons

cally all the nuclear dust is "supplied" with radioactive isotopes, part of which are long-lived.

In the case of the 5000-megaton scenario the largest part — 2850 megatons (57%) — falls on ground bursts, including 1000 megatons on cities and industrial objects. The yield of individual warheads ranges between 0.1 and 10 megatons, and the explosions total 10,400 megatons. In the scenario of maximum yield (10,000 megatons), the share of ground bursts was increased to 63%, with 15% falling on urban and industrial targets.

There are two other scenarios which (from numerical modeling results) lead to heavy stratospheric pollution. The total yield of each of these scenarios amounts to 5000 megatons, but the yield of individual warheads and the number of explosions differ greatly. For example, in the scenario with a small number of explosions on the surface (10%), 33% of the total yield fall on cities. Charges are relatively small (0.1 to 10 megatons), but the total number of explosions is very large (22,500). In Fig. 3.7 this scenario is

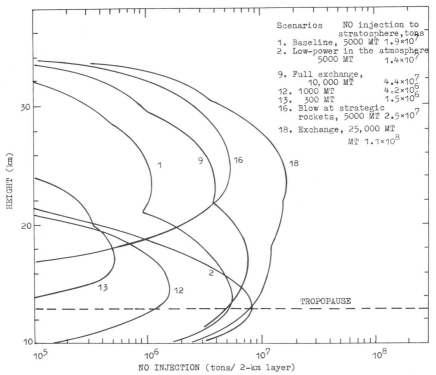

FIGURE 3.7. Vertical profiles and total amounts of NO injected into the stratosphere according to some nuclear war scenarios. Scenario number 18 (TTAPS) illustrates the impact of a possible nuclear conflict in the future in case of a continued arms race and nuclear weapon accumulation.

marked number 2. The second (of these two) scenario is marked number 16. According to this scenario, the whole yield (5000 megatons) is directed to continental rocket-launching silos (e.g., 100% ground bursts). The yield of individual warheads is very large, on the order of 5 to 10 megatons. The number of explosions totals 700. The pollution of the middle stratosphere (20 to 35 km) by the oxides of nitrogen and by small-sized dust is very heavy (number 16, Fig. 3.7). Twenty-six teragrams of NO is injected into the stratosphere, an amount that exceeds by five times the background value of NO.

In the case of maximum yield (10,000 megatons), the NO injection reaches 44 Tg. Table 3.2 shows the basic results of model estimations of the direct impact of nuclear war on the atmosphere and climate. All of the parameters change drastically. For example, the solar radiation in the NH midlatitudes will not exceed 1 to 3% of its usual values for 45 days. This means that the so-called "nuclear winter" will last for one and a half months.

TABLE 3.2   Long-Term Impact of a 10,000-Megaton Nuclear Exchange on the NH Biosphere

Parameter	Changed Value[a]	Possible Range of Changes	Duration of the Impact	Geography of the Impact
Total solar radiation	×0.01	×0.003–0.03	1.5 month	Midlat. NH
flux	×0.05	×0.01–0.15	3 months	Midlat. NH
	×0.25	×0.1–0.7	5 months	NH
	×0.50	×0.3–1.0	8 months	NH
Surface air	−43°C	−53° – −23	4 months	Midlat. NH
temperature (°C)	−23°C	−33° – −3	9 months	NH
	−3°C	−13° – +7	1 year	NH
UV solar radiation	×4	×2–8	1 year	NH
(290–320 nm)	×3	×1–5	3 years	NH
Dosage of radioactive	≥500 R		1 hour–	30% of midlat. NH
fallout	≥100 R	3 times	1 day	50% of midlat. NH
	≥10 R		1 day–	50% of midlat. NH
			1 month	
			1 month	
Concentration of radioisotopes				
^{131}I	4×10⁵ MCi	—	8 days	Midlat. NH
^{106}Ru	1 × 10⁴ MCi	—	1 year	NH
^{90}Sr	400 MCi	—	30 years	NH
^{137}Cs	650 MCi	—	30 years	NH

[a]×0.01 (etc.) means a hundredfold decrease of the background (normal) value.

Source: Ref. 48.

The surface temperature will drop to -23 to $-53°C$, independent of the season.

In the process of atmospheric cleansing of smoke, dust, and various gaseous pollutants, the UV carcinogenic radiation will rapidly increase in the interval 290 to 320 nm, and its excess will destructively influence the eye. An increase in UV solar radiation is associated with a supposed biggest decrease of TOC after a nuclear conflict. The UV radiation increase, as seen from the table, is supposed to be substantial (three to four times) as well as long term (1 to 3 years).

In this scenario the total NH stratospheric ozone decrease is estimated at 70%. A still greater decrease was estimated by Obukhov and Golitsyn (49)—down to 90%. An additional destruction of ozone, in their opinion, is connected with the heterogeneous destruction of ozone by the stratospheric explosive aerosol whose concentration should greatly increase from nuclear explosions.

The estimates made by Turco et al. of the ozone content decrease due to the Tunguska meteor fall (discussed later) were based on erroneous ideas about a negligible NO_2 absorption in the visible (500 to 750 nm) spectral region. Therefore, doubts arise about the reality of the system of photochemical processes used in modeling the phenomena in the ozonosphere, in studying both the Tunguska meteor (TM) fall and a nuclear conflict. This conclusion was experimentally verified from the TOC temporal change following the 1961–1962 nuclear tests. By 1970, TOC had increased by 8% compared to the pretest period—not decreased as Turco et al. believed.

The estimates given in Table 3.2 again confirm that the NH midlatitude belt can be subject to such a high level of radioactive pollution that lethal doses will become a global phenomenon. Postexplosion survivors and those who were fortunate enough to avoid the direct impacts will face the insoluble problem of protecting biological organisms against radioactive isotopes for years to come.

Now let us analyze in more detail the temporal scales of the nuclear war impact on climatologically important atmospheric parameters such as the atmospheric optical thickness τ and air temperature in the troposphere and stratosphere. Figure 3.8 shows model temporal changes of the atmospheric optical thickness at $\lambda = 550$ nm for a set of scenarios. Naturally, maximum atmospheric turbidity results from the maximum yield of explosions (10,000 megatons), both in short (e.g., weeks) and long time periods. The scenario with a relatively small yield (100 megatons) refers to a group of scenarios that produce very large atmospheric pollution ($\tau = 3$ to 6). Here big impacts on the atmosphere are connected with the production of large amounts of soot aerosols from numerous urban fires.

The impact of the 4 April 1982 El Chichón (Mexico) large-scale eruption,

40 km have to be catastrophic (Fig. 3.9). The lowest model surface temperature drop on day 30 is 40°C for scenarios numbers 9 and 2 and 35°C for numbers 14 and 1 (baseline). A relatively strong effect of soot aerosols on the optical thickness of the upper troposphere creates conditions for a similar surface temperature decrease, even for a scenario with a minimum yield (number 14). Comparing the temperature changes in scenarios numbers 14 and 4, one can evaluate a model contribution to the thermal regime of the lower troposphere by the soot and dust (submicron) aerosols. Apparently, the soot aerosol determines the temperature during the first 2 to 3 months, and the dust aerosol in the 6 months following. A temperature deficit of 3 to 4°C remaining by day 300 is rather substantial, since it concerns the entire Northern Hemisphere.

Stratospheric temperature changes (Fig. 3.10) are even more dramatic. Model calculations give almost unimaginable changes in temperature profiles — a 90°C increase at a height of 18 km. A comparatively slow recovery of temperature in the stratosphere indicates that the normal circulation regime of the upper troposphere as well as the lower and middle stratosphere

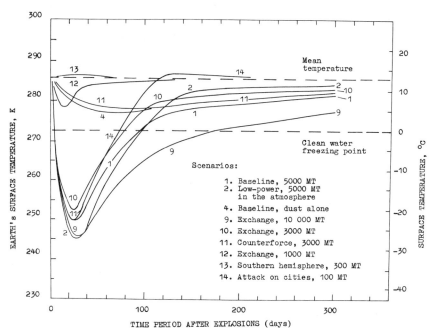

FIGURE 3.9. Temporal variations of mean hemispheric land surface temperature for the baseline scenario and some others. Scenario numbers 4 and 11 ignore the consequences of fires.

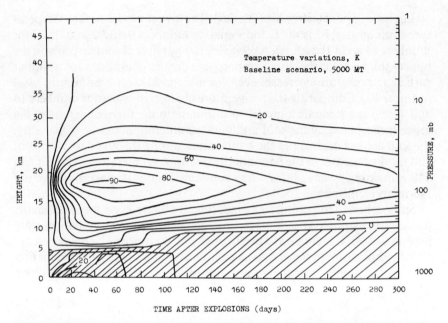

FIGURE 3.10. Isolines of air-temperature variations in the troposphere and stratosphere of the Northern Hemisphere after 300 model days for the basic scenario of nuclear war. The shaded part of the graph is for negative temperatures.

will be totally destroyed in case of a nuclear conflict (according to the basic scenario). Enormous vertical and horizontal temperature gradients will create large-scale vortices that will lead to chaos in the ozonosphere. With large amounts of dust in the stratosphere, the ozone layer may even be completely destroyed.

Modeling the impact of multiple nuclear explosions (partly presented in Figs. 3.7 through 3.10) makes it possible to identify similar events in the past, such as during the 1958–1962 nuclear tests. These are the events observed between 1963 and 1970:

1. Before 1965 a continuous increase had taken place in the meridional temperature lapse rate between the tropics and NH midlatitudes, as well as an increase in the vertical gradient over Asia. A surface temperature decrease constituted 0.2°C in the southern extratropical NH latitudes and 0.4°C in the northern ones.
2. The decrease in mean global temperature for 1958–1965 reached 0.3°C.

3. In January 1963 one of the largest temperature increases occurred in the stratosphere, and the production of kinetic energy exceeded an average level by a factor of 6.

An important study of the possible consequences of a nuclear war was undertaken in the Computing Center of the USSR Academy of Sciences (CC). Numerical modeling performed by Aleksandrov and Stenchikov (50) was aimed at an estimate of the climatic implications of post–nuclear war fires which will strongly increase the smoke-laden atmosphere turbidity. Calculations were made using an improved version of a two-level Mintz-Arakawa model which simulates the atmospheric general circulation, and a thermodynamic model of the oceanic upper quasihomogeneous layer. At a horizontal resolution of $12° \times 15°$ latitude-longitude, the geographical distribution of the ocean and continents is prescribed taking into account orography, sea ice cover, continental glaciers, and snow cover.

Heat flux divergences due to radiation, latent heat of phase transformations of water in the atmosphere, and sensible heat exchange were taken into account in the interactive consideration of the formation of cloudiness (the tropopause being at a level of 200 mb). Calculations were made for mean annual conditions (a mean annual extra-tropospheric insolation was prescribed).

Numerical modeling (50) was aimed at calculations of a quasiequilibrium climate (the state of the atmosphere-oceans-continents system) after an instantaneous change in the properties of the NH atmosphere north of $12°$ N as a result of its dust loading after nuclear explosions at and near the surface, as well as under the influence of soot particles injected into the atmosphere as a result of various fires (in forests, cities, at oil and gas pipelines, etc.). The troposphere is assumed to be only soot polluted, and the stratosphere only dust polluted (the effect of these aerosols on longwave radiation transfer is neglected). A temporal variation is prescribed for the aerosol optical thickness of the atmosphere in the shortwave spectral interval, which decreases to 7 immediately after the explosions and to zero in 360 days.

Beginning with day 100 after a nuclear conflict, stratospheric pollution prevails, 50% of the absorbed stratospheric solar radiation being assumed to be reemitted to the troposphere as longwave (thermal) radiation, and 50% to space. Calculations showed that immediately after the explosion, the land surface and the near-surface atmosphere cool rapidly in the Northern Hemisphere, but on the whole, the atmosphere is warmed for approximately 6 months.

During the first three postexplosion months, the radiation budget of the surface-atmosphere system is positive, but then it changes its sign, dropping to −10 W/m² (this value characterizes the mean rate of cooling the thermally inert ocean). Global atmospheric heating reaches approximately 20°C, after which the temperature slowly decreases. The mean global air temperature near the land surface decreases rapidly by 15°C and then, still more slowly, increases. A sea surface temperature decrease during the first 10 months is about 1.2°C. A maximum decrease in the mean zonal air temperature near the land surface takes place in the NH midlatitudes, reaching 23°C and 10°C on days 40 and 243, respectively. A temperature increase of up to 25°C on day 378 is caused not only by tropospheric cleansing a year after the explosion, but also by the residual heating of the upper troposphere, which intensifies the atmospheric thermal emission. The cooling of air in the Southern Hemisphere constitutes only 3 to 4°C.

A sharp decrease in the surface air temperature in the Northern Hemisphere changes the sign of the temperature lapse rate (the inversion), which leads to suppressed convection processes, the cessation of rainfall, and prolongation of the time for the tropospheric cleansing of soot aerosols. The formation of a large interhemispheric temperature contrast radically changes the regime of meridional circulation (Fig. 3.11). In the equatorial belt, instead of the usual Hadley circulation cells, a new interhemispheric cell is formed which promotes a strong intensification of the mass and pollutants exchange between the hemispheres. A considerable intensification of precipitation takes place in the SH tropics and in the belt at 20–30°N.

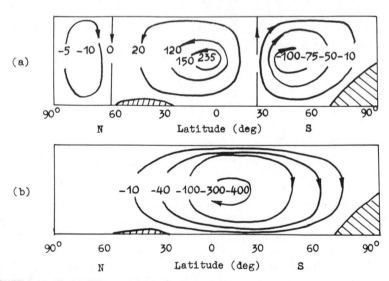

FIGURE 3.11. Meridional atmospheric circulation (air mass transport in 10¹² g/s) *(a)* before and *(b)* on day 297 following a nuclear exchange.

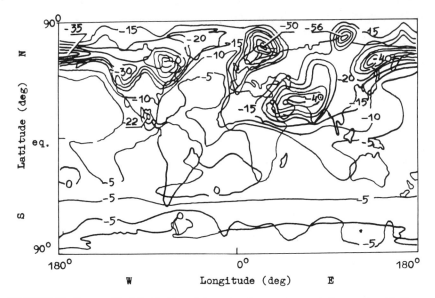

FIGURE 3.12. Geographical distribution of air surface temperature variations on day 40 after a nuclear exchange.

The geographical distribution of the nuclear explosion impacts on climate is nonuniform (Fig. 3.12). An air temperature decrease near the land surface in 40 days reaches 36°C (Alaska), 34 and 40°C (the central and eastern regions of North America), 41°C (Kamchatka), 51°C (central Europe), 56°C (Kolsk Peninsula), and 51°C (Arabian Peninsula). A much weaker ocean surface cooling determines the appearance of strong ocean-land gradients in the Northern Hemisphere. On day 378 day after the postcleansing warming of the atmosphere the air temperature over the continents of the Northern Hemisphere exceeds the normal one by 25 to 35°C. The Southern Hemisphere turns out to be 2°C cooler than normal. Nuclear explosion-induced climatic perturbations do not go farther than 30°N, on the assumption that the products of fires and explosions do not reach the Southern Hemisphere.

Computer Centre scientists (51) also consider two extreme scenarios with total explosive yields of 10,000 and 100 megatons. Calculations of the evolution of a disturbed atmosphere were made for the hemispheric and global versions of 3-D climate modeling.

Despite the two-order difference in the yields of the scenarios assumed in calculations, the behavior of the model system remained homogeneous for the impacts of such different scales. A major climatic effect, as in the basic scenario, remains to be a rapid and drastic drop in the temperatures of the surface and lower tropospheric layer over the continents. Maximum temperature drops for these scenarios differ negligibly.

A major difference is in the duration of the cooling period. In the opinion of Aleksandrov et al. (51), the characteristic time of cooling for the 10,000-megaton scenario is about 1 year, and for 100 megatons about 3 months.

The formation of disturbances in the thermal state of the atmosphere over the continents and an appearance of enormous interhemispheric temperature gradients in the upper troposphere and lower stratosphere will lead to a breakup of the usual system of meridional circulation and the creation of conditions for a rapid interhemispheric air masses exchange. It will take 3 to 4 weeks for the SH atmosphere to become polluted with nuclear dust and smoke almost to the same extent as the NH atmosphere (51). Quite unexpectedly, processes of the thermal transformation in the case of the minimum-yield scenario are more intensive than for the 10,000-megaton scenario.

Large temperature gradients (and hence fast motions of air masses) will appear between interior continental areas and the oceans (seas) because SST changes will be negligible. For the 10^4-megaton scenario, the Computing Centre model gave a decrease of mean global SST of 1 °C for 10 months. An intensified air mass exchange between the continents and the oceans will result in enormous rainfall in coastal regions.

Due to a strong air temperature increase in high mountain regions, snowpacks and glaciers can begin to melt, resulting in catastrophic snowmelt runoff through mountain streams to the valleys.

Before undertaking a critical discussion of results obtained in studies on nuclear winter, we shall mention the considerations by Thompson (52) on the status of model studies. He says that some experts are sceptical because model predictions of a postnuclear catastrophic air surface temperature drop (by 30°C and more) are controversial and hence insufficiently reliable. In solving this complicated and important problem, a critical analysis of the results obtained is needed with the participation of experts in various fields. Although the threat of nuclear war becomes increasingly obvious, an increasing concern has appeared about the reliability of quantitative estimates of the possible climatic impacts of a nuclear war, to the extent that the applied 3-D climate models were not intended to take into account large-scale impacts on the atmosphere, such as its extreme smoke loading. No doubt of extreme importance here is a sufficiently adequate consideration of the nonlinear interactivity of radiative processes and the propagation, transformation, and fallout of aerosols, the concrete features of which are difficult to predict. One of the most serious uncertainties is the amount of smoke aerosols that can form a day after a nuclear attack. Available estimates of smoke production are based on suppositions of such almost unpredictable

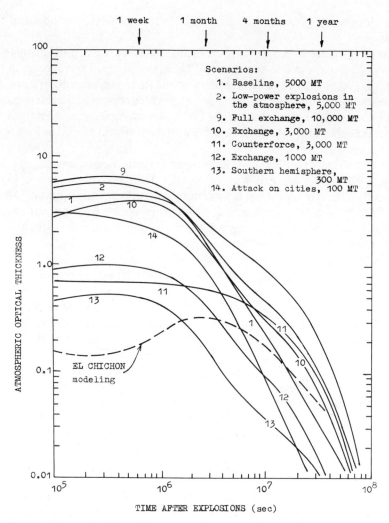

FIGURE 3.8. Temporal variations of atmospheric optical thicknesses averaged over the Northern Hemisphere ($\lambda = 0.55$ μm) for some nuclear war scenarios. A model temporal change of transformations of the optical parameters of a volcanic aerosol cloud (El Chichón, Mexico) is given for comparison.

modeled taking into account gas-to-particle conversion of aerosols in the stratosphere, turns out to be compatible with respect to atmospheric pollution with the second scenario in Fig. 3.8 (5000 megatons) in half a year, and due only to the specific production of stratospheric aerosols.

Temperature changes at the surface level and in atmospheric layers up to factors as the area of fire, the amount of burning material and smoke produced, and the properties, processes of diffusion and fallout of aerosols of the smoke produced, and so on. These factors are difficult to describe, especially during the first hours and days of fire in the presence of very dense smoke veils, at a time when the coagulation processes must be very active. Here, an analysis of conditions for the production of smoke from large-scale forest fires of the past can be useful.

Even estimates of the direct radiative effects of aerosols are hindered by the limited data base on the optical characteristics of smoke. Recent calculations confirm the importance of the consideration of diurnal change of insolation. Even more substantial are the effects of stratospheric aerosols not previously considered, including the maintenance (and intensification) of cooling near the surface and the considerable reduction (by half) of tropospheric heating.

Initial calculations using a 3-D model and taking into account only tropospheric aerosols revealed a radical change in the meridional circulation expressed in the formation of a giant circulation cell in low latitudes, covering both the hemispheres; with the addition of stratospheric aerosols, the meridional circulation turned out to be closer to that usually observed, and two Hadley cells were formed. Results of these new calculations illustrated the difficulty of adequately modeling the climatic consequences of a nuclear war. Note, however, that although "accurate" simulation experiments are impossible, even approximate estimates of the potential ecological catastrophe are extremely important. What is needed is to avoid either the extremes of alarmism or the underestimation of the possibilities for ecological disaster.

Admittedly, so far there are no 3-D mathematical models ready which would reliably reflect huge disturbances in a multilink and cross-correlated climate system. Nevertheless, nuclear war modeling is gradually broadening with an improved physicomathematical foundation.

Using a 2-D climate model, MacCracken (53) obtained estimates of the possible climatic implications of nuclear war, based on considerations of related changes in the gas and aerosol composition of the atmosphere, especially of ejected soot aerosols resulting from urban and forest fires. Comparative estimates of the mass of various gas and aerosol components ejected to the atmosphere by volcanic eruptions and nuclear explosions are given in Table 3.3. Figures for tropospheric aerosols correspond to the accumulated aerosol mass but not to its instantaneous content in the troposphere. As seen from Table 3.3, the supposed postnuclear ejections of submicron aerosol into the stratosphere are comparable with large-scale volcanic ejections which have caused a marked (about 1 °C) air surface temperature drop for several years after the eruption. Nuclear explosions also cause a considerable

TABLE 3.3   Mass of Various Components (Tg = 10^{12} g) Ejected to the Atmosphere by Volcanic Eruptions and Multiple Nuclear Explosions

Ejection Components	Stratosphere	Troposphere
Aerosols (including dust)		
Nuclear explosions	118	5000
Volcanic eruptions		
Tambora (1815)	Hundreds	2×10^5
Krakatoa (1883)	50	2×10^4
Agung (1963)	10–20	$(4-8) \times 10^3$
Mount St. Helens (1981)	~0	2×10^3
El Chichón (1982)	10–20	$(2-4) \times 10^3$
Background aerosol (NH)	~1	5–10
Oxides of nitrogen		
Direct ejections by nuclear explosions	~7	~1
Fires	0	26
Background content (NH)	~4	~0.2
Ozone		
Postnuclear production	~–600	40–100
Background content	~1700	~20
Postnuclear fire soot		
Cities	0	150
Forests	0	57
Gas ejections from natural forest fires	~0	a few
Background soot aerosol	~0	0.1–1

Source: Ref. 53.

(as compared to background values) increase in the NO_2 content (which absorbs solar radiation in the visible) and a decrease in the content of ozone, optically active in the UV and IR regions.

The most serious consequences of nuclear explosions are, however, forest and urban fires, which eject huge amounts of soot into the atmosphere. This particular aerosol was assumed (53) to have a lognormal size distribution with a mode radius of 0.06 μm and a dispersion of 2.21; the optical characteristics were taken from the data of Turco et al. (46). The size of aerosol particles was assumed to be constant and small and therefore does not affect the thermal radiation transfer.

For initial estimates of the effect of soot on the radiative regime and mean hemispheric (NH) air surface temperature, a 1-D radiative-convective model (RCM) was used, taking into account the heat capacity of the soil surface layer at fixed cloud conditions and rates of aerosol fallout from the atmosphere. The aerosol is considered to mix instantaneously up into the global atmosphere, which, no doubt, leads to an overestimation of the mean hemispheric impacts on the radiative regime and climate. A most serious limitation is, however, neglect of the horizontal inhomogeneity of the surface (the distribution of land and oceans), because, under conditions of extreme effects on the atmosphere, sharp land-ocean contrasts are formed which will affect much of the general circulation of the atmosphere.

The scale of disturbance is prescribed on the assumption that 150 Tg of soot is ejected to the troposphere by urban fires during the first day after the explosions and that 57 Tg of aerosols resulting from forest fires is also ejected into the troposphere. In this case the vertical optical thickness of the atmosphere in the visible will constitute about 4 with contributions by scattering and absorption being approximately equal.

For 2-D calculations, background data on the meridional aerosol profile were selected from numerical experiments with the use of the 3-D large-scale diffusion model GRANTOUR, on the assumption that the presence of soot aerosols does not affect the processes of diffusion and removal of aerosols from the atmosphere (the development of an interactive model is planned for the future). As a result, the rate of removal is overestimated.

Calculations showed that even in a month the distribution of aerosols remains patchy, but in 2 months only several weak centers remain, with an optical thickness $\tau > 1$. The mean hemispheric optical thickness decreases by that time to 0.5. However, with a slower removal of aerosols from the atmosphere the optical thickness will be larger and the time of soot residence in the troposphere will be longer.

Analysis of the results of calculations with a radiative-convective model (53) shows that during the first 2 months, the temperature will rapidly decrease, with a maximum decrease of more than 30°C being reached in 2 weeks and subsequent gradual (and slower in time) increase of temperature with the rainout of soot from the troposphere, with a characteristic time of 2 to 3 weeks. The temperature decrease is much slower over the ocean because of the ocean's thermal inertia, which also must act as a buffer for the cooling over the continents. Naturally, the mean hemispheric value of ΔT should be closer to that for the ocean. The radical decrease of ΔT takes place if a contribution by soot is excluded. The small total effect of nitrogen dioxide and ozone can be explained by their mutually compensating radiative impacts. Preliminary results of calculations with a 2-D model (a nine-layer model with a 10° latitude resolution for a diagnostically calculated cloudiness as a function of relative humidity) revealed a qualitatively similar pattern for the hemisphere on the whole, but much smaller maximum temperature decreases (about 7 to 8°C) in the case of land (depending on the rate of aerosol removal) and a slower temperature recovery. This results from taking into account both the effect of the ocean and the reduced effect of aerosols (with consideration of its meridional profile). Naturally, the midlatitude temperature decrease is much stronger than the hemispheric decrease. Of great importance is the consideration of variations in the regime of cloudiness and precipitation (e.g., a decrease in the rain rate) under conditions of the transformed atmospheric circulation.

Applying the baroclinic parameterization of the eddy transport to a 2-D model, MacCracken studied the horizontal meridional motion of the air and found that heated smoke clouds move rapidly to the equator. This probably happens because of large changes in the vertical and horizontal temperature gradients during the "nuclear winter" period.

Using the National Center for Atmospheric Research (NCAR) 3-D numerical climate model, Covey et al. (54) performed simulation experiments to analyze the possible climatic impacts of urban and forest fire smoke resulting from multiple nuclear explosions in the atmosphere. The applied model is spectral (wavenumbers up to 15 are considered) and provides the spatial resolution 4.5° (latitude) by 7.5° (longitude). Parameterization of the physical processes responsible for the formation of climate makes it possible to describe the formation of clouds interacting with the fields of relative humidity and convection. The ocean surface temperature is prescribed (taking into account the annual change) as a lower boundary condition (on land its analogue is the heat-balance equation). Fixing the SST (leaving out its decrease following a smoke loading of the atmosphere) means that there will be some overestimation of the input of the latent and sensible heat to the atmosphere. With a zero heat capacity assumed for the soil surface layer, the initial temperature drop is overestimated. The annual course of the sun's elevation, snow and ice cover extent, and ozone content in the atmosphere are all considered.

The simulation modeling proceeds from the supposition that immediately following nuclear explosions, all the solar radiation in the NH midlatitudes is practically absorbed in the middle troposphere (without reaching the surface) by an optically thick, purely absorbing smoke layer. The thermal emission of the smoke layer determining its contribution to the greenhouse effect is neglected, because the absorptivity of the smoke aerosol in the infrared is 10 times less than in the visible.

The optical parameters of the smoke aerosol layer were found according to the basic scenario of nuclear war with a cumulative explosive yield of 6500 megatons. In this case the layer from 1 to 10 km in the band 30–70° contains 2×10^{14} g of smoke particles. The coefficient of absorption for particles is assumed to be 1.8 m²/g, which determines the optical thickness of the layer at 0.5 μm to be 3. The initial spatial inhomogeneity of the smoke layer is characterized by the total content of aerosols in the latitude band 31–67°N 1.62 g/m², and half as much in the intervals 27–31°N and 67–72°N (initially, there is no smoke over the rest of the globe). The aerosol concentration is assumed to be independent of longitude (bearing in mind a strong zonal transport), and the vertical distribution of smoke is characterized by its equal contents in four atmospheric layers, with their thickness changing within 870 to 260 mb. Initial conditions are prescribed from results of the numerical integration of the undisturbed annual change of climate for a period of about 20 years.

Three versions of the nuclear war–induced climate change are considered: (i) summer ($t = 0$, 30 June); (ii) winter ($t = 0$, 27 December); (iii) spring ($t = 0$, 22 March).

Since the insolation in the summer is maximum, the strongest effect of the atmospheric smoke loading can be determined. Immediately upon the formation of a smoke layer, a 20°C heating of the upper troposphere takes place (in 10 to 20 days the stratospheric heating at the level 30 mb near 60°N reaches 80°C), whereas near the Earth's surface the process of cooling begins. In the band 50–70°N the mean zonal temperature drop exceeds 10°. At an altitude below 2 km, cloudiness intensifies, while in the upper troposphere the clouds almost disappear (because of strong heating).

Large temperature gradients in the middle troposphere cause a considerable intensification of the zonal circulation. In the NH stratosphere (200 to 50 mb), the west-east transport is intensified on the polar side of the zone of heating and the east-west winds on the equatorial side. A substantial intensification of zonal winds takes place near the surface which may favor heat transport from the warm ocean to the cold continents.

The geographical distribution of the continental surface temperature is characterized by a temperature drop below zero by the tenth day over vast regions of Eurasia and North America (except along the coastlines). An average temperature decrease in "summer" constitutes 15 to 20°C.

The mean zonal circulation changes radically; the Hadley cell crossing the equator (especially in the Northern Hemisphere spring) is strongly intensified.

Instead of two Hadley cells, one large cell is formed, like a jet stream, which during a fortnight can transport warm clouds of nuclear dust and smoke to the Southern Hemisphere, thereby favoring the formation of a "nuclear winter" or, at least, a "nuclear fall." Such changes in the circulation cells take place only in "summer" and in "spring." A more detailed pattern of the wind field disturbance can be obtained from analysis of its geographical distribution. While variations in the mean zonal wind speed do not exceed 5 m/s, at individual locations these changes can be 10 times greater.

A continuation of numerical modeling is planned, with various prescribed scenarios and with an interactive account of radiation and aerosol dynamics and transport. The problem of scattering solar radiation by smoke aerosols and of the contribution of aerosols to the atmospheric greenhouse effect requires more detailed consideration. A thorough analysis of the parameterization of physical processes in conditions of a heavily disturbed atmosphere is needed. The subgrid processes of the vertical transport, under conditions of increased atmospheric stability, are changing.

Analysis of the climatic consequences of changes in the gas composition of the atmosphere has led to the conclusion that the effects of variations in the gas composition on surface air temperatures (SAT) do not exceed several degrees (47). In most cases, however, the effect of gas components on the temperature regime is opposite in sign and as yet cannot be estimated with the necessary accuracy, since some physicochemical connections require specification and continued study. Therefore, let us discuss a work by Luther (55) which deals with the effects of the postnuclear oxides of nitrogen.

Production of the oxides of nitrogen (NO and NO_2) in the high-temperature postnuclear cloud in the atmosphere triggers reactions of catalytic destruction of ozone taking place at a temperature above 2000 K (a temperature decrease below 2000 K in the zone of explosion takes place about 1 min after the explosion). Estimates of the amount of the postexplosion oxides of nitrogen vary, but (in the recalculation for NO) we assume a value of 6.7×10^{31} molecules of NO per megaton (55). With this estimate calculations are made of the possible impact of 1- and 20-megaton nuclear explosions on the atmospheric ozone content over various regions. In the first case a postexplosion cloud is stabilized in the lower stratosphere (its lower boundary is near the tropopause), and in the second case the postexplosion cloud is stabilized near the level of maximum ozone concentration.

The following scenarios were prescribed: (i) explosions of three hundred 1-megaton bombs over an area of 25×10^3 km²; and (ii) explosions of eighteen 20-megaton bombs over an area of 16×10^3 km². The total content of NO_x in an undisturbed atmosphere is $(1-2) \times 10^{16}$ molecules/cm², and its postexplosion content in the stratosphere is, respectively, 8×10^{19} and 1.5×10^{20} molecules/cm²; that is, the content of the oxides of nitrogen is several orders of magnitude greater than that of the background levels.

A numerical modeling activity used the Lawrence Livermore National Laboratory's (LLNL) 1-D photochemical model of the atmosphere (taking into account the kinetics of chemical reactions as well as vertical diffusion) to analyze the possibility of the formation of "ozone holes" in the atmosphere (a 44-layer atmosphere between 0 and 56 km is considered, with 40 components and 170 chemical and photochemical reactions). Calculations, according to the first scenario (1-megaton bomb explosions), showed that with the coefficient of horizontal diffusion K in the stratosphere taken between 3×10^9 and 2×10^9 cm²/s, at $K = 3 \times 10^9$ cm²/s a maximum decrease (25%) in the ozone content occurs 2 days after the explosions. With a rapid diffusion of the exploded products ($K = 1 \times 10^{10}$ cm²/s), however, this would decrease by only 18%. It is important to remember that although the amount of oxides of nitrogen rapidly decreases, the decrease in ozone remains almost unchanged for several days. The relatively low level of the decrease of the ozone content is because the cloud of material ejected into the atmosphere is located below the maximum of the ozone layer. Within the cloud the ozone is almost completely destroyed.

Under conditions of an undisturbed atmosphere, NO_2 absorbs only 0.3% of solar radiation, but the postexplosion absorption increases to 40 to 50% and takes place largely in the stratosphere. This must cause a strong heating of the lower stratosphere, which can stimulate stratospheric convection ("overturn"), which will take NO_2 to great heights. With $K = 3 \times 10^3$ cm^2/s and the emitted products supposed to occupy the layer 12.8-30.3 km, the maximum ozone decrease would be 71%. This points to the possibility of a strong regional decrease of TOC in the case of 1 Mt bomb explosions. However, this requires a thorough analysis before reliable conclusions can be made.

A serious impact on the regional radiative regime and climate can also be caused by nuclear ejections of dust aerosol. Calculations for the second scenario (20-megaton explosions, $K = 5 \times 10^9$ cm^2/s) show that within several hours the regional decrease in ozone reaches 70% and remains at that level for several days. The larger area encompassing the zone of a 40 to 50% decrease in ozone is several millions of square kilometers. The intensity of hard UV radiation strongly increases, 5 to 12 times, respectively, within this zone; however, during the first several days its influence on biological systems must be reduced by dust aerosol.

Emphasis was also placed (56) on the impact of postnuclear gas products on atmospheric climatic parameters. Based on a 1-D radiative-convective climate model, the conclusion was drawn that the combination of solar radiation strongly reduced by upper troposphere aerosols, the intensified greenhouse effect, and the heat of fires attest to possible intensive initial heating and subsequent cooling, which may be manifested by major climatic instability.

The nuclear exchange-produced climatic impact is not confined to aerosols. Hydrocarbon fuels burn giving off large amounts of CO_2, SO_2, CO, and NO. The greenhouse effect is directly produced by CO_2 and SO_2. In the presence of NO, through chemical reactions, CO is transformed into CH_4 and O_3, which make the greenhouse effect of the troposphere even more pronounced. The oxides of nitrogen injected into the stratosphere enrich it with NO_2 and N_2O_5. The climatic effect of NO_2 is manifested mainly through its effect on the albedo of the surface-atmosphere system, which decreases with increasing NO_2 concentrations.

The influence of NO_2 on the greenhouse effect is indirect and shows itself through changes in the stratosphere-troposphere radiation budget, because of solar radiation absorption. N_2O_5 affects climate mainly through the greenhouse effect. Of importance is the greenhouse effect of water vapor ejected to the stratosphere, the concentration of which may be very high in a warmer stratosphere.

Along with the atmospheric heat budget, a huge amount of heat released from fuel combustion must be considered. The postexplosion greenhouse effect on the atmosphere should be estimated with due regard to an internal heat source, which in the presence of a strongly absorbing troposphere will cause its heating and will in part compensate for a decrease of the shortwave radiation influx into the lower troposphere (caused by atmospheric aerosols).

Estimations using a 1-D model of the radiative-convective heat exchange showed that during the first stage the atmospheric greenhouse effect increases and the surface temperature T_s rises at the locations of nuclear explosions. This is determined by the combined effect of (i) a decrease of the surface-atmosphere system albedo due to shortwave radiation absorption by atmospheric NO_2 and tropospheric soot aerosols: and (ii) an increase in the greenhouse effect of tropospheric CH_4, SO_2, CO_2, O_3, and soot aerosols. The development of the greenhouse effect is also enhanced by the impact of the internal heat source (fires) on the radiative heat exchange. The antigreenhouse effect of the stratospheric dust clouds at this stage is compensated by other factors that determine what global or regional climate will be.

At the second stage, when the process of fossil fuel burning has come to an end, the surface temperature T_s decreases due to the downward motion of the soot aerosols from the atmospheric boundary layer, when the internal heat sources dissipate. The temperature of the stratosphere and upper troposphere will rise due to eddy-diffused transport of part of the soot aerosols into the upper troposphere and stratosphere.

A temperature decrease was obtained, assuming an equilibrium between radiative temperatures of the gas and aerosol components of the stratosphere. A consideration of nonequilibrium conditions will lead to some additional temperature decrease, $\Delta T_s \simeq -3$ K. With the process going on, the stratospheric temperature decreases because of the sinks of stratospheric aerosols and NO_2 and because of changes in the ozone concentration. The surface temperature T_s will, in time, rise again and could exceed the surface temperature of the initial equilibrium state of the planet.

These results reveal the importance of the need for a thorough account of changes in the chemical composition as well as in the structural and optical characteristics of the atmosphere when modeling the climatic impacts of a nuclear war. The major consequence of a nuclear war will be a strong instability of Earth's climate for 1 to 3 years following the war.

Three-D modeling studies of nuclear war consequences for the atmosphere and studies based on 1-D radiative-convective models (RCMs) (52–

54,56) show that "nuclear winter" will not be as severe as suggested in some of the research cited above. The temperature decrease will probably constitute from one-third to one-half that of earlier RCM results ($\sim40°C$).

For one thing, the probable rapid transport of nuclear dust and smoke to the Southern Hemisphere points to an accelerated dispersion and dilution of nuclear pollutants in the Northern Hemisphere. These results indicate that the postnuclear aerosol pollution of the atmosphere cannot last as long as had been supposed. Therefore, the importance of considering variations in the gas composition of the atmosphere during the postnuclear disturbance of its thermodynamical regime becomes increasingly obvious.

We, the authors, believe that a nuclear war can be the last step toward the final destabilization of the climatic system, not only because the composition of the atmosphere will be drastically altered, but so will be its structure and dynamics. No doubt the state of the cryosphere will also be drastically changed.

We also think that the climate system have already surpassed its natural variability. It can be illustrated by the curve of deviations of the averaged NH surface air temperature from a multiyear average (Fig. 3.13).

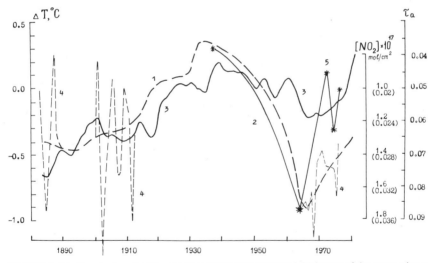

FIGURE 3.13. Secular trend of the mean annual residual optical thickness of the atmosphere (minus Rayleigh scattering); NO_2 content and air surface temperature variations ΔT in the Northern Hemisphere. 1, Residual optical thickness τ_a; 2, total NO_2 content in an air column; 3, air surface temperature deviations from an average over the three last decades; 4, variations in τ_a during large-scale atmospheric disturbances; 5, NO_2 content from measurement by Vassy, the authors, Brewer et al., Kulkarni, and Pommereau. The optical thickness of an air column containing 1.8×10^{17} molecules of NO_2 per square centimenter is 0.036 (for $\lambda = 520$ nm). (After Ref. 2.)

After 1950, temperature variations began to oscillate (an account should be taken of a low value of ΔT for 1984, absent in the graph).

3.3 IMPACT OF THE TUNGUSKA METEOR FALL ON THE OZONE LAYER AND CLIMATE

On 30 June 1908, between 7 and 8 o'clock in the morning, a large explosion was heard by inhabitants of a sparsely populated region in eastern Siberia. A controversy has developed over the origins of this explosion, which took place under a clear sky. Although it is sometimes referred to as the Tunguska Meteor Fall, suggestions about its possible origin also include a comet, a black hole, or antimatter. The explosion, heard and felt for hundreds of kilometers away from the epicenter (see Fig. 3.14), apparently occurred about 8.5 km above the Earth's surface. A superheated jet of air felled trees for many kilometers from the epicenter. Close investigation by scientists from Moscow showed that while the bark of the trees had been seared, the wood underneath had remained untouched (98). Today, speculation about the Tunguska event continues unabated.

Analysis of the impact of the 1908 Tunguska meteor (TM) fall on the atmosphere is not only interesting but is necessary because the TM fall has been considered in many Soviet and foreign studies as an analogue to a large-scale nuclear explosion (the estimate of the explosive energy of this meteor fall in nuclear equivalent varies between 10 and 40 megatons). Therefore, the scientific community believed that changes in the composition of the lower stratosphere and even in the Earth's climate could have been caused by such an explosion.

In studies by Turco et al. (57,58) the TM fall was used to check a numerical model that simulated major impacts on the atmosphere, such as a nuclear war. Numerical modeling, even in instances of relatively close correspondence to actual processes, can be used to forecast results of big impacts on the environment and ecosystems. The modeling of the TM event was pursued to verify the reliability of numerical experiment results based on the Tunguska meteor fall data. Turco et al. (57,58) concluded that this modeling provided a correct description of the basic features of the TM event and of subsequent changes in the concentration of minor optically active gaseous components of the stratosphere (e.g., ozone, the oʻides of nitrogen, etc.). However, probably as a result of our critique at the Leningrad SCOPE ENUWAR Workshop (May 1984), Turco removed the TM event from the calibration list for his numerical model (59).

Naturally, the results of numerical forecasts depend heavily on background data put into the model. In this section we try to specify the input

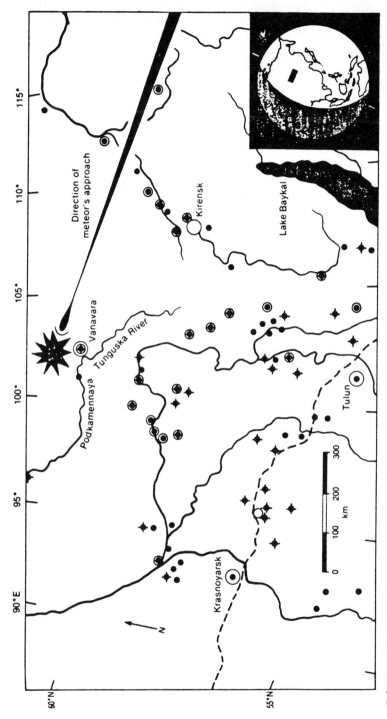

FIGURE 3.14. *Inset:* The Earth is shown for the time of the Tunguska meteor's early morning fall, with the sunrise line crossing central Asia. *Above:* This map shows the epicenter of the explosion over central Siberia at 65° 55'N, 101° 57'E (star), as well as places where witnesses described seeing (○), *hearing* (◉), and feeling (+) earth tremors from the meteor's arrival. Reprinted by permission from *Sky and Telescope* January 1984.

parameters for the TM event, to make it possible for modelers to improve the TM numerical models.

Some new data and ideas about the nature of TM and the processes following its fall have been obtained, based on observations of the atmospheric spectral transparency at the Smithsonian Astrophysical Observatory Mount Wilson station for the years 1905, 1906, and 1908–1911, and on evidence by the TM eyewitnesses reinterpreted from a different point of view. These data have opened up new possibilities for understanding the real scale of the TM event and for verifying its cometary origin.

3.3.1 The Nature of the Tunguska Object: Composition, Structure, and Size

Despite some work [e.g., Ganapathy (60) and Sekanina (61)] creating doubts about the cometary nature of the Tunguska meteor, an analysis of available observations of this event taking place more than 75 years ago has made it possible to substantiate sufficiently reliable ideas about the Tunguska object parameters and to support the view about the consequences of this event as having been the result of invasion by a small comet.

Ganapathy (60) supposed that increased concentrations of iridium in the Antarctic ice cores (dated 1909–1910) and in microscopic spherules of cosmic matter found at the TM explosion site (in the layer of moss referenced to the year 1908) point to a single source—a meteor with a mineral composition corresponding to that of well-known meteors of carbonaceous chondrite. Assuming that the TM material has been uniformly scattered over the Earth's surface, Ganapathy estimated the mass of TM at 7 megatons. However, from results of recent studies within the Complex Amateur Expedition (CAE) (62), the total mass of solid cosmic material in the form of spherules of (presumably) TM origin, found in the region of the fallen trees, does not exceed 2 tons. The area is estimated at 20,000 km^2; it has the form of an oval, stretched along the trajectory of the meteor fall. This oval "is sown" with spherules most heavily as compared to adjacent areas and with the area of the globe approximately equal to 5×10^8 km^2. It becomes clear that the ratio of areas does not confirm to Ganapathy's calculations.

The uniform settling of spherules over the Earth's surface, as suggested by Ganapathy, cannot be valid because of the very nonuniform areal distribution of spherules even at the site of the fall and because of the specific downward motion of air over the Antarctic, which creates conditions for freezing and sedimentation on the glacier of both moisture and aerosol. Thus the Antarctic can be represented as a pump with a filter on which the atmospheric pollutants are forced to settle. These facts point to a very roughly estimated (that is, overestimated) mass of TM mineral constituents

— all the more so because the spherules found in the Antarctic are not of TM origin but are derived from a chondrite meteor of May 1908.

Among stone meteorites most numerous are the so-called chondrites, containing round-shaped formations — chondres — which usually occupy most of a meteorite. The substance of chondres and the substance filling the interchondres space — matrix — are most often identical. Depending on the content of iron and silicates, chondrites are subdivided into classes (HH, H, L, LL, C). Carbonaceous C-chondrites contain much iron, but the iron is almost totally in silicate minerals. Further subdivision of C-chondrites is connected to their changing composition, densities, and amounts of chondres (C1, C2, C3). Carbonaceous chondrites are so called due to the dark (like carbon) color of major minerals, containing much fine-dispersed magnetite. However, the carbon content is not very high but increases with the transition from C3 to C1. If we extrapolate further to the densities of meteorite substance below unity, we may find the so-called pre-C1 group, a generalized type of the comet substance.

As a comparative analysis of the mineral composition of carbonaceous chondrites shows, a relative content of bound water grows from type C3 to type C1. The dependence of the H_2O content on the density of meteor matter turns out to be linear (Fig. 3.15), which makes it possible to extrapolate to type pre-C1, that is, the comet matter. Taking (from the literature) 0.6 to 0.75 g/cm^3 as the density of the comet substance, we determined that the mineral component of the comet contains 45 to 50% of bound water.

The Tunguska cosmic body can be supposed to have been the core of an old microcomet consisting of mineral formations as embryos (probably, these had been lumps of alkaline metals compounds) bound by water-and-gas-condensate ice into a solid conglomerate. It can also be supposed that bound water had been an element of the comet's mineral core, since carbonaceous chondrites of the types C1, C2, and C3 are solid mineralized formations. Minerals such as gypsum, calcite, glauberite, myrabilite, tenardite, slavikite, and xylomelane contain 30 to 60% of bound water (by mass). These data suggest that the embryo of the core of the comet largely consists of crystal compounds of light elements with added compounds of heavier elements [e.g., lead and uranium (in such minerals as uraninite and otenite)]. The presence in the TM debris of the compounds containing the oxides of uranium was confirmed by the Evenk people, who had observed a strong luminescence of boulders showing from craters at the site of the TM debris fall; also, increased concentrations of lead of cosmic origin have recently been found in that region. It is well known that compounds of the uraninite type are very unstable in the face of direct environmental impacts (e.g., precipitation, sharp temperature gradients, and so on).

Studies of anomalies of the chemical composition of soil surface and peat

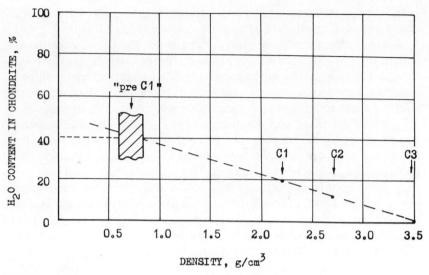

FIGURE 3.15. Bound water content in carbonaceous chondrites C1, C2, C3, and in the substance of cometary meteors (pre-C1). The dashed zone determines the upper limit of the densities of cometary meteors, rather than an effective mean (Ref. 80). In TM-related calculations Park (79) used a very low value, $\rho = 0.002$ g/cm³, apparently for a disintegrated body. Extrapolating the dependence for such density values gives the H_2O content in loosely bound comets up to 50% of their total mass.

bogs referenced to the year 1908 point to an excess amount of compounds of volatile elements — sodium, potassium, calcium, and so on. From spectral studies of the luminescence of comets close to the sun, Millman (63) found the same sequence of light elements. Evidence of the TM fall eyewitnesses, results of studies showing an unusual taste and poisonous properties of water in the South Bog, discoveries of specific remnants of gypsum crystals, and firsthand accounts of glittering boulders in craters all help to get an idea of the physicochemical properties, structure, and composition of the Tunguska comet. Probably, less stable salts of sodium and potassium had been washed out by the environment's gradual disintegration of clods of the mineral-ice conglomerate. The more stable compounds of calcium were found by the CAE participants. Eyewitnesses attributed large crystal gypsum directly to the TM fall.

Golonetsky and Stepanok (64) assumed that the mineral part of TM could have constituted not more than 0.1% of its weight. Presumably, this estimate does not concern alkaline metal compounds but characterizes the content of the compounds of zinc, bromine, tin, hydrargyrum, antimony, lead, and selenium, among others. An abundance of volatile elements, small amounts of settled spherical particles, a small ejection to the atmosphere of dust aerosols, as well as other TM features, such as (in particular) the specific character and multiplicity of explosions, favor the conclusion by Wipple, Fesenkov and others regarding the cometary nature of the Tunguska meteor.

Now consider the firsthand account of the TM fall (65,66) in order to clarify the scale as well as the features of the comet meteor explosion when it slowed down and "stopped" in the upper troposphere. Eyewitnesses at a distance of 65 km from the site of the TM fall said that at the moment of the TM fall the skies "had broken, opened;" they spoke about a very tall fiery column (the body of the explosion). No doubt, they had seen not the postexplosion release of hot gases, but the explosion itself (the body of the explosion), since the stinging light sensed by people (about 1.5 W/cm^2) must have been emitted from an extended object, most of which had not been near the Earth's surface but had been in a weakly absorbing atmospheric layer (e.g., the lower stratosphere between 12 and 30 km).

What was the size of the body of the explosion? Presumably, as TM penetrated the atmosphere, the rate of ablation of the external shell of the cometary meteorite led to the growth of the sizes of the TM gas shell and trail.

Important generalizations of evidence from the region of the Upper Lena were made by Epiktetova (67). Tracing of the TM path, following the descriptions given by eyewitnesses, showed that the trajectory had an azimuth of $95°-110°$ and crossed the Lena near the Mironovo village; that is, the flight of the TM could be traced for about 450 km. The firsthand account showed that the outline of the body (together with the gas tail) had been continuously changing during the flight. At high-altitude parts of the path, the TM had been seen as a globe-shaped body (a "red lump"), and then a "sheaf" with sparks in the tail had formed, with some of the sparks reaching the Earth's surface and breaking trees. By the end of the flight the silhouette of the body had stretched into an unbroken band ("log"), before hiding behind the ridge of the left bank of the Lena River. Thus we get a clear pattern of the gradual destruction in the atmosphere of the Tunguska meteor. This, of course, is natural when a cosmic body, penetrating the atmosphere, brakes.

To date, no adequately substantiated calculations of TM trajectory parameters have been made, because basic TM parameters have not as yet been established, including its initial mass and velocity, as well as the density and structure of its body. If we estimate (from firsthand accounts) the visible diameter of the object, the result will show not only the huge size of the gas shell but also its rapid growth upon entry into the lower stratosphere. Here one may use the results of calculations by Sekanina (61) of the parameters of the motion of large bolides in the terrestrial atmosphere.

Figure 3.16 shows the paths of bolides in the atmosphere, both calculated

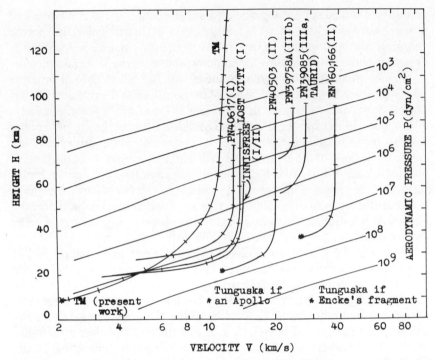

FIGURE 3.16. Changes in the velocity of brightest bolides at their entry to the atmosphere. Sloped curves are isolines of dynamic pressure growing at the meteor's "immersion" to the atmosphere (dyn/cm²). Bars on the trajectories are second-long periods of the path. Abbreviations of the trajectories: PN 40,617(I), a bolide photographed at the U.S. Prairie network; EN160,166(II), a bolide registered at the European network (160,166 is the catalog order number of the bolide); II, the type of bolide according to the density of the substance [I, $\rho \sim 3.7$ g/cm³, σ (ablation coeff.) = 0.032 s²/km²; II $\rho \sim 2.1$ g/cm³; $\sigma = 0.050$ s²/km²; ablation parameter = $\theta = \sigma V_0^2$; IIIa, $\rho \sim 0.6$ g/cm³; $\sigma = 0.13$ s²/km²; IIIb, $\rho \sim 0.2$ g/cm³; $\sigma = 0.2$ s²/km²]. IIIa, Taurid—the type of bolide and the fact that it belongs to the meteor stream Taurid; Innisfree and Lost City, individual names of bolides that had fallen as meteors. Asterisks mark the points of final bolide flares with complete destruction of the body; TM-A, the point of TM explosion for the case of low entry velocity (the body belongs to the family of Apollo asteroids); TM-B, for the case of TM being a fragment of Encke's comet. The extreme left TM trajectory is drawn from the authors' estimates of the TM parameters. (From Ref. 61.)

and observed (photographing by the bolide network). As seen from the figure, the velocity of low-speed bodies in the upper stratosphere and above is practically constant, which is explained by the thin air density at these altitudes. Only from a height of 30 to 40 km does their marked braking begin.

The height/speed dependencies shown in Fig. 3.16 refer to most-interesting bolides known for their relatively slow motion in the atmosphere (due to large masses) and their exceptionally intensive final flares. For the TM case,

Sekanina gives only two terminal points, preceeding from extra-atmospheric velocities of about 14 and 31 km/s. The height/speed dependence (extended curve) for TM was drawn by Nikolsky and was based on published information and estimates.

According to Sekanina, flares occur at the moment when the pressure and respective internal stress encountered pass a critical limit (these differ for each bolide) and when the atmosphere becomes like a wall to the meteorite. The body very rapidly disintegrates and falls into fragments, probably due to deformation shifts. Gorazdovsky (68) in a laboratory experiment shows that with a deformation shift (at a tangential stress in the meteorite body of about 2×10^{11} dyn/cm^2) an explosion occurs, with dispersal and ejection of 5 to 20% (by volume) of the cosmic body's solid-core matter. The size of particles ranges between 10 and 100 μm.

The deformation explosion is followed by emission in the visible, UV, and X-ray intervals, as well as by intense electrification of the surfaces of the disintegrating fragments with resulting potentials of hundreds of thousands of volts and subsequent lightning discharges. Examination of the curves in Fig. 3.16 shows that against the isolines of internal stresses, growing in the bolide's body as it enters the denser atmospheric layers, the terminal parts of the curves of three low-speed bolides rise slightly. Here, according to Sekanina's calculations, the internal stresses start decreasing, which does not favor the final rapid fragmentation and flare of the bolide and can be explained by the gradual disintegration of the bolide at the time. As for the TM, its supposed (by analogy) quasiconstant velocity in the upper stratospheric layers is also connected with the high-speed ablation of the ice gas–condensate shell.

In the case of the TM the point of "stopping" is at the very bottom, which, again, testifies to a huge loss of mass during its flight and to its relatively slow entry into the atmosphere ($V_0 < 14$ km/s); V_0 is probably close to a minimum of about 11.2 km/s. One of the terminal points of the calculated TM path [according to Sekanina (61)] refers to very fast entry into the Earth's atmosphere ($V_0 > 30$). Such must be the velocity of the meteorite if it is a fragment of Encke's comet, having an orbit's inclination of about 10°. This comet has specific features. Thus Sekanina gives evidence for Encke's comet's having ejected large amounts of dust particles, mainly of millimeter and centimeter sizes. These two features, however, are in sharp contrast to the TM data.

There is some evidence (audible and visual events) that before "stopping," the TM body had started dividing into two parts. As Voznesensky (69) states, this probably took place at altitudes near 20 km. At the point of stopping, the TM core had rapidly fragmented, with resulting powerful electric discharges between the fragments.

Since the hydrocarbon components of the ice core (CH_4, C_2H_2, C_2H_6,

C_3H_8, H_2, etc.) had largely been converted into gases, disintegrated, and mixed with the air by the moment of stopping, a detonating gas-air mixture formed which at the moment occupied a cylindrical volume larger than $350 \ km^3$. Powerful electric discharges triggered the explosion of the entire gas trail, whose size could be estimated from the eywitness accounts (the diameter of the head of the explosion body $d \geq 5$ km, length 25 to 30 km).

Naturally, the explosion of such a column of gas mixture had not been instantaneous; it had taken several seconds. By thorough calculations one can estimate several of the important parameters of the explosion, such as the radiative effect on people and vegetation. Compared with real events, one can try to determine the geometry of the explosion.

Our estimates suggest that the length of the explosion body was 25 to 30 km. The literature suggests that the explosive projection was about 20 to 22 km. The traced path of TM is about 570 km, with the angle of slope of the middle part of the trajectory about 25° [a value close to that found by Zotkin (70)] and in the end 50°. The latter means that the Tunguska meteor ought to have been seen at an altitude of about 250 km. Note that at first the eyewitnesses had seen TM as a sphere with a size of about half the moon and without a noticeable tail. Hence the size of TM (with coma) ought to have been about 1 km. According to Kovalevsky and Potapov (71), such a body could have been seen at altitudes above 500 km. The time of the TM flight through the middle atmosphere (120 to 10 km) constituted about 25 s at a speed of 11.2 km/s. These data are in sharp contrast to results calculated by Sekanina (61). Apparently, he did not establish an adequate data base from firsthand accounts and therefore used erroneous background characteristics of TM. For example, Sekanina believed that TM had not suffered any fragmentation until the final explosion, at which time the TM material was totally transformed into small-sized dust particles.

In fact, much information exists about the explosions that occurred during the TM flight along the trajectory (67) and about subsequent (three or four) secondary explosions (the so-called aftershocks) that followed the major explosion, which resulted in the breaking up of the TM into several huge boulders up to 30 m in diameter. Sekanina, however, referred to the lack of secondary explosions, which, according to Gorazdovsky (68), should have been associated with the delayed TM breakup, a delay due to the reaction in the polycrystal structure of the meteorite's mineral core. The delay parameter is very important for an understanding of numerous bolide events, but to date, we know very little about the limits of their variability in real cometary meteorite bodies. Probably, the effect of delay depends to a large extent on the size of the body. Therefore, the TM body and its accompanying effects should be modeled thoroughly.

Changing the orbital parameters of asteroids and short-period comets, Sekanina chooses the most probable (in our opinion) asteroid origin of the TM and, comparing most reliable firsthand accounts, takes the following parameters of the TM motion in the atmosphere: the radiant azimuth $A_R = 110°$ and the zenith distance of the radiant $Z_R = 85°$ (the radiant is a point in the skies from which the apparent trajectory of a body's motion originates). Calculating (using these values) the projections of the TM flight onto the celestial sphere for several observation locations, Sekanina uncovers an obvious problem with respect to the angular heights of the TM flight over the following locations: Kezhma, Kirensk, and Nepa, among others. In Kezhma, for example, the TM fall eyewitnesses attested to a height of TM in the skies of about $25°$. Yet the calculations by Sekanina suggest only $7°$.

Many uncertainties were introduced into the analysis by Sekanina's unawareness of additional factors, such as the firsthand account of two events from the village of Kamenskoye, that is, the TM fall and, after some time, the fall of another meteor. Despite an abundance of observational data and results of model calculations, Sekanina could not reconstruct the direct "trail" of the TM (the parameters of its body and its orbit).

3.3.2 On the Mass of the Tunguska Meteor

In the works by Fesenkov (72) and Turco et al. (58), estimates are given of the TM's initial kinetic energy E_0. Values of E_0 in these works are within $(1-4) \times 10^{25}$ erg. We take $E_0 = 4 \times 10^{25}$ erg (58), since below we shall use the results of the study by Turco et al. The Turco et al. study is one of the most representative studies of the TM problem.

An attempt was made earlier (Section 3.3.1) to support a low entry speed V_0 to the Earth's atmosphere. Presumably, the TM orbit was of dynamic origin (i.e., an interaction of several bodies), which in some cases provides conditions for the reduction of the usual velocities of comets to about 10 km/s and, perhaps, even lower.

Analysis of the parameters of the largest bolides observed at the Prairie Network in the United States and at the observation networks in the Federal Republic of Germany and Czechoslovakia revealed a correlation between the entry speeds of meteors and their masses. The character of the correlation is reliably represented by the parabolic curve in Fig. 3.17. Because data about large bolides are scarce, the number of points in the graph is very limited. Data taken from Ceplecha and McCrosky (73) (solid circles) ideally fit the curve begun by the solid line and continued by the dashed line, which in the energy interval of interest to us $[(2.5-4) \times 10^{25}$ erg] points to low entry velocities for the larger bodies (e.g., 6 to 8 km/s) and to their huge masses of about 150 megatons. Although data from McCrosky et al. (74) (circles with

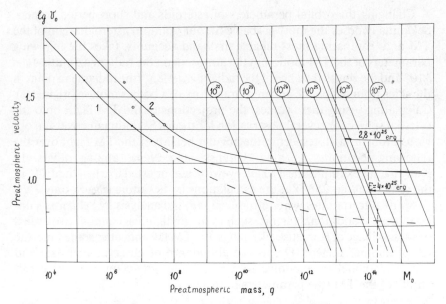

FIGURE 3.17. Dependence of the entry velocity on extra-atmospheric mass of bolides from data of the U.S. Prairie network and European network for largest bolides. The sloping kinetic energy isolines cover only the range of the multiorder extrapolation of the dependence $\log V_0 = f(M_0)$ of a parabolic character. The upper curve is from data of Ref. 74, the bottom curve from data of Ref. 73. The dashed-line continuation of curve 1 corresponds to an accurate analytic curve crossing the observation points. A deviation from the analytic dependence is explained by the formally required minimum entry velocity $V_0 = 11.2$ km/s.

dots) have a systematic upward shift on the velocity axis, the shape of curve 2 points to a similar mass/velocity dependence for large meteors. Because of the lack of data for model calculations of possible variations of the TM dynamic orbit, both dependencies had to be referenced for large masses to $V_0 = 11.2$ km/s, despite a pronounced change in the parameters of dependence found in the initial (left-hand) part.

A weak substantiation of the obtained dependence $V_0 = f(M_0)$ and its possible ambiguity do not permit an estimation of the mass M_0 to an accuracy better than the estimate of its lower limit, because of this multiorder extrapolation. Thus we arrive at $M_0 \geq 70$ megatons.

Another possibility for estimating the initial TM mass appeared when considering variations in the spectral optical thickness of the atmosphere as determined from observations of the spectral transparency at the high-altitude Smithsonian Astrophysical Observatory—SAO (Mount Wilson, California) before and after the TM fall (e.g., the summer of 1908) (75). Observations at Mount Wilson began in 1905 in order to estimate the total solar radiation flux beyond the atmosphere (e.g., the solar constant S_0), its variability, and the contribution of the S_0 spectral constituents to the observed variations in S_0. Observations were also designed to study the forcing mechanisms of solar activity on weather and climate. In 1906 the SAO scientists, headed by Abbot, started monitoring the total solar radiation flux and its spectral composition, supplementing these measurements with those of the aureole radiance and of the water vapor content in air column. Daily (May to November) spectral measurements covered the wavelength interval 0.35 to 2.8 μm. Publications provide processed data in the form of S_0 daily means and average values of spectral transparency. The monitoring generally continued until 1954.

Upon scrupulous and detailed studies of the behavior of the atmospheric spectral optical characteristics and of the variability of the total water vapor content over Mount Wilson, some new circumstances have been uncovered. These new circumstances clarify the real nature of the Tunguska body as well as the cause-and-effect pattern of variations in the optical characteristics of the atmosphere with regard to the Tunguska meteor fall.

From numerous works on the TM event by Fesenkov (76,77) it follows that the TM was a microcomet with an initial mass of several megatons. It was assumed that the cosmic object, constituted largely of frozen dust material, had totally disintegrated into particles about 1 μm in size and that these aerosols had spread over the Northern Hemisphere, causing a sharp decrease in transparency at all these wavelengths at Mount Wilson. As Fesenkov (77) supposed, a reduction of transparency in California had been caused by disintegrated TM material and depended little on wavelength. The initial marked transparency minimum was observed [from the initial unsmoothed data (75)] on 17 July, with the next major minimum occurring only on 4 August. Fesenkov believed that the first minimum of transparency was a fluctuation against growing turbidity of air masses moving from the east. This explanation, however, does not fit reality.

With the "point" character of the TM explosion, an income of air masses should have been expected with a maximum concentration of explosive products shortly after its initial signs (not later than half the time of the path, i.e., 8 days). According to Fesenkov, the initial products of the explosion came in 2 weeks (from smoothed data) and maximum concentrations were observed 20 days later (4 August). Such estimates are unacceptable, although the conclusion about the 24 July minimum transparency as a major TM manifestation would apply. Therefore, the 4 August maximum of τ (Fig. 3.18)* does not apply directly to TM but is only a second appearance of an aerosol cloud of pre-TM origin.

---

*Optical thickness of the atmosphere $\tau = -\ln P$, where P is the atmospheric transparency in the vertical direction.

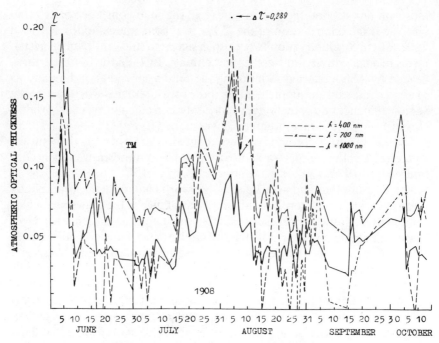

FIGURE 3.18. Temporal change in the total optical thickness of the atmosphere (for $\lambda = 700$ and 1000 nm) and residual optical thickness (for $\lambda = 400$ nm) over Mount Wilson (1740 m) for the period June–October 1908. The maximum residual optical thickness for $\lambda = 400$ nm is given in the upper part of the graph (0.289), beyond the graph. There are several points under the abscissa axis for τ_{400} values. Their relativity slight transition to the negative area illustrates the reliability of the chosen value of Rayleigh optical thickness and relatively high accuracy of observational data. Sloped curves show the rate of τ maximum value decrease for the aerosol cloud circling the globe. The formation of this cloud is related to a large store bolide that entered the atmosphere in early May, at an entry velocity of 35 to 40 km/s, and exploded at an altitude of about 22 km. The vertical line indicates the moment of the TM entry.

This conclusion is not the final evidence for the interference of an incidental event (with respect to the TM) with the development of the optical phenomena over Mount Wilson but only a preamble to it. Apparently, sufficient evidence for an impact of TM is a reappearance of the aerosol cloud exactly 60 days after its first appearance on 4 June 1908 (Fig. 3.18).

Observations of the transport of the erupted material of the Mount St. Helens volcano (1980), located in approximately the same section of the United States, showed that the erupted products, ejected to heights of 21 to 27 km, circled the globe twice within a period of 60 days. Examination of the curves of variations in the optical thickness at 400, 700, and 1000 nm given in Fig. 3.18 (from the SAO data) shows that for 700 and 1000 nm, the atmospheric optical thickness on 4 August and 4 October is indicated by the solid line that connects the values of τ for these dates and data for 4 June. The optical thickness at 400 nm (the Rayleigh thickness is extracted from the total optical thickness) has no such drops of τ maxima, but there is a peak ($\tau = 0.289$) on 4 August. Hence at 400 nm the combination took place of the optical thicknesses of the TM cloud and the dust-loaded masses at altitudes of 21 to 27 km, circulating around the globe within a period of 60 days.

During the 1908 observations the first working wavelength 400 nm was located at the very end of the shortwave interval of the optical range used at SAO to study the transparency variations. Data on the optical thicknesses for the second (450 nm) and third (500 nm) working wavelengths illustrate a rapid decrease in the optical thickness of the TM air masses, with increasing wavelength. In other words, the residual optical thickness of the TM air mass (minus the Rayleigh and ozone ones) turned out to be very small in the longwave spectral interval ($\lambda > 600$ nm) and sharply increasing at $\lambda = 400$ nm. This characteristic feature of the TM air mass is connected with the inevitable presence in the TM air mass of large amounts of nitrogen dioxide, with a broad and sufficiently deep absorption band between 350 and 600 nm. After a thorough analysis of the spectral change of the residual optical thickness, it was divided into components according to their origins and the content of nitrogen dioxide in an air column of a unit cross section was calculated.

A question arises about the origin of the dust-loaded air mass with strong scattering properties in the near infrared (700 to 1600 nm) that first appeared over Mount Wilson on 4 June 1908. Its initial occurrence was determined from the dependence of the rate of increasing atmospheric turbidity at each of the three appearances of this air mass over Mount Wilson. A distinct pattern was obtained of the temporal change of the optical thickness, whose extrapolation for May 1908 made it possible to determine the timing of this dust loading in the stratospheric layer from 20 to 30 km.

Since the obtained dependence had a very smooth and regular course, it was possible to determine the onset of stratospheric dust loading: early May 1908. It was not difficult to identify the geographical region where the dust loading of the stratosphere had taken place. For example, judging from the motion of the Mount St. Helens effluent, whose speed of motion was about 35 km/h, one can point to the place of the May 1908 event — Tibet. Based on this and taking into account Ganapathy's data on atmospheric pollution by the aerosols of meteor origin in 1908–1909 (carbonaceous chondrite containing rare soils, including iridium in respective layers of Antarctic ice), the real cause of the dust loading can be reliably established — the intrusion of a high-speed cosmic object ($V_0 \geq 35$ km/s) in which, upon aerodynamic braking (Fig. 3.16), critical stresses up to 0.5×10^8 dyn/cm^2 appeared. As a result, a deformation explosion occurred in the form of a huge bolide flare, with the dispersion of a large part of the object (up to 30%), whose mass apparently reached 10^5 tons. Data from recently begun continuous monitoring of atmospheric turbidity at some optical sounding stations confirm the possibility of such an event. Cases of heavy stratospheric dust loading of unknown origin have been observed. Apparently, they were not very exceptional events. Such penomena should be ascribed to large-scale bolide events taking place outside the observers' field of view, such as, for example, over deserted regions of the Pacific, among other isolated locations.

Yet another possibility has appeared to estimate the TM mass (through the ratio of trace components in the cloud containing the products of explosion and ablation of the TM body). It is necessary, therefore, to specify the differences between temporal variations of the optical thicknesses at wavelengths 400, 450, and 500 nm, and variations at $\lambda \geq 700$ nm. The optical thickness τ_{400} (Fig. 3.18) during the period July–August 1908 has an anomalous course compared to τ_{700} and τ_{1000}, with maxima for 4 June, 4 August, and 4 October fitting straight lines. The observed law of linear temporal decrease of maximum optical thickness points to a homogeneous diffusion and obviously to sedimentation of meteor matter from a circumglobal cloud of products of the May ("pre-TM") ejection (e.g., a bolide over Tibet).

Applying the observed linear dependence of decrease to τ_{400}, one can distinguish between the total contribution by two air masses (each with the products of only one of the two intrusions) to τ_{400}. In the upper part of Fig. 3.18 a maximum for τ_{400} is given ($\tau_{400} = 0.289$) which is beyond the graph. Subtracting from this value $\tau_{400} = 0.108$ obtained from the linear dependence for 4 August, we obtain $\tau_{400}{}^{TM} \approx 0.18$ only for TM.

This technique for dividing the total optical thickness into components is justified because the products of respective intrusions were concentrated in different air masses at different levels (the TM products in the layer of 13 to 19 km, and the pre-TM products between 21 and 29 km) and were transported at different velocities and in different directions. The TM products in air masses moving with the circulation systems of the midlatitude Ferrel cells are rapidly transformed and mixed with local air masses. Note that the TM products are mainly water vapor and nitrogen oxides, with small amounts of small-sized meteor aerosols.

Taking the TM aerosol component to be 1.5 times greater than the background level determined for $\lambda = 400$ nm as $\tau \approx 0.04$ (78), we have $\tau_{aerosol}{}^{TM} \approx 0.06$. Then $\tau_{NO2}{}^{TM} = 0.12$ remains for nitrogen dioxide. With an absorption cross section $\sigma_{400}{}^{NO2} = 5.5 \times 10^{-19}$ cm^2, we obtain the NO$_2$ content in air column $N = \tau/\sigma \approx 2.2 \times 10^{17}$ molecules of NO$_2$ per square centimeter.

Calculations with photochemical models and measurements in the stratosphere showed that in the daytime the NO and NO_2 concentration ratio varies between 2 and 5. Taking $NO/NO_2 = 3.5$, we have an NO_x content of about 10^{18} molecules per square centimeter. Note that during 17 days, the NO removal through reactions with HO_x could reach two orders of magnitude. Taking this loss into consideration, we can determine the NO amount remaining after the diffusion of TM products from the first cloud (without photochemical removal); it is about 10^{20} molecules of NO_x per square centimeter.

Park's (79) calculations of the post-TM NO amount ejected to the atmosphere were based on an initial velocity and density of the TM body, selected in accordance with the hypothesis of the cometary origin of TM. Park found it impossible in choosing the direction of the TM orbital flight to make an exception for TM and therefore neglected several specific features of the TM entry into the atmosphere that pointed to a low entry speed of TM, typical of objects captured by the Earth. Park performed calculations with $V_0 = 40$ km/s and efficient density $p = 0.002$ g/cm^3.

Initially [$V_0 = 11.2$ km/s; $\rho_{cover} = 0.03$ g/cm^3 (79)], the production of NO should be reduced, at the least five to six times. Park believed that its estimate of 6×10^{35} molecules could be overestimated 10-fold. We shall use the value 1.0×10^{35} molecules.

Next, correction of the NO amount is needed because the cloud moving toward California has only a part of the NO formed during the TM flight at altitudes below 20 km and partly during the explosion (without NO ejected by the explosion to the middle stratosphere). Let these shares constitute 20% of the total NO production. Then, in order to calculate the coefficient of the TM products diffusion from the cloud moving toward Mount Wilson during a 17-day period, we use two estimates of NO: 2×10^{34} (initial) and 10^{20} (final) molecules. Dividing the first value by the second, we find the diffusion coefficient to be 2×10^{14}. Applying it to a water vapor content increase of about 0.5 g of precipitated water over Mount Wilson at the moment of the arrival of the TM mass, we calculate the water mass thrown off by TM below 20 km to be about 10^{14} g.

When estimating the initial TM mass, one should take into account the loss of TM mass (ablation) on the slant path (estimated at 400 km) from the level of the initial loss (about 160 km) to an altitude of 20 km. It follows from the data of Bronsten (80) that because of ablation, a meteor body can lose on this trajectory a considerable portion (about 3/5) of its initial mass containing water ice and frozen CH_4 (2:1) and a small part of NH_3. Thus the initial TM mass constituted about 2.5×10^{14} g. Obviously, this estimate of M_0 is very rough, but it leads to a value of M_0 close to that obtained from the dependence $\log V_0 = f(M_0$ (dashed line in Fig. 3.17), $V_0 \approx 7$ km/s at $E_0 = 4 \times 10^{25}$ ergs.

Since we accept the hypothesis about a TM chemical explosion caused by

an electric discharge in an explosive gas-air mixture formed by the evaporation of the TM gas-condensate shell, it is important to specify characteristic
parameters of the TM explosion. Methane is taken as a major "working"
component of the gas-air mixture, although it should be remembered that at
the ejection of volatile hydrocarbons across the zone of ablation, the pyrolysis of methane could have taken place creating a hydrogen-air mixture
(e.g., a fire damp). The efficiency of the process of pyrolysis was confirmed by
a find in the peat layer of diamond-graphite-crystal amalgams whose origins
were dated at 1908 (81).

Taking methane as a major component and an agreed-upon estimate of
5×10^{23} ergs for the explosive yield, one can estimate the mass of methane in
the gas-air mixture constituting the body of the explosion. Since the reaction
$CH_4 + O_2$ gives about 213 kcal/mol, to obtain 5×10^{23} ergs it would take
about 0.9 megatons of CH_4 and about 9 megatons of O_2 contained in
350 km^3 of air at an average pressure of 100 mb.

No doubt, not all of the methane that evaporated in the lower stratosphere
from the gas-condensate ice shell of TM had taken part in the explosion.
Most likely, only about one-tenth of the mass of methane in the gaseous tail
of TM had exploded.

Thus, according to our estimates, the initial mass of TM varies some-
where between 70 to 250 megatons. Then the diameter of the TM body
before its entry into the atmosphere was $600 \leq D_0 \leq 920$ m. (For an assumed
value of the efficient density of the TM body, $\rho_{eff} = 0.6$ g/cm^3.) An estimate
of the mass of the remaining part of the TM that entered the atmosphere,
falling in three to four fragments in the environs of the South Bog, gave about
10^5 tons ($\rho \sim 1.5$ g/cm^3), with an initial diameter of about 23 m.

With all the necessary data on parameters of the TM entry known, one
can estimate the loss of mass by the TM body. Based on the studies mentioned earlier and using our estimates, we compiled a summary of TM
parameters (Table 3.4).

Calculations are made using the equation of mass losses to calculate the
characteristics of bolides (80):

$$dm/dt = -K(S\rho V^3/2Q),$$

where K is the coefficient of heat transfer, S the area of the bolide's cross
section, ρ the air density at an altitude for which calculations are made, V the
bolide's velocity, and Q the specific heat of evaporation of the bolide's body.

For convenience, we take $K = \sigma 2\Gamma Q$, where σ is the parameter of ablation
and Γ is the coefficient of resistance of the medium to the intruding body:

$$dm/dt = -\sigma\Gamma S\rho V^3.$$

TABLE 3.4 A Summary of the Tunguska Meteor Parameters

TM Parameters	Units	Extreme Values from Basic Studies	After R. Turco	After Z. Sekanina	After Kondratyev and Nikolsky	Remarks
Initial energy, E_0	$(10^{25}$ erg)	0.02–4	4	—	4	1 Mt
Initial mass, M_0	$(10^{12}$ g)	0.4–4.9	5	8	$70 < M_0 < 250$	$\approx 4 \times 10^{22}$ erg
Initial velocity, V_0	(km/s)	20–47	40	30	$6.5 \leq V_0 \leq 11.2$	
Initial diameter, D_0	(m)	50–850	850	—	606–927	Body of explosion 350–450 km^3
Explosion yield, ϵ_e	$(10^{23}$ erg)	0.1–5	5	5 (8.5 km)	5	12.5 Mt
Trajectory slope, α	(° altitude)	11–40	30	5	25	
Trajectory azimuth, ϕ	(° N)	100–120	115	110	95	
Mass of debris, M	(tons)	4×10^6 (dust)	—	—	10^5	
Effective TM density, ρ_{eff}	(g/cm^3)	0.001–1	0.002	—	0.6–0.7	$(\rho_{cover} = 0.02–0.03)$
Effective density of the TM fragment, ρ_{fr}	(g/cm^3)	—	—	—	1.5	
Size of debris, d	(m)	—	—	—	10–30	

Mt = megaton; TM = Tunguska meteor

In calculations we take $\sigma = 0.13 \times 10^{-10}$ s^2/cm^2, $\Gamma = 2$, $S = 2.9 \times 10^9$ cm^2; $\rho = 1.225 \times 10^{-5}$ g/cm^3, and $V = 11.2 \times 10^5$ cm/s. Introducing these values to the equation of mass losses, we find that for 22 s—the time of the flight in the atmosphere to an altitude of 20 km—the TM body looses $\Delta m = 28.5$ megatons or 40% of $M_0 = 70$ megatons.

Of course, one can question the technique used for estimating mass losses (the calculation of losses for some "middle" height and subsequent distribution of the obtained value, for unit time, along the entire working trajectory), but it should be remembered that in order to restrain the upper limit of the estimate, the value $\Gamma = 2$ was taken for the coefficient of resistance instead of the calculated value 3.5.

3.3.3 Optical Phenomena in the Upper Atmosphere

The explanation of the nature of the Tunguska Meteor fall as an intrusion of an ice body with a diameter greater than 0.5 km and a mass more than 70 megatons which threw off about half its mass (consisting of mineral fragments covered with a thick layer of the water-methane-ammonia ice) when flying through the atmosphere, suggests numerous possibilities of optimal combinations of all (still inadequately related) aspects of the enormous event into a well-balanced chain of cause-and-effect events of unusual scale.

In our opinion, the Tunguska comet is not an aggregate of loose lumps of snowflakes and dust, but rather, a sound ice boulder (i.e., a TM is the core of a comet), with small additions of mineral matter constituting, apparently, only parts of a percent of the total mass.

As follows from the analysis of the optical phenomena at Mount Wilson, the contribution of the dust component to the optical thickness of the TM cloud is negligible compared to that by gaseous components and, of course, water vapor. It is very difficult to isolate the contribution by the small-sized aerosol component of the TM from the residual optical thickness (wavelengths 400 and 450 nm). It would be logical to believe that in the shortwave spectral interval its contribution does not exceed the background level, since in the near IR (0.7 to 1.6 μm) the TM aerosol is not detected when one tries to determine its individual contribution or any characteristic features in the spectral attenuation.

High light and color displays of the 30 June – 1 July sunset phenomena and their rapid termination during the following nights indicates a heavy but short-term pollution of the upper atmosphere with the substance of the gas-condensate ice shell of the Tunguska meteor body. An appearance in the mesosphere and stratosphere of large amounts of water vapor, nitrogen oxides, small particles of comet ice, and carbon and ammonia compounds is most probable.

An excess amount of water vapor in the summer cold mesosphere favored the formation of anomalously numerous luminous clouds appearing as air masses, rich in water vapor, nitrogen oxides, and so on, moving westward when the TM was passing the mesosphere. Indeed, during this period a high-speed east-to-west airstream resides between 55 and 75 km (in the latitudinal belt 55–65°N). The Tunguska comet had a small gas-dust tail which entered the upper atmosphere in the morning of 30 June over the zone of Eurasian midlatitudes (west of the site of the TM fall). We believe that the tail was probably bent in such a way that its middle part pointed toward the southwest and the end toward the northeast. The TM tail consisted of the same substances, except for the oxides of nitrogen formed during the flight of the TM through the atmosphere.

An unusual range of colors of the twilight segment could be connected with selectivity of the optical properties of substances which got into the mesosphere and formed there from initial TM products. A short duration of the afterglow radiance anomalies (three nights) is easily explained by the rapid photodissociation of water vapor molecules and photochemical reactions between HO_x, NO_x, and O_x families, yielding solid nitrate and ammonium compounds. Special attention should be paid to anomalous twilight phenomena at the southern boundary of their areal extent, encompassing Tashkent, Stavropol, and Sevastopol.

In Tashkent the night-sky lightening was so strong that only the brightest stars could be seen. This phenomenon cannot be explained in the same way as for the northern regions of the areal extent, because at midnight the sun's rays pass above Tashkent at an altitude of about 700 km. Hence, an abnormally high lightening of nocturnal skies in these regions cannot be ascribed to sunlight scattering. Most probably, it was caused by the strong luminescent radiance of settling TM aerosol particles containing the oxides of uranium. The major source of the luminous aerosols must have been the tail of the Tunguska comet, where it accumulated as the substance gradually left the TM body. The oxides of lead found in TM debris at the site of the fall confirm the possible presence of unstable uranium compounds in the TM body.

There is one more (58), better substantiated explanation of the intensive luminescence of skies (10^{-4} of the daytime sky radiance) in the southern regions of the 30 June–1 July nocturnal optical phenomena. During the flight of the NO-enriched TM cloud, near its lower and upper boundaries (~20 and 40 km), the NO molecules reacted with stratospheric ozone. Approximately 1 out of 40 molecules of NO_2 will be excited. One may assume that one of the five excited molecules will luminate in the visible. Data of Turco et al. (57) show that with the assumed amounts of ejected NO, luminescence could be observed across an area of 1500 by 500 km. The cloud probably stretched from southeast to northwest, moved westward, and was 1500 km long and 500 km wide. This ensured the luminescence of the skies from central Asia to central Europe for at least one night.

3.3.4 Variations in the Aerosol Optical Thickness and in the Ozone Content between 1905 and 1911

Results of recent studies (78) on the atmospheric spectral transparency under high-altitude conditions enable us to reconsider data by Abbot et al. (75) for the period 1905–1911 as well as analysis of these data (76,77).

In some cases, at high atmospheric transparency, the spectral course of the optical thickness of the aerosol component τ_a acquires the specific shape of a smooth wave with a peak at about 520 to 540 nm (78). The authors explain this by the selectivity of the optical properties of stratospheric aerosols, to which volcanic eruptions contributed little.

Data obtained one October morning in 1980 at Cheget Peak in conditions of high total transparency ($H = 3.1$ kilolumens; transparency coefficient $P = 0.825$) and low humidity ($\omega = 0.16$ cm of precipitated water) show a pronounced maximum near 0.55 μm and a sharp decrease near 0.45 μm in the spectral course of τ_a (Fig. 3.19, curve 1).

An almost similar spectral course of τ_a (curves 4, 5) in the interval 0.45 to 0.65 μm was obtained at Mount Wilson (75) on 4 and 6 June 1908. The aerosol optical thickness on these days exceeded by three to four times that obtained at Cheget. However, it referred to the aerosol cloud, which circled the globe several times, returning to Mount Wilson with a period of 2 months. Hence this aerosol belonged in the stratosphere. Even in the spectral course of τ_a (curve 3) obtained by Turco et al. (58) by averaging over the first decade of June 1908 (the period of the first passage of the stratospheric cloud over Mount Wilson), a maximum was observed near 600 nm and a decrease of the aerosol optical thickness to between 400 and 500 nm. The spectral change of τ_a for these days differed considerably from λ^{-1} in this spectral region and followed this pattern beginning with 700 nm to the longwave spectral interval.

Under low-altitude conditions the aerosol optical thickness is seldom that small. But even if that is so (82) (curve 2), τ_a exhibits selectivity, with the same features in the spectral course as for the high-altitude aerosols (curve 3).

Thus a tentative conclusion (because of inadequate observations) can be drawn that when the contribution by volcanic aerosols is small and tropospheric aerosols do not change specific features of the spectral course of the stratospheric aerosol thickness, it is manifested through a maximum at 530 nm and decrease in the shortwave interval. Such stratospheric aerosols

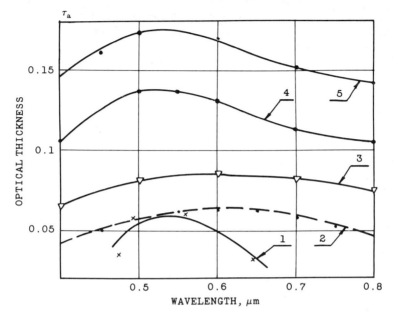

FIGURE 3.19. Spectral change in residual optical thicknesses of the atmosphere. 1, From measurements at Cheget Peak (Caucasus, 3100 m; Ref. 78); 2, from measurements by Niki-tinskaya et al. near Leningrad (Ref. 82); 3, from measurements at Mount Wilson (averaged over a decade; from Ref. 58); 4 and 5, from the 4 and 6 June 1908 Mount Wilson data.

consist mainly of meteorite aerosols coming continuously from the meteor zone and accumulating in the lower stratosphere.

Present monitoring of the spectral transparency and Lidar soundings of the atmosphere show that sharp changes in the aerosol optical thickness do sometimes happen in the stratosphere. Each case of growth of the strato-spheric aerosol mass probably results from either a volcanic eruption or the fall of a large meteor.

Observations of volcanic aerosol transformations following recent erup-tions of Mount St. Helens and El Chichón showed that an ejection of solid particles and gas products to the stratosphere changes its optical characteris-tics for many months in the latitudinal belt covering the source as well as outside that belt. In 6 months the effect becomes pronounced over the hemisphere, and in 1 to 1.5 years over the entire globe. The removal of aerosols continues for several years as a result of gravitational settling and various kinds of washout.

The optical characteristics of volcanic stratospheric aerosols, known since the 1912 Katmai volcano eruption, were studied thoroughly following the two last eruptions. The characteristic feature of stratospheric aerosols of that period was that the spectral dependence of the aerosol thickness varied in the visible in inverse proportion to wavelength, pointing to the presence in the stratosphere of a large amount of small-sized aerosols.

The disintegration of a large meteor, as seen from observations, takes place in the stratosphere in the "stopping" layer, the height of which depends on the velocity and angle of entry into the atmosphere, and, of course, on the solidity of the body of the meteor. Thus additional masses of submicron aerosols formed with complete (bolide) or even partial disintegration of the meteor body in the stratosphere in one region or another are sporadically superimposed on the continuous flow of meteor matter moving from the lower mesosphere. This explanation best fits the case of stratospheric turbidity variations over Mount Wilson on 4 June 1908 (Fig. 3.18), when the maximum amplitude of the optical thickness daily mean was observed in all intervals of the measured spectrum (400 to 1600 nm). The spectral course of the aerosol optical thickness determined by this event is shown in Fig. 3.19 by curves 3, 4, and 5, of which dependencies 4 and 5 refer to concrete days with maxima in the optical thickness, and dependency 3 was obtained by averaging over the entire decade.

Note that in this period the air humidity over Mount Wilson did not follow an increase in the optical thickness but remained at a background level (4 to 5 mm of precipitated water). The behavior of humidity was quite opposite, however, when the products of the TM explosion reached the vicinity of the station. The correlation coefficient between the humidity and the turbidity of the atmosphere between 17 July and 25 August exceeded 0.9. In other periods of observations in 1908 and for the following several years, no correlation was observed.

At the moment of the TM's entry into the atmosphere and its explosion, about 10^{35} molecules of NO_x were formed. It was mainly the oxide of nitrogen NO which, being in the lower stratosphere, reacted with ozone, giving NO_2. As a result of photochemical reactions with O_x, about half of the NO molecules must be transformed into nitrogen dioxide, optically active in the UV and in the visible with maximum absorption near 400 nm. The content of NO_2 and O_3 at Mount Wilson on 4 August was estimated from the Smithsonian Institution spectral data. It turned out that an air column of 1 cm^2 contained about 2×10^{17} molecules of NO_2 and 9×10^{18} molecules of O_3 (2,78).

These values contradict the conclusions from and results of photochemical modeling of the TM entry into the atmosphere (58). It turns out that under certain conditions, an injection of oxides of nitrogen can lead to an increase in the ozone content or, at the least, not change its content. The uncertainties of modeling are probably connected with the omission of reactions with the HO_x group.

With each intrusion of a meteor body into the atmosphere, in the zones of the shockwaves with temperatures about 10,000 K and pressures of tens and hundreds of atmospheres, a heating of the air up to temperatures of 5000 to 6000 K takes place and, as a result, ionization and dissociation of N_2 and O_2 molecules occur.

In the eddy zone of the meteor's trail (tens of kilometers in cross section in the case of large bodies), the cloud of dissociated atoms of nitrogen and oxygen rapidly cools, and molecules of N_2 and O_2 form, with part of the atoms of nitrogen and oxygen combining into the molecules of nitrogen oxide (NO). Maximum amounts of NO are formed when the temperature drops to 3400 K. With a further decrease of temperature, part of the NO molecules disintegrate, and the NO concentration decreases. At 1800 K NO molecules cease to disintegrate, and the remaining molecules of NO are a final product of the process of intrusion.

According to estimates (57,58,79), the production of NO at the TM entry ranged between 19 and 29 million tons. With new values of entry velocity (11 to 14 km/s), initial TM mass ($70 < M_0 < 250$ megatons), and a higher efficient density of the TM body ($\rho_{eff} \approx 0.6$ g/cm^3), the estimate of NO production must be decreased to 5 to 6 megatons. As follows from stratospheric photochemical models, half the NO molecules must be transformed into molecules of nitrogen dioxide through the reaction of ozone destruction. The catalytic role of NO in these reactions should markedly reduce the ozone content. Such a pattern of photochemical process was supposed by Turco et al. (58) and Park (79), who estimated the effect of TM on stratospheric ozone. A similar approach to photochemical phenomena can be found in studies on the atmospheric effects of nuclear tests (7,13) and on the consequences of possible nuclear war (7,46).

In 1964 Hampson outlined the possible changes in photochemical interactions of stratospheric ozone with NO_x and HO_x. Further development of this idea can be found in the works by Hesstvedt (83), Hunt (84), Crutzen (85), and Nicolet (86).

Based on model studies of photochemical processes associated with the TM entry, it was found (58) that during the first year following the TM fall, the ozone content in the NH stratosphere can decrease by 35 to 45%. This decrease is a result of a single injection to the stratosphere of 6×10^{35} molecules of NO. This is comparable to an annual production of 2×10^{35} molecules of NO_x by aurora, with the background NO_x content in the stratosphere estimated at about 10^{35} molecules.

Keeping in mind an overestimated amount of the TM-produced molecules of NO and an underestimated NO_x production resulting from high-altitude nuclear tests, one can compare these events: the TM entry (about 10^{35} molecules of NO), aurora (2×10^{35} molecules per year), nuclear tests in 1958 and 1961–1962 (1.5×10^{35} molecules of NO). Thus one can speak of the importance of nuclear tests in NO_x production and possible climatic impacts, although as Crutzen and Birks (7) show, none of the most thorough studies (10,19,20) mention the possibility of a global ozone depletion resulting from NO_x ejected by nuclear tests.

Observational data (35) show that in the period following nuclear tests the ozone content in the NH stratosphere increased by 8%, as noted at the beginning of this chapter.

Returning once again to the events of 1908, we draw the reader's attention to observations of the spectral transparency in the Chappuis band (600 nm) at Mount Wilson. Turco et al. (57,58) state that according to variations in the optical thickness for the 3 years following the TM fall, the total ozone content has increased by 30% ± 15%. In one of their works (57) the TOC could not be estimated directly from the 1908 Mount Wilson data because of strong fluctuations of the spectral transparency of an unknown origin. An extrapolation was made of mean seasonal data for 1909–1911 at the moment of TM entry and the estimate of TOC from the extrapolation curve was compared to that found for the year 1911. The TOC change ascribed to the TM event was then estimated from the ratio of these values (Fig. 3.20).

As our analysis showed, the 1908 observational data can be used to estimate TOC, but only with a thorough accounting of the NO_2 content, since the NO_2 absorption band covers the spectral interval between 350 and 650 nm, and thus totally includes the ozone absorption band. Based on estimates of the NO_2 content when the TM cloud was above Mount Wilson, the total ozone content was estimated; so, too, were the possible errors caused by leaving out the contribution by NO_2 to the optical thickness at 500 and 600 nm. Thus, when air masses containing gas and aerosol components ejected by the TM passed over Mount Wilson, the TOC was about 8×10^{18} molecules of O_3 per square centimeter. With the use of the linear relationships proposed in Turco et al. (57,58), neglect of the NO_2 contribution can lead to an imaginary 30% decrease of TOC.

To understand the controversy between the TOC estimates obtained from our data and those from the data of Turco et al. (57,58), we considered the mean annual spectral transparencies for the years 1905–1906 and 1908–1911, as given in Turco et al. (58) and Abbot et al. (75). The differences between the mean spectral optical thicknesses for 1905 and 1906 and for 1908, 1909, 1910, and 1911 were examined.

Figure 3.21 illustrates well a gradual decrease in these differences from 1908 to 1911. Clearly, although these differences do not contain any information about molecular scattering, they do characterize interannual variations in the attenuation by the aerosol component as well as similar variations in the absorption by gaseous components (ozone and nitrogen

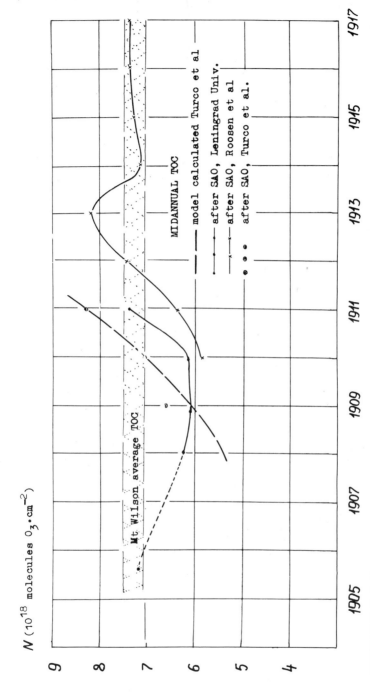

FIGURE 3.20 Probable variations in the mean seasonal ozone content over Mount Wilson between 1905 and 1917. N = total ozone content in units of 10^{18} molecules/cm².

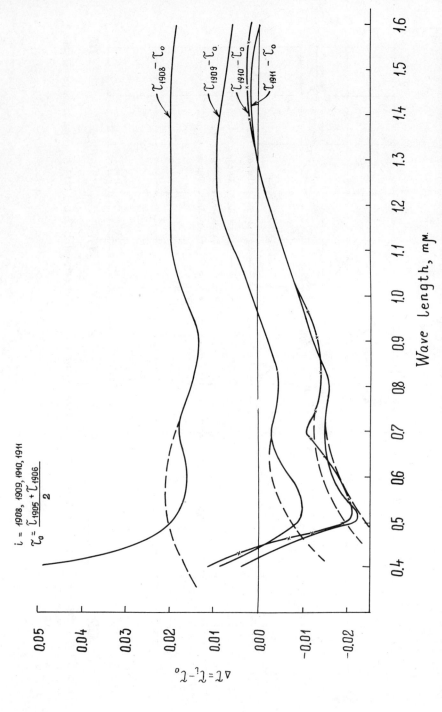

FIGURE 3.21. Spectral change in the differences in optical thicknesses for 1908, 1909, 1910, and 1911, and an average for 1905–1906. Dashed-line continuations of the difference curves to the spectral interval with wavelengths shorter than 0.8 μm show a probable change (in the first approximation) in the spectral thickness of an aerosol addition to an average for 1905–1906.

dioxide). Dashed curves and their extrapolations to the IR characterize the spectral course of excess aerosol attenuation as compared to average aerosol attenuation for 1905–1906.

The transition of the curves for 1910 and 1911 across zero difference between optical thicknesses (to negative values) indicates that the "optical" weather in 1905–1906 was supposed to be undisturbed and close to the background one and in fact preserve the traces of disturbance at wavelengths shorter than 1.2 μm. An important feature of the dependencies for 1910 and 1911 is their close course in the entire spectral interval and crossing the zero difference between 1.4 and 1.6 μm. It indicates that the atmosphere in 1910 and 1911 had practically cleansed itself of the aerosol component ejected by the two meteors, and mainly by the pre-Tunguska bolide (May 1908).

Crossing the zero difference between the optical thicknesses points to their coincidence between 1.4 and 1.6 μm in 1905, 1906, 1910, and 1911, a coincidence which means that the thicknesses correspond to their background values. The close course of the curves for 1910 and 1911 indicates that the optical thicknesses for these years are equal to the background values over the entire spectral region under consideration. But then the question arises that perhaps the TM-ejected products (e.g., water vapor) promoted a more complete cleansing of the atmosphere of the aerosol component, such as that found in 1905–1906.

Of particular interest are bends near wavelengths 700 and 400 nm. The bend at 700 nm is apparently indirectly connected to variations in ozone, and the bend at 400 nm is directly related to the NO_2 absorption band.

The bends on all four dependencies can be explained by the following two reasons:

1. A substantially increased NO_2 content between 1908 and 1911 compared to 1905–1906, and a decrease (by 15%) of TOC in 1908–1910 with the TOC in 1911 being greater than in the preceding years.

2. The possible underestimation of the NO_2 content and the overestimation of TOC between 1905 and 1906 compared to 1908–1911. This contradicts the work of Neftel et al. (87), which addresses the maximum content of nitrates in the Greenland ice layer referenced to 1906.

Calculations for 1905 and 1906 using mean annual spectral thickness data (88) gave normal TOC values (e.g., 7.2×10^{18} molecules of O_3 per square centimeter).

Figure 3.20 shows the temporal changes of TOC between 1905 and 1917. The lack in the "SAO Annals" of data on the spectral transparency for 1907 makes it impossible to verify stepwise variations in TOC and to attribute without doubt a 14% decrease of the mean annual TOC to the year 1908. Nevertheless, an attempt should be made to interpret these data from the perspective of pre-TM and TM effects on the atmosphere.

Consider variations in the mean annual TOC values (Fig. 3.20) calculated from the SAO data by Kondratyev and Nikolsky (this volume), by Turco et al. (58), and by Angione et al. (89). TOC values calculated at Leningrad State University fall approximately between the values calculated by the other two groups. A correct estimate of the aerosol spectral optical thickness and an account of the contributions by nitrogen dioxide to solar radiation extinction and by scattered light in the spectrobolometer play an important role in the calculations. The determination of the spectral change of all the constituents of extinction in the working spectral interval (400 to 1200 nm) is, apparently, obligatory for TOC calculations. Different calculation techniques used by American scientists produced TOC values that differed by 18% for 1910 and by 24% for 1911, with the background data being the same.

The model-calculated curve obtained by Turco et al. does not fit the experimental points. This is probably connected to the strongly linearized correlations when modeling photochemical processes in a 1-D model. We think that the TOC decrease from 7.2×10^{18} molecules of O_3/cm^2 in 1905 and 1906 to 6.25×10^{18} in 1908 should be referred to the summer of 1908. The TOC decrease in 1908 must not be attributed to the TM event. Our studies have shown that an injection of dry meteor matter in May 1908 from a pre-TM high-speed bolide had a major adverse impact on the ozone layer. Since the pre-TM products appeared three times over Mount Wilson, including the period of the simultaneous passing of the TM products, a conclusion can be drawn that the TM had not destroyed ozone but rather had compensated for the TOC decrease by maintaining it at a level of 6.1×10^{18} molecules of O_3/cm^2 in 1909, then promoting a TOC increase in 1911–1913. This conclusion is based on TOC values obtained for the first passage of the pre-TM products (4 June 1908) over Mount Wilson when the maximum TOC decrease reached 30%. During the second passage of the pre-TM cloud and the first (and the last) passage of the TM cloud (4 August), the TOC reached $(8 - 10) \times 10^{18}$ molecules of O_3/cm^2; that is, the average TOC level was exceeded by about 18%.

Neftel et al. (87) found a maximum of nitrates in the ice layer (Greenland) referenced to the year 1906. From our studies the mean annual NO_2 content in 1905 and 1906 amounted to 5×10^{16} molecules of NO_2 per square centimeter. For 1906 the NO_2 content should have been larger, about 6.5×10^{16} molecules of NO_2 per square centimeter. However, an analysis of the data of Neftel et al. revealed a 2-year delay of fallout with respect to the timing of the events. Such a long delay can be associated with the great distance (in latitude and longitude) between Greenland and the source of release.

Summarizing the discussion of ozone variations, an injection of 150 megatons of water vapor, 80 megatons of methane, and ammonia into the NH stratosphere must sharply change the course of photochemical processes. Probably, under "wet" conditions, the photochemical processes of ozone destruction in the presence of large amounts of NO changed in such a way that high-speed reactions of NO with HO_x have to some extent prevented ozone from destruction.

The presence of additional amounts of water vapor, methane, and ammonia in the stratosphere must also alter (at the expense of an intensified greenhouse effect) the thermal regime of not only the stratosphere but the troposphere as well. Probably, because of the long lifetimes of methane (7 to 11 years), the NO_x group (3 to 4 years), and water vapor (1.5 to 2 years) in the stratosphere and because of the TOC increase during the years following the TM fall, the climate warming [started during the previous century (Fig. 3.6)] should have accelerated first in the Northern and then in the Southern Hemisphere. The enormous 1912 eruption of the Katmai volcano, however, delayed and changed the trend of the climatic warming.

An attempt was made (58) to show that the Tunguska catastrophe had promoted a temporary climate cooling in the Northern Hemisphere. Estimates by Turco et al. were based on ideas about TM as containing an enormous amount of small-sized dust particles like those of Encke's comet. In fact, the climatic effect of TM has been the opposite, favoring the compensation for the cooling effect of the 1907 eruption (Stiubelya Sopka) and the effect of the 1908 pre-Tunguska stone meteorite. One may suppose that the effect of the Katmai eruption on climate in the Northern Hemisphere has also been partially compensated for by the TM contribution to the greenhouse effect.

REFERENCES

1. K. Ya. Kondratyev and G. A. Nikolsky, 1984. *Possible Impacts of a Nuclear Conflict on the Atmosphere and Climate,* Moscow: GKNT USSR.

2. K. Ya. Kondratyev, S. N. Baibakov, and G. A. Nikolsky, 1985. Nuclear war, atmosphere and climate, *Sci. USSR* **2,** 2–13; **3,** 2–11.

3. F. M. Luther, 1976. A parameterization of solar absorption by nitrogen dioxide, *J. Appl. Meteorol.* **15**(5), 479–481.

4. G. C. Reid, 1977. Influence of ionization on the neutral atmosphere, in *Dynamic and Chemical Coupling between Neutral and Ionized Atmosphere.* Dordrecht, The Netherlands: D. Reidel, pp. 191–202.

5. M. I. Budyko, 1979. *The Problem of Carbon Dioxide.* Leningrad: Gidrometeoizdat.

6. M. W. Carter and A. A. Moghissi, 1977. Three decades of nuclear testing, *Health Physics,* 33(July), 55–71.

7. P. J. Crutzen and J. W. Birks, 1982. The atmosphere after a nuclear war: twilight at noon, *Ambio,* 11(2–3), 114–125.

8. L. Machta, 1975. Comments to the report of E. Bauer and F. R. Gilmore, The effect of atmospheric nuclear explosions on total ozone, *Proc. 4th Conf. on CIAP,* DOT-TSC-OST-75-38. Washington, D.C.: U.S. Department of Transportation.

9. A. J. Miller, A. J. Kruger, C. Prabhakara, and E. Hilsenrath, 1974. Nuclear weapons tests and short-term effects on atmospheric ozone, presented at AIAA/AMS Meeting, San Diego, Calif., July.

10. A. D. Christie, 1975. Atmospheric ozone depletion by nuclear weapons testing, *Preprint APRB 11, No. 1,* Atmospheric and Environmental Service, Ontario, Canada.

11. J. K. Angell and J. Korshover, 1973. Quasi-biennial and long-term fluctuations in total ozone, *Mon. Weather Rev.* 101, 426.

12. J. London and J. Kelly, 1974. Global trends in total atmospheric ozone, *Science* 184, 987.

13. H. S. Johnston, G. Whitten, and J. Birks, 1973. Effects of nuclear explosions on stratospheric nitric oxide and ozone, *J. Geophys. Res.* 78, 6107.

14. Yu. A. Izrael, V. N. Petrov, and D. A. Severov, 1983. On the impact of atmospheric nuclear explosions on the stratospheric ozone content, *Meteorol Hydrol.* 9, 5–13.

15. Yu. A. Izrael, 1984. *Ecology and Environmental Control.* Moscow: Gidrometeoizdat.

16. J. London, R. D. Bojkov, S. Oltmans, and J. J. Kelley, 1976. *Atlas of Global Distribution of Total Ozone, 1957–1967,* Boulder, Colo.: NCAR.

17. B. G. Mendonca, K. J. Hanson, and J. J. DeLuisi, 1978. Volcanically related secular trends in atmospheric transmission at Mauna Loa Observatory, Hawaii, *Science* 202, 513.

18. K. R. Peterson, 1970. An empirical model for estimating worldwide deposition from atmospheric nuclear detonations, *Health Phys.* 180, 357.

19. E. Bauer and F. R. Gilmore, 1975. Effect of atmospheric nuclear explosions on total ozone, *Rev. Geophys. Space Phys.* 13(4), 451–458.

20. H. S. Johnston, 1977. Expected short-term local effect of nuclear bombs on stratospheric ozone, *J. Geophys. Res.* 82(21), 3119–3124.

21. J. Hampson, 1974. Photochemical war on the atmosphere, *Nature* 250, 189–191.

22. K. Ya. Kondratyev and G. A. Nikolsky, 1970. Solar radiation and solar activity, *Quart. J. R. Meteorol. Soc.* 96, 509–522.

23. K. Ya. Kondratyev, G. A. Nikolsky, D. G. Murcray, J. J. Kosters, and P. R. Gast, 1971. The solar constant from data of balloon investigation in the USSR and the USA, in *Space Research XI.* Berlin: Akademie-Verlag, pp. 695–703.

24. J. Hampson, 1977. Surface insolation and climatic potential effect of atmospheric contamination, *Preprint,* University of Western Ontario, London, Canada.

25. M. N. Markov, Ya. I. Meerson, and M. R. Shamilev, 1966. Emission from ionospheric layers in the infrared, *Dokl USSR Acad. Sci.* **167**(4), 803–806.

26. V. F. Gordiets, M. N. Markov, and L. A. Shelepin, 1970. On the mechanisms for infrared emission in the upper atmosphere," *Kosm. Issled.* **8**(3), 437–448.

27. Yu. M. Kondratyev, Yu. A. Martynov, V. G. Sochnev, and S. G. Yakovlev, 1980. Instrumentation, techniques, and results from infrared upper atmospheric radiation measurements from the Meteor-25 satellite, *Tr. Gos. NITs IPR* **11**, 15–24.

28. K. Ya. Kondratyev and G. A. Nikolsky, 1970. Direct solar radiation up to 30 km heights and solar constant, in *Meteorological Studies,* Vol. 17. Moscow: Nauka, pp. 23–33.

29. K. Ya. Kondratyev and G. A. Nikolsky, 1978. Solar activity and climate, *Dokl. USSR Acad. Sci.* **3**, 607–610.

30. M. N. Markov, 1983. The earth's thermal corona, *Sci. Life* **6**, 25–32.

31. S. Glasstone, 1964. *Effects of Nuclear Weapons.* Washington, D.C.: U.S. Atomic Energy Commission.

32. G. A. Nikolsky, 1981. On the role of some anthropogenic factors in present climate changes, *Proc. All-Union Symp. on Physical Basis for Present Climate Changes.* Moscow: Nauka, pp. 225–251.

33. J. K. Angell and J. Korshover, 1977. Estimate of the global change in temperature surface-100 mb between 1958 and 1975, *Mon. Weather Rev.* **105**(4), 375–385.

34. A. W. Brewer and A. E. Wilson, 1966. Measurements of solar ultraviolet radiation in the stratosphere—discussion, *Quart. J. R. Meteorol. Soc.* **92**, 267–274.

35. A. D. Christie, 1973. Secular or cyclic change in ozone, *Pageoph.,* **106–108** (V–VII), 1001–1009.

36. J. K. Angell and J. Korshover, 1978. Recent trends in total ozone and ozone in the 32–46 km layer, in *Papers at WMO Symp. on Geophysical Aspects and Consequences of Changes in the Composition of the Stratosphere, June, Toronto.* Geneva: WMO, pp. 107–114.

37. J. K. Angell and J. Korshover, 1978. Recent rocketsonde-derived temperature variations in the Western Hemisphere, *J. Atmos. Sci.* **35** (9), 1758–1764.

38. J. E. Walsh, 1978. Temporal and spatial scales of the arctic circulation, *Mon. Weather Rev.* **106**(11), 1532–1544.

39. K. Telegadas, 1973. *The Seasonal Stratospheric Distribution and Inventories of Excess Carbon-14 from March 1955–July 1969,* Health and Safety Laboratory Report 243, Washington, D.C.: U.S. Goverment Printing Office.

40. R. G. Oliver, 1976. On the response of hemispheric mean temperature to stratospheric dust: An empirical approach, *J. Appl. Meteorol.* **15**(9), 933–950.

41. J. J. DeLuisi, 1977. Estimation of solar radiation absorption by volcanic strato-spheric aerosols from Agung using surface-based observations, *Proc. Symp. on Radiation in the Atmosphere, Garmisch-P., FRG, 1976*, pp. 247–249.

42. B. G. Mendonca, K. J. Hanson, and J. J. DeLuisi, 1978. Volcanically-related secular trends in atmospheric transmission at Mauna Loa Observatory, Hawaii, *Science* **202**, 513.

43. Ch. H. Duncan, R. C. Harrison, J. R. Hickey, J. M. Kendall, Sr., M. P. Thekae-kara, and R. C. Willson, 1977. Rocket calibration of Nimbus-6 solar constant measurements, *Appl. Optics* **16** (10), 2690–2697.

44. K. Ya. Kondratyev and G. A. Nikolsky, 1970. Solar constant variations from the 1962–1968 balloon measurements, *Izv. Akad. Nauk. SSSR Fiz. Atmos. Okeana* **6**(3), 227–238.

45. A. A. Kmito and Yu. A. Sklyarov, 1981. *Pyrheliometry.* Leningrad: Gidrometeorizdat.

46. R. P. Turco, O. B. Toon, T. P. Ackerman, J. B. Pollack, and C. Sagan, 1983. Nuclear winter: global consequences of multiple nuclear explosions, *Science* **222**(4630), 1283–1292.

47. R. P. Turco, O. B. Toon, T. P. Ackerman, J. B. Pollack, and C. Sagan, 1984. Climatic consequences of nuclear war, *V. mire nauki* **10**, 4–16.

48. P. R. Ehrlich, J. Harte, M. A. Harwell, P. H. Raven, C. Sagan, G. M. Woodwell, J. Berry, E. S. Ayensu, A. H. Ehrlich, T. Eigner, S. J. Gould, H. D. Grover, R. Herrara, R. M. May, E. Mayr, C. P. McKay, H. A. Mooney, N. Myers, D. Pimentel, and J. M. Teal, 1983. Long-term biological consequences of nuclear war, *Science* **222**(4630), 1293–1300.

49. A. M. Obukhov and G. S. Golitsyn, 1983. Potential atmospheric consequences of a nuclear conflict, *Earth Universe* **6**, 7.

50. V. V. Aleksandrov and G. L. Stenchikov, 1983. On the modeling of the climatic consequences of the nuclear war, *Proc. on Applied Mathematics, The Computing Centre of the USSR Academy of Sciences, Moscow.*

51. V. V. Aleksandrov, N. N. Moiseev, S. L. Skorokhodov, and G. L. Stenchikov, 1984. On studies of climatic consequences of nuclear war in the Computing Centre of the USSR Academy of Sciences, report at the SCOPE ENUWAR Meeting on Climatic Consequences of a Nuclear War and Their Influence on the Biosphere, May 14–16, Leningrad.

52. S. L. Thompson, 1984. An evolving 'nuclear winter'—guest editorial, *Clim. Change* **6**(2), 105–108.

53. M. C. MacCracken, 1983. Nuclear war: preliminary estimates of the climatic effects of a nuclear exchange, *Preprint UCRL-89770*, Lawrence Livermore National Laboratories, Livermore, Calif.

54. C. Covey, S. H. Schneider, and S. L. Thompson, 1984. Global atmospheric effects of massive smoke injections from a nuclear war: results from general circulation model simulations, *Nature* **308**(5954), 21–25.

55. F. M. Luther, 1983. Nuclear war: short-term chemical and radiative effects of

stratospheric injections, *Preprint UCRL-89957,* Lawrence Livermore National Laboratories, Livermore, Calif.

56. K. Ya. Kondratyev, N. I. Moskalenko, and S. V. Gusev, 1985. Climatic consequences of a nuclear war from 1-D radiative-convective heat exchange model, *Dokl. Akad. Nauk. SSSR* **280**(2), 321–324.

57. R. P. Turco, O. B. Toon, C. Park, R. C. Whitten, and P. Noerdlinger, 1981. Tunguska meteor fall of 1908: effects on stratospheric ozone, *Science* **214**(45126), 19–24.

58. R. P. Turco, O. B. Toon, C. Park, R. C. Whitten, J. B. Pollack, and P. Noerdlinger, 1982. An analysis of the physical, chemical, optical, and historical impacts of the 1908 Tunguska meteor fall, *Icarus* **50**, 1–52.

59. R. P. Turco, O. B. Toon, J. B. Pollack, and C. Sagan, 1984. Nuclear winter to be taken seriously, *Nature* **311**(5984), 307–308.

60. R. Ganapathy, 1983. The Tunguska explosion of 1908: discovery of meteoritic debris near the explosion site and the South Pole, *Science* **220**(4602), 1158–1160.

61. Z. Sekanina, 1983. The Tunguska event: no cometary signature in evidence, *Astronom. J.* **88**(9), 1382–1413.

62. N. V. Vasilyev, 1983. Fantastic hypotheses without cover of fantasy, *Sci. USSR* **3**, 88–983, 110–111.

63. P. M. Millman, 1974. Interplanetary material, *JRAC Can.* **68**(1), 13–22.

64. S. P. Golenetsky and V. V. Stepanok, 1983. Cometary matter on the Earth (on studies on the Tunguska cosmic anomaly), in *Meteoritic and Meteor Studies.* Novosibirsk: Nauka, pp. 99–122.

65. E. L. Krinov, 1949. *The Tunguska Meteorite.* Moscow: USSR Academy of Sciences.

66. V. G. Konenkin, 1967. First-hand account of the 1908 Tunguska meteorite, in *The Problem of Tunguska Meteorite,* issue 2. Tomsk: TSU, pp. 31–35.

67. L. E. Epiktetova, 1976. New evidence of the Tunguska meteor fall eyewitnesses, in *Problems of Meteoritics.* Tomsk: TSU, pp. 20–34.

68. T. Ya. Gorazdovsky, 1976. Explosive dynamics of the Tunguska meteor in the light of laboratory reological explosion, in *Problems of Meteoritics (The Problem of Tunguska Meteorite).* Tomsk: TSU, pp. 74–82.

69. A. V. Voznesensky, 1925. The 30 June 1908 meteor fall in the Upper Khatanga river, *Mirovedenie* **14**(1), 25–36.

70. I. T. Zotkin, 1966. The Tunguska meteorite trajectory and orbit, *Meteoritics* **27**, 109–118.

71. A. F. Kovalevsky and I. N. Potapov, 1983. On the angle of inclination of the Tunguska meteorite trajectory, in *Meteoritic and Meteor Studies.* Novosibirsk: Nauka, pp. 161–166.

72. V. G. Fesenkov, 1961. On the cometary nature of the Tunguska meteorite, *Astron. J.* **28**(4), 577–585.

73. Z. Ceplecha and R. E. McCrosky, 1976. Fireball and heights: A diagnostic for the structure of meteoric material, *J. Geophys. Res.* **81**(35), 6257–6275.

74. R. E. McCrosky, Shao Tsi, and A. Posen, 1978. Bolides of the Prairie Network, I. General information and orbits, *Meteoritics* **37**,44–68.

75. C. G. Abbot, F. E. Fowle, and L. B. Aldrich, 1913. *Annals of the Astrophysical Observatory of the Smithsonian Institution,* Vol. III. Washington, D.C.: U.S. Government Printing Office.

76. V. G. Fesenkov, 1966. A study of the Tunguska meteorite fall, *Sov. Astron.* **10**, 195–213.

77. V. G. Fesenkov, 1949. Increase in the turbidity of the atmosphere due to the fall of the Tunguska meteorite on June 30, 1908, *Meteoritics* **6**, 8–12.

78. G. A. Nikolsky, M. M. Safronova, and E. O. Shults, 1984. On the optical thicknesses of the aerosol component of attenuation in high-mountain conditions, *Proc. 12th Symp. on Actinometry "Application of Actinometric Information to National Economy," Irkutsk,* pp. 87–88.

79. C. Park, 1978. Nitric oxide production by Tunguska meteor, *Acta Astronaut.* **5**, 523–542.

80. V. A. Bronsten, 1975. The nature and origin of meteor bodies, in *Problems of the Origin of the Solar System Objects.* Moscow: VAGO, pp. 265–301.

81. E. V. Sobotovich, V. A. Kvasnitsa, and N. N. Kovaliukh, 1983. New evidence for the Tunguska object reality, in *Meteoritic and Meteoric Studies.* Novosibirsk: Nauka, pp. 138–141.

82. N. I. Nikitinskaya, O. D. Barteneva, and L. K. Veselova, 1973. On the availability of the spectral thickness of the atmosphere in high-transparency conditions, *Izv. Akad. Nauk. SSSR Fiz. Atmos. Okeana* **9**(4), 437–442.

83. E. Hesstvedt, 1971. *A Time-Dependent Photochemical Model of the Upper Stratosphere.* Oslo: Institute of Geophysics, University of Oslo.

84. B. G. Hunt, 1966. Photochemistry of ozone in the moist atmosphere, *J. Geophys. Res.* **71**(5), 1385.

85. P. J. Crutzen, 1966. Determiniation of parameters appearing in the 'dry' and the 'wet' photochemical theories for ozone in the stratosphere, *Tellus* **21**(3), 368–388.

86. M. Nicolet, 1970. Ozone and hydrogen reactions, *Annal. Geophys.* **26**(2), 531–546.

87. A. Neftel, J. Beer, H. Oeschger, F. Zürchert, and R. C. Finnel, 1985. Sulphate and nitrate concentrations in snow from South Greenland, 1895–1978, *Nature* **314** N6013, 611–613.

88. C. G. Abbot, F. E. Fowle, and L. B. Aldrich, 1908. *Annals of the Astrophysical Observatory of the Smithsonian Institution,* Vol. II. Washington, D.C.: U.S. Government Printing Office.

89. B. J. Angione, E. Medeiros, and R. G. Roosen, 1976. Stratospheric ozone as viewed from the Chappuis band, *Nature* **261**(5554), 289–290.

90. E. Teller, 1982. Deadly myths about nuclear arms, *Readers Digest* **121**(Nov.) 139–146.

91. E. Teller, 1984. Widespread after-effects of nuclear war, *Nature* **310**, 621–624.

92. J. Blamont, J. P. Pommereau, and G. Souchon, 1975. Observations crepusculaires du NO_2 stratospherique, *C.R. Acad. Sci. Paris Ser. B* **281**, 247–252.

93. J. E. Harris, D. G. Moss, N. R. Swann, G. F. Neill, and P. Gildwarg, 1976. Simultaneous measurements of H_2O, NO_2 and HNO_3 in the daytime stratosphere from 15 to 35 km, *Nature* **259**, 300–302.

94. A. W. Brewer, J. B. Kerr, and C. T. McElroy, 1973. Nitrogen dioxide concentrations in the atmosphere, *Nature* **246**, 129–133.

95. R. M. Bloxam, A. W. Brewer, and C. T. McElroy, 1975. NO_2 measurements by absorption spectrometer: Observations from the ground and high altitude balloon, Churchill, Manitoba, July 1974, *Proc. 4th Conf. on CIAP*, DOT-TSC-OST-75-38. Washington, D.C.: U.S. Department of Transportation, 454–457.

96. E. A. Brun, 1976. Principal activities of COVOS in 1974, *Proc. of the 4th Conf. on CIAP (1975)*. Washington, D.C.: U.S. Department of Transportation, pp. 45–54.

97. J. F. Noxon, 1975. Nitrogen dioxide in the stratosphere and troposphere measured by ground-based absorption spectroscopy, *Science* **189**,547–549.

98. G. Greenstein, 1985. Heavenly fire, *Science* **6**(6), 70–77.

Chapter 4

Recent Studies and Conclusions

The results of studies on the potential impacts of a nuclear war on the atmosphere and on climate testify to the catastrophic character of such impacts. A major result of numerical modeling and analyses of the consequences of nuclear tests is a conclusion about the unavoidable variability of the climate, which, in turn, will adversely affect the biosphere and will bring everyday industrial and agricultural activities to a halt. To obtain reliable quantitative estimates about such a global-scale ecological catastrophe, additional serious research efforts are needed.

First, an analysis of the principal uncertainties of the available theoretical estimates is needed. MacCracken (1, Appendix) compiled a useful summary of previous assessments of the effect of smoke aerosols and the reliability of those assessments (see Tables 4.1 through 4.5 for the results of his analysis). No doubt, to overcome such uncertainties, an extensive program of laboratory and field observations as well as simulation numerical experiments must be accomplished. Such a program was outlined by MacCracken (1, Appendix). From the viewpoint of the "nuclear winter" concept, of major importance are studies of the physical and chemical properties as well as of the characteristic features of the mesoscale and large-scale transport of smoke aerosols of various origins. This must be done taking into account aerosol transformation at different stages of transport caused by the gas-to-particle conversion processes, coagulation, sedimentation, washing out, interaction with cloudiness, and other processes.

The simulation numerical models used to estimate the climatic effects of

TABLE 4.1   Historical Summary: Estimates of Smoke Amount

Factor	Value Used Most Often	Range	Depending on:
Number of explosions over combustible targets	6000	From several to many thousands and more	War scenario
Share of unburned material	20%	From several to 100%.	Strategy of the selection of targets, coverage, and probability of burning
Density of combustibles	3 (g/cm²) in cities, 0.5 (g/cm²) in forests	0.1–10 g/cm²	Type of the target, season (forest)
Area under fire	200 km²	10–1000 km²	Power and height of explosions, type of the target, the weather, probability of propagation, local topography, conditions of burning
Share of the aerosol particles (<1.0 μm) suspended in the air	3×10^{-2}	10^{-3}–10^{-2}	Specific properties of fuel and fire, the share of burnt fuel
Total mass of aerosol	2×10^{14} g = 200 Tg	10–1000 Tg	Current ejections due to forest fires in the USA are about 10 Tg/yr, and the global soot aerosol content totals <1 Tg

nuclear explosions in the atmosphere require further improvement. Here a major difficulty relates to the lack of an adequate account of numerous feedback mechanisms responsible for the formation of climate under conditions of a strongly disturbed gas and aerosol composition of the atmosphere. A correct description of the altered conditions for shortwave and longwave radiation transfer (an account of the dynamics of the atmospheric greenhouse effect) is undoubtedly needed. Hence reliable information about variations in the gas composition of the atmosphere as well as an account of the gas-aerosol interactivity is required.

Results given in recent publications convincingly testify to an urgent need for the continued improvement of simulation models. The observed laws and causes of present climate changes are still poorly understood because the problem is an extremely complicated one (2).

Even new results of analysis of the now observed climate changes revealed serious controversies between available estimates. For example, Kelly et al. (3) obtained specified data on the secular trend of mean annual surface air temperatures (SAT) for the NH continents for the period 1851–1984 in the form of SAT anomalies against average values for 1946–1960. The basic difference between the earlier estimates and the new ones (the available information is of limited representativeness) is the conclusion about a not-so-strong cooling in the 1880s, as had been discovered earlier (the difference in the amplitudes of cooling was 0.07 °C). The stronger SAT variability in the

TABLE 4.2 Historical Summary: Height and Power of Smoke Ejections

Factor	Value Used Most Often	Range/Remarks	Depending on
Height of the smoke veil	Up to 10 km	From several to 20 km	Type of fires (urban, forest, etc.), their area and intensity, air temperature, and winds
Share of smoke aerosol ejected to the stratosphere	5%	0–20%	Fire intensity, temperature, and wind
Transformation due to:			
Chemical reactions	Negligibly small	Changes in the composition	Size distribution, chemical composition, number density, atmospheric conditions (relative humidity, temperature, etc.), clear air involvement
Coagulation	Slow	May be faster in the beginning	
Aerosol–cloud interaction	Negligibly small	Changes in the size distribution and optical properties of clouds are possible	
Washing out	25%	From several to 50%	

TABLE 4.3 Historical Summary: Effect on Shortwave and Longwave Radiation Transfer

Factor	Value Used Most Often	Range/Remarks	Depending on:
Smoke-induced extinction of global radiation in the Northern Hemisphere (optical thickness is 3)	Up to 90%	From zero in smoke-free regions up to 99%	Composition, structure, size distribution, and shape of particles
Size distribution	Lognormal with a maximum near 0.1 μm	Temporal variations of the size distribution (from initially nonlognormal)	Processes of formation, coagulation, and washing out
Smoke particles single-scattering albedo	0.5	0.5–0.8	Composition, size, shape, etc.
Composition of smoke aerosol	Soot particles covered with hydrocarbons	Broad range depending on specific fuel mixture	Processes of formation and chemical transformation
Shape of particles	Spherical	Can be very nonspherical (axes ratio 1 : 10)	Processes of formation and coagulation
Effect of smoke on the IR radiation transfer	Negligibly small	Can be strong in case of thick smoke clouds	Size distribution, shape, and composition of smoke particles

beginning of this period could be explained by an inadequate observational data base.

The absence of a long-term SAT trend at the end of the nineteenth century (the temperature remained almost unchanged until the late 1910s) is a most important feature of the new data. There is a clearly pronounced difference between the temperature levels before and after 1920. During the 5-year periods 1915–1919 and 1920–1924, a warming of about 0.3°C took place. This constituted about 75% of the total warming in the periods before and after 1930. The warming in the early 1980s proved to be the most intensive.

An important result from a new analysis of the SAT data is the conclusion about a stepwise (discrete) climate warming. This runs counter to the view that the SAT secular trend was caused by continuously increasing CO_2 concentrations or by such factors as volcanic eruptions and solar activity.

Based on results of processing a more complete SAT data base for the

1949–1972 period, for both hemispheres, averaged over a $1° \times 1°$ latitude-longitude grid (for latitudes north of $75°N$ and south of $50°S$, only fragmentary data are available), Chen (4) found estimates of mean monthly and mean annual SAT anomalies (against averages for the whole period under

TABLE 4.4 Historical Summary: Diffusion and Effect of Smoke Aerosol on Thermal Regime

Factor	Typical Estimates	Range/Remarks	Depending on:
Mesoscale transformation and the processes of cleansing	Left out	Can be very important	Weather conditions and dynamical disturbances
Mesoscale propagation (diffusion) of smoke	Undisturbed vertical diffusion, very fast horizontal propagation	Both the vertical and horizontal strong diffusion is possible, depending on specific dynamical interaction	Weather conditions, self-induced heating, conditions of smoke diffusion and its patchy distribution
Hemispheric transformation and the processes of cleansing	Undisturbed	Can be either accelerated or slowed	Mesoscale disturbances (for instance, land–ocean contrast), specific global dynamics, and vertical structure of the atmosphere (convection)
Hemispheric diffusion	Instant	Most likely slow, especially in the Southern Hemisphere (in the absence of the self-induction processes)	Season, wind speed, disturbances in the AGC structure
Temperature variations	Strong (rapid cooling over the continents)	From small to very strong	Patchy smoke, buffer effect of the ocean, AGC disturbances, season, cloudiness, height of explosions, optical properties of smoke, etc.

TABLE 4.5 Historical Summary: Potential Impact on Climate

Factor	Value Used Most Often	Range/Remarks	Depending on:
Extinction of the incoming solar radiation	90%	From several to 100%	Rate of diffusion properties and patchy distribution of aerosol
Temperature variations on the midlatitude land	From −30° to −40°C	From 0 to −30°C and more; stronger in summer than in winter	Season, latitude of the extent of the oceanic buffer effect, the rate of the vertical and horizontal diffusion of smoke, climatic feedbacks
Temperature variations in the tropics and on the Southern Hemisphere land	From −20° to −40°C	From 0 to −20°C	Specific diffusion and washing out of smoke, the buffer effect of the ocean
Duration of temperature changes	Several months at instant appearance	From several days to a year, at possible rapid appearance (under certain conditions)	Distribution, rate of diffusion, and lifetime of soot particles in the atmosphere
Variations in rainfall	Decrease	Either increase or decrease	Geographical coordinates (coastline, inner part of the continent, etc.), variations in atmospheric stability, possible early washing out due to convection
Surface winds and storms	Intensification	Either intensification or ceasing	Dynamic reaction to temperature contrasts and disturbances

consideration) as well as respective trends for different latitudinal belts, with separate analysis of data for land and ocean (Table 4.6).

Data for the trends in the Northern Hemisphere agree well with those obtained earlier using considerably different processing techniques. How-

TABLE 4.6   SAT Trends (°C/year) for the Period 1949–1972

Latitudinal Belt	Land and Ocean	Land	Ocean
90°S–90°N	−0.002	−0.006	−0.003
0–90°N	−0.010[a]	−0.011[a]	−0.005
0–90°S	0.004	0.004	0.003
20°N–90°N	−0.014[a]	−0.014[a]	−0.010[a]
20°S–20°N	0.001	−0.003	−0.001
20°S–90°S	0.009[a]	0.012[a]	0.004

[a]Data with an 85% statistical significance.

ever, an analysis of the time series for the Earth as a whole showed that a SAT decrease was very small and cannot be considered to be statistically significant. Only on regional scales are statistically significant trends revealed. A statistically reliable temperature decrease took place, for example, in the extratropical latitudes of the Northern Hemisphere (80–90°N). In the same latitudinal belt in the Southern Hemisphere, however, a positive SAT trend was observed, with almost complete compensation of surface air temperature changes in both the hemispheres. There were practically no variations in temperature in the equatorial zone. The SAT variations on land and in the ocean are qualitatively similar. Only in the 20–90°S belt did the warming over the land exceed a temperature increase over the ocean. That, however, might be explained by the urban effects on SAT observations. Estimates of the coefficient of correlation between the SAT over land and ocean gave maximum values for the tropics (0.804 for mean annual values). The interannual SAT variability is most pronounced in the polar regions, both on land and in the ocean.

The observed interannual SAT variability is governed by numerous factors which can be classified into three categories: external forcing, internal regular oscillations, and random processes (with the latter caused partly by the dynamic instability of the atmosphere). The most probable external factors are large-scale volcanic eruptions and an increasing CO_2 concentration. The ENSO phenomenon also contributes to the formation of the interannual variability over North America and, apparently, affects the global-scale internal coherent oscillations.

As Robock (5) noted, understanding the nature of interannual variability as well as any hope for long-range weather forecasts are possible only with reliable estimates of the contribution by each of these significant factors and with the recognition of the factor-induced "signals" in the observed SAT variations. As a result, an analysis of the secular trend and annual change of the SAT global fields was undertaken which showed SAT variability during the last century as having been caused by both external and internal factors.

Since the characteristic time of the reaction of SAT to an increase in the CO_2 concentration and a phase shift due to the oceanic thermal inertia still remain unknown, recognition of the CO_2 signal remains difficult. Considerable changes caused by other factors can bring about the SAT trends, which may erroneously be interpreted as a CO_2 signal.

As MacCracken (6) noted, although an impression may emerge that the development of numerical modeling of the anthropogenic effects on climate in the last decade was characterized by an increasing consistency of the estimates, a more thorough analysis suggests that, in fact, the degree of uncertainties of the empirical and theoretical estimates of the present climate changes have increased (e.g., "present' climate refers to the period 1850–2100). This is a natural consequence, because for such a complex multifaceted phenomenon as climate with numerous feedbacks, traditional, statistically strict estimates of the reliability of the diagnosis and forecasts are practically impossible. This applies also to the calculations of a mean global warming with an assumed doubling of the CO_2 concentration.

With the CO_2 content and the SAT assumed to change in phase, with a CO_2 increase from 280 ppm (1850) to almost 345 ppm (now), a warming by 0.5, 0.9, or 1.35 °C should have taken place with a temperature increase (due to a CO_2 doubling) of 1.5, 3, or 4.5 °C. With a 0.5 °C warming observed in the last 100 years and a logarithmic extrapolation (depending on the CO_2 concentration), the lower limit of the estimates seems to be most probable.

The controversy surrounding the available estimates of the anthropogenic impact on the climate is due to four major factors: (i) the inadequacy of the available observational data base; (ii) the inadequate techniques for analysis of such data; (iii) the multifaceted nature of climatic changes; and (iv) the inadequacy of numerical models, which brings about contradictory estimates of the sensitivity of the climate to various factors (how one takes into account the buffer effect of the ocean plays an important role).

Available observational data do not reveal any quasihomogeneous, latitude-intensified, monotonic temperature increase near the surface and in the troposphere, as had been predicted by theoretical calculations of the climatic impact of an increased CO_2 concentration. The data for the Northern Hemisphere reveal a temperature field that varies with time and location, and with variability that increases with increasing latitude. This intensification of temperature variability with latitude suggests that the high latitudes will contribute to the formation of any hemispheric temperature trend. It is important to note, however, that these data refer to the continents and do not take into account the 0–20° equatorial belt, where temperature variations are out of phase with those observed in the Arctic and can therefore compensate for them.

Ellsaesser (7) emphasizes that the interpretation of the climate warming trend observed in the last 100 years, as produced by the impact of an increased CO_2 concentration, is acceptable only under the following conditions:

i. that the warming is not connected with the natural process of a return to a normal, milder climate, such as that observed during the Little Optimum Period (900–1300), following a temperature decrease during the Little Ice Age (about 1550–1850);

ii. that there is evidence that a biospherically induced CO_2 release of 100 to 200 gigatons of carbon (e.g., as a result of forest cutting and agriculture) took place before the year 1938, whereas the CO_2 ejections due to fossil fuel combustion (about 175 gigatons of carbon) has occurred mainly after 1938;

iii. that there are other climate-forming factors (more powerful than CO_2) which have prompted warmings and coolings lasting for 30 to 50 years and that suppressed a persistent trend of warming caused by an increased CO_2 concentration;

iv. that the warming caused by a CO_2 doubling does not exceed $1.5°C$, and that the time shift of the warming with respect to the change in the CO_2 concentration constitutes less than 50 years.

Because the first two conditions cannot be guaranteed, and the problem of the contribution by other factors remains unclear, there are no grounds for persuasively ascribing the climate warming trend to the impact of an increased CO_2 concentration.

To date, attempts to filter out from the observed interannual temperature variability the contribution by external factors have not led to any convincing results, because of arbitrarily prescribed correlations with forcings due to solar activity and volcanic eruptions, among other factors. Hoffert et al. (8) analyzed the role of internal, free (self-induced) oscillations in interannual climate changes. These appear when a system of interactive nonlinear differential equations with a destabilizing feedback manifested near the basic state is considered. A stabilizing feedback starts functioning when the amplitude of fluctuations becomes large. In this case a cyclic variability can form in the absence of external factors and a more complicated regime of the climate in the presence of such factors.

Hoffert et al. (8) made calculations using a model of SST variability (T_a), taking into account a mixed layer $h_m = 100$ m thick, with an eddy-mixing coefficient $k = 2000$ m² per year, and with a fixed temperature of the polar ocean $T_p = 1.5°C$:

$$dT_a/dt = (T_e - T_a)/\tau + 1/h_m[k(\partial T/\partial z) + w(T - T_0)]_0,$$

where $T_e(t)$ is the equilibrium temperature, and the relaxation time for the mixed layer $\tau = 4$ years; w is the varying velocity of the upwelling. The expression in brackets determines the total heat flux at the mixed layer–thermocline interface, equal to zero under equilibrium conditions, when diffusion compensates for the upwelling.

With the destabilizing feedback between the upwelling and T_a expressed as

$$w = w_0 + \alpha\,\Delta T \quad (\alpha > 0,\ \Delta T = T_a - T_e),$$

the stabilizing feedback is given by adding the third nonlinear term $\rho(\Delta T)^3$. The resulting feedback turns out to be unstable at "small" $\Delta T (\partial w/\partial T = \alpha > 0)$, but stable at "large" $\Delta T (\partial w/\partial T = -3\beta(\Delta T)^2 < 0)$, both at positive and negative ΔT. At $T_e = $ const the given model illustrates the formation of free interval oscillations of T_a in a time scale of about 80 years, at realistic values of the parameters w_0, α, and β, and favorably reproduces the change of the globally averaged surface air temperature for the last 100 years, taking into account the forcing caused by an increased CO_2 concentration. This points to the importance of an adequate consideration of the atmosphere-ocean nonlinear feedback.

Gates and Potter (9) used an interactive climate model (a combination of the models of atmospheric general circulation, mixed-layer dynamics, and a seasonal thermocline) to assess climate changes caused by a CO_2 doubling. A controlled experiment showed that the model adequately describes the present climate but underestimates the SST because of the insufficiently warmed oceanic surface layer and because of excess mixing.

When the CO_2 concentration doubles, the mean global SST rises by about $0.1\,°C$ per year (mainly in the Southern Hemisphere), and the SAT rises by about $0.3\,°C$ per year. Precipitation practically does not vary (0.01 mm per day for annual average). The climate warming depends largely on latitude and season, increasing three times in the winter toward the pole, as a result of both the effect of the feedback and the intensified poleward latent heat transfer. During the first 6 years there is no regular spatial structure of warming over the continents.

Meehl and Washington (10) analyzed the impact of a CO_2 doubling on the wind field and precipitations in the tropics, based on calculations using the NCAR climate model and a simple model of a 50-m oceanic mixed layer, enabling one to consider the annual course of the heat budget leaving out the dynamics of advection and diffusion. Calculations revealed a nonuniformly distributed increase of mean seasonal precipitation connected with the warming of the tropical ocean by about $2\,°C$. Both in Winter (December to February) and in summer (June to August) precipitation is maximum in the region of the Pacific Ocean. Over India (in summer and in winter) and Africa (during the western African monsoon) the rains cease.

Schlesinger et al. (11) calculated the function of the reaction of climate F_c that characterizes the buffer effect of the ocean on the climatic impact of a CO_2 doubling, using an interactive climate model, a combination of a two-level model of the atmospheric general circulation developed at Oregon State University (OSU) and a six-layer model of the oceanic general circulation (at a fixed SST).

Analysis of results of numerical modeling for the past 20 years showed that the atmospheric warming near the surface intensifies from the tropics to the subtropics, decreases near the middle latitudes, and increases in the high latitudes. If in the tropics and subtropics the warming grows with height, the remaining part of the globe exhibits an opposite trend. In both hemispheres an SST increase is more intensive in the midlatitudes than in the tropical belt.

The ocean in the subtropics and midlatitudes is warmed down to deeper layers than in the equatorial zone, with maxima in the layer from 250 to 750 cm near 65°N and 60°S. The parameter F_c is a ratio between the current warming ΔT and its equilibrium value T_{eq} (the latter is calculated using a simpler energy-balance model for a period of 200 years, since a period of 20 years is insufficient to reach an equilibrium; it is 2.82°C). Its calculations have led to the conclusion that the surface air is warmed by 30% in less than a year. Then the rate of warming decreases and approaches that observed in the upper layer of the ocean.

Due to a slow heat exchange with deeper layers of the ocean, a 1°C increase of the upper-layer temperature takes place in about 75 years. Such an increase, corresponding to $\Delta T/\Delta T_{eq} = 1 - 1/e = 0.63$, occurs in the atmosphere in 50 years. The mean global sensitivity of the ocean-atmosphere system to the forcing is 0.72°C/(W/m²).

Apparently, detection of any climatic "signal" (including that caused by volcanic eruptions or by an increased CO_2 concentration) is possible only by averaging observational data over large regions. Most rational is an averaging with a weight taking into consideration numerical modeling results that reveal the geographical regions and the seasons to which substantial climate changes correspond.

Bell and Abdullah (12) considered an example of SAT observations averaged over the period 1958–1981 (as background data) and over a later period as a source of information about a signal of the El Chichón eruption. They took into account numerical modeling results according to which a maximum posteruption SAT decrease was to be observed in 1983, reaching 0.3 to 0.5°C in the tropics and 0.7 to 1.2°C in polar regions. However, data for 1983 revealed a warming (exceeding 2σ), which can be explained apparently by the almost simultaneous effect of an intensive El Niño – Southern Oscillation (ENSO) event.

Of considerable interest are new estimates of the role of various feedbacks in the formation of the climate. Based on the use of an improved radiative-convective climate model proposed earlier by Wang et al., Gutowsky et al. (13) analyzed the role of feedback determined by the meridional transport of latent and sensible heat, with the contributions by the heat transport in the atmosphere and the ocean being considered separately. The model foresees a prescribed total meridional heat transport (MHT) across 30°N latitude in such a way that the heat flux in the ocean at this latitude is considered constant (24×10^{14} W), while in the atmosphere it varies, making it possible to isolate the effect of the feedbacks due to heat transport in the atmosphere.

Calculations showed that the relationship between the following two counteracting factors is of paramount importance: (i) the feedback between the MHT and SST meridional gradient; and (ii) an intensification of the ice-albedo feedback determined by the water content in the atmosphere. The dynamic reaction to a change in the radiative forcing depends on this relationship, since this relationship largely determined the formation of the SST field. Therefore, to assess the temperature variations, it is not enough to know only the reaction of the MHT to the radiative forcing, because similar MHT changes can be associated with substantially different variations in the SST meridional gradient, which depends in turn on the relationship between latent and sensible heat fluxes. The latter testifies to the importance of reliable parameterization of the latent and sensible heat transport.

Somerville (14) analyzed the role of the feedback mechanism in the formation of the climate, manifested through the dependence of the cloud water content on temperature, based on calculations with the use of an improved 15-layer 1-D radiative-convective climate model. The specific character of this model consists of using (i) a wet-adiabatic convective adjustment (instead of a fixed vertical temperature lapse rate); and (ii) the empirical dependence of the cloud water content on temperature (optical thickness of clouds, respectively).

Calculations for the control experiment gave a CO_2 doubling-induced mean global SAT increase of 1.74 K. Such an underestimation of a climate warming (as compared to 3-D climate model data) is typical for the radiative-convective models, which do not consider the albedo and some other feedbacks. An account of the cloud feedback gave a halved CO_2 climate warming due to increased cloud water content (hence the optical thickness and albedo of clouds) when the temperature rises, as well as a prevalence of the albedo effect on the radiation budget over the greenhouse effect (the latter holds for all types of clouds except cirrus clouds).

Clearly, the need to consider cloud feedback in 3-D climate models becomes obvious. This will be difficult, however, because, no doubt, the temperature dependence of the cloud water content is specific to different conditions.

Watts (15) made calculations based on the use of an approximate 1-D (globally averaged) model of the oceanic thermal regime. This showed that the variability of the rate of formation of the circulation and temperature field for deep water may cause SST fluctuations which are close in value to the SAT trends observed during recent centuries. Results of this numerical modeling also agree with data of recent deep-water temperature observations in the Northern Atlantic. It follows then that variations of the large-scale characteristics of deep-ocean circulation can be a factor of climate change on the time scale of centuries.

Although SST variations resulting from varying solar radiation coming to the ocean surface can be substantial, they cannot penetrate the ocean deep enough to cause the same effect as do variations in the rate of upwelling. It means that the observed deep-water temperature variations cannot be explained by the variability of insolation. If they are caused by varying thermohaline circulation, then, consequently, this variability can contribute much to variations in SST and, possibly, the climate.

The use of an NCAR nine-level spectral GCM with a rhomboidal cutoff at wavenumber 15 has shown that this model satisfactorily reproduces the basic features of the general circulations. However, it has three serious drawbacks: (i) it systematically underestimates land-surface air temperature; (ii) it poorly reproduces thermal troughs over the deserts of North Africa and North America; and (iii) it overestimates mean rainfall and unreliably outlines arid regions and zones of moderate rainfall. Since all these drawbacks can at least in part be determined by prescribed constant soil humidity, Black and Pitcher (16) undertook numerical experiments taking into account the dependence of soil moisture on the type of vegetation in deserts and on the high-mountain plains of North America and North Africa. The results obtained show that numerical modeling results agree better with observations when variations in the soil moisture are taken into account (calculations were made for January and July).

With regard to an assessment of major impacts on the atmosphere, an account of the aerosol-induced changes in cloud-cover properties is of particular importance. The effects of aerosols on the optical properties of clouds can be manifested through the following three mechanisms: (i) as a result of the presence of aerosol particles inside the cloud (interstitial aerosols); (ii) as unsoluble nuclei in water droplets (this concerns particularly the particles of graphitic carbon); and (iii) by increasing the content of condensation nuclei, which leads to changes in the size distribution of cloud particles (an increase in the small-sized fraction of droplets).

Ramaswamy (17) obtained estimates of the effect of the first two mechanisms. When the volume concentration of graphitic carbon in droplets grows from 0 to 2 μm/m^3, the values 1 to w (w is the single-scattering albdeo) grow from 10^{-7} (pure water) to 10^{-3}, depending on the concentration and optical properties of the interstitial aerosol. Naturally, with a strongly absorbing interstitial aerosol, the single-scattering albedo decreases considerably. Estimates based on a 1-D radiative-convective climate model showed that with the cloud albedo decreased by 3% the SAT rises by 1 K. With a soot aerosol concentration in the global cloud cover of 0.5 μm/m^2, the climate will warm by 2 K.

Of interest are recent estimates of the contribution to the formation of the atmospheric greenhouse effect by various optically active atmospheric trace gases. Previous estimates have led to the conclusion about the comparability of the contributions by numerous optically active trace gases and CO_2 to the formation of the greenhouse effect. Data of Table 4.7 illustrate the calculated and observed trends of the mixing ratio for various trace gases for 1850–1984.

It follows from Table 4.7 that the mixing ratio for all the trace gases observed (especially CFCs) increases. Based on the use of a 1-D radiative-convective model, Wuebbles et al. (18) obtained estimates of the contribution by various trace gases to variations in SAT between 1850 and the present. Total warming must have constituted 1.0 K, with the following contributions by the various trace gases (in parentheses): 0.67 K (CO_2); 0.21 K (CH_4); 0.20 K (N_2O); and 0.08 K (CFC-11 and 12). As is seen, the total

TABLE 4.7 Probable Mixing Ratios of Various Trace Gases in the Preindustrial Period (1850) and at Present (1984), as Well as the Recently Observed Rate of Trace Gas Increases

| Component | Mixing Ratio Near the Surface (ppm) | | Present Rate of Increase (%/yr) |
	1850	1984	
CO_2	270	341	0.4
CH_4	1.0	1.73	1–2
CO	0.11	0.13	1–6
N_2O	0.285	0.302	0.25
CFC-11	0	1.9×10^{-4}	5–8
CFC-12	0	3.5×10^{-4}	5–8
CCl_4	0	1.5×10^{-4}	1–3
CH_3Cl	6.5×10^{-4}	6.5×10^{-4}	0
CFC-22	0	0.5×10^{-4}	7–10
CFC-113	0	0.3×10^{-4}	15–17
CH_3CCl_3	0	1.3×10^{-4}	5–8

contribution by trace gases (compared to CO_2) amounts to almost 50%. Since the warming observed during the last century constituted only 0.5 K, one may suppose that the difference with theoretical estimates should be explained by neglecting several feedbacks and the oceanic thermal inertia.

Having commented on some complications of the present-day climate theory, we shall now discuss the difficulties in assessing major impacts on the atmosphere. Covey et al. (19) emphasized the necessity to continue numerical modeling with different prescribed scenarios and with an interactive account of radiation, dynamics, and transport of aerosols. The problem of solar radiation scattering by smoke aerosols and the contribution by aerosols to the atmospheric greenhouse effect must be studied. A thorough analysis of the adequacy of the parameterization of physical processes in conditions of a strongly disturbed atmosphere is needed. The processes of formation of the smoke layer from urban and forest fires are still unclear. Studies of the optical properties of the smoke aerosols are needed.

In view of numerous uncertainties in assessments of the possible climatic impact of nuclear war, Cess et al. (20) calculated the sensitivity of climate to various factors using the Oregon State University (OSU) two-level climate model, with the total postnuclear smoke aerosol assumed to reside in the troposphere. The factors considered were: diurnal change, threshold aerosol optical thickness (AOT), which determines a large effect; vertical profile of smoke concentration; single-scattering albedo of the dust particles (all calculations were made with prescribed "July" sun elevation and with fixed SSTs).

Calculations for large AOT values confirm previous conclusions about a strong climate cooling. However, results for the threshold AOT turned out to be quite different. AT AOT ≈ 1 the climate can widely change, depending on the specific vertical distribution of smoke and the single-scattering albedo value (0.7 or 0.5 at $\lambda = 0.55$ μm). A consideration of the diurnal change (instead of a fixed sun elevation) plays an important role.

Robock (21) emphasized the importance of considering the albedo feedback in calculations with the use of an improved 2-D (zonal) seasonal version of the energy-balance climate model developed by Sellers. Preliminary estimates of the possible climatic impact of nuclear war scenarios (the dynamics of the incoming solar radiation at the surface level) have been suggested by Turco et al. (TTAPS). The major purpose of numerical modeling was to take account of the annual course and feedback determined either by the onset or the melting of the snow (ice) cover.

Calculations of SAT averaged over $10°$ latitudinal belts with a time step of 15 days were made for land and ocean, separately. A detailed account was taken of radiative processes and (in a parameterized form) of the horizontal transport of energy in the atmosphere and ocean. A simplified model was taken of the ocean as a mixed layer of a fixed thickness (from 90 m at the equator to 40 m at the poles).

To consider a nonuniform distribution of the solar radiation transmission T' with the atmosphere loaded with dust and smoke, this quantity was parameterized as $T' = T^f$, where T is the transmission after TTAPS, and f is the parameter dependent on latitude and in inverse proportion to the area of the globe covered with smoke (in percent), with respect to the Northern Hemisphere area (TTAPS assumed the smoke and dust to be uniformly distributed over the entire Northern Hemisphere).

Naturally, at a uniform global distribution, $f = 0.5$. In the case of a gradual input of dust and smoke to the Southern Hemisphere, in 3 months $f = 0.667$ (0–90°N) and $f = 0.333$ (0–90°S). Three "scenarios," with cumulative explosive yields of 5×10^3 megatons (basic case), 100 megatons (attack of cities), and 10^4 megatons (maximum yield), were modeled according to the following versions: the war begins in winter (1 January), in spring (1 April), in summer (1 July), and in fall (1 October). For control, the basic case is taken with summertime explosions (between 30 and 70°N) and $f = 2.273$.

Also, it is assumed that the small-sized dust and smoke are concentrated in the upper troposphere and stratosphere, which excludes the necessity of considering their effect on the longwave radiation transfer (the greenhouse effects).

Calculations made for nine versions of prescribed input parameters (different yield, varied seasons, and distribution of dust and smoke) suggest that in 4 years the reaction of the Northern Hemisphere atmosphere appears only as slow temperature oscillations, with preserved spatial distribution (characteristic of the fourth year) and in a gradual reduction of the effect (a similar situation is observed after large volcanic eruptions). Maximum temperature drops vary between -8.1 and -22.9°C.

Naturally, the effect of explosions is largest over land because of its lower thermal inertia. In a control case maximum cooling (21.4°) takes place in the 50–60°N band 45 to 60 days after the beginning of the war. The Northern Hemisphere averaged temperature drop over land is 11.5°C. Over the ocean the cooling is weaker and slower.

Due to the lower thermal inertia of the sea ice cover, a maximum temperature drop (10.7°C) takes place in the highest latitudes 90 to 105 days after the beginning of the war. Even over the midlatitude oceans a substantial cooling happens, one not revealed by calculations made by TTAPS. The mean zonal maximum temperature drop is 15.4°C.

During the first year, the effect on the climate can be described as a linear reaction to a decreasing insolation (21). During the second year, however, the nonlinear feedbacks start to appear because of the changes in the extent of the snow and ice cover. So, for example, the formation of a vast zone of cooling on the land (by more than 5 °C) in summer mid- and high latitudes is determined by the snow-albedo feedback (by increasing snow cover extent under conditions of lowering temperatures and by a respective increase in albedo). Maximum cooling over the ocean takes place at the poles in winter (during the second year) and results from the ice-thermal inertia feedback (the growth of the amplitude of the temperature annual change): an increase in the sea-ice cover extent leads to a decrease in the oceanic thermal inertia. This type of feedback prevails over the well-known ice-albedo feedback (however, variations in the snow and ice albedo caused by their contamination by settled dust and smoke particles must be taken into account).

By the end of the second year the distribution of cooling over the continents becomes similar to that over the oceans, in view of the dominating influence of the oceans due to horizontal mixing. In 18 months the cooling over the oceans becomes similar to that over the land and, later, even exceeds it since the oceanic thermal inertia slows down the recovery of the undisturbed state.

Due to cryospheric feedbacks, the reaction of climate to postnuclear disturbances turns out to be stronger and of longer duration than was suggested by TTAPS. As for the effect of the timing of the beginning of the war, calculations showed that the effect of atmospheric smoke loading in the fall and winter is not as large as in the control case, due to a lower level of insolation. Nevertheless, it is substantial. For example, despite a small initial effect in the winter, the cooling in the summer of the following year reaches 8 °C and spring conditions are nearly equivalent to those in summer.

With a maximum explosive yield (10^4 megatons), the first-year cooling remains, as in the control case, but because of cryospheric feedbacks it becomes much stronger (exceeding 16 °C) during the second year and remains for several subsequent years (the temperature field for the fourth year is similar to that for the second year in the control case). In accordance with the TTAPS data, even with the 100-megaton explosive yield, the effects are similar to those in the control case.

Robock (21) emphasizes the preliminary nature of his results (explained by the approximate climate model used), which does not take into account the numerous interactive effects: the dynamic and radiative interaction between the atmospheric circulation and smoke; the possible patchy spatial distribution of smoke; the effect of radiative heat exchange in the atmosphere; the circulation in the ocean and in the mixed layer of varying depths; and the water cycle in the atmosphere (the processes of formation and dissipation of clouds, and aerosol rainout).

Warren and Wiscombe (22) obtained estimates of variations in the snow cover albedo due to the sedimentation onto its surface of soot particles formed after nuclear explosions in the atmosphere. With the mean Northern Hemisphere mass of deposited aerosol assumed to be 7.1×10^{-5} g/cm^2, and rain rates of 13 g/cm^2 per year (the Arctic Ocean) and 37 g/cm^2 per year (Greenland), the relative mass concentration of aerosols in the snow cover will be, respectively, 65×10^{-5} and 2.2×10^{-5} (i.e., an accumulation of aerosols during a 1-month period). In this case the portion of solar radiation absorbed by snow cover will increase by a factor of 2.8 (for fresh snow) and 2.7 (for old melting snow) in the Arctic, and, respectively, 2.2 and 2.4 in Greenland. Because the soot aerosol penetrates the snow cover, its effect on albedo can be of long duration. A decrease in albedo can markedly affect the climate. Estimates by H. Lettau (1977) show, for example, that at the South Pole a decrease in albedo from 85% to 65% raises the temperature by 16 K.

The numerous assumptions with regard to "nuclear war" scenarios made in several studies have caused a critical reaction to the results obtained. Thus Barton and Paltridge (23), for example, obtained estimates of the optical characteristics of urban and forest fire smoke that could result from multiple nuclear explosions in the atmosphere. The estimates testify to the existence of somewhat overstated conclusions about the climatic effects of nuclear war. If, according to Crutzen and Birks, the maximum mass of smoke particles ejected to the atmosphere is 4×10^{14} g, smoke covers half the Northern Hemisphere, the time for particle sedimentation is 5 to 10 days, the aerosol content in the vertical air column is 5×10^5 g/(cm^2 m), and the average particle radius is 0.05 μm, then the aerosol optical thickness at 0.5 μm wavelength will be about 5. This means that less than 1% of the solar radiation will reach the Earth's surface, and therefore intensive cooling would take place.

However, results of such calculations depend largely on the prescribed size of particles. Yet Crutzen and Birks did not take into account such an important fact as a rapid coagulation of particles whose rate is in proportion to the square of their number density. Independent of an initial value, several hours later their concentration must decrease because of coagulation to about 10^4 cm^{-3} and, during subsequent days, by an order of magnitude more. Therefore, the maximum number density of particles in 5 to 10 days cannot exceed 10^3 cm^{-3}, which (with constant total mass of aerosol) indicates, at the least, a doubling of the particle radii. One can assume that in 5 days the radius corresponding to a maximum of number density will be about 0.1 μm (instead of 0.05 μm for fresh smoke) and total number density of particles will decrease to 3.8×10^8 cm^{-2} (e.g., by a factor of 8).

Table 4.8 compares estimates of aerosol optical thickness and its components determined by absorption and scattering (indices a and s, respectively), as well as estimates for new and old aerosols (calculations were made with the prescribed complex refraction index 1.3 to 0.31). The ratio F/B is that of the backscattered and forward-scattered radiation.

TABLE 4.8 Optical Characteristics of New and Old (5–10 days) Smoke Aerosols

Type of Aerosol	Optical Characteristics			Transmission		F/B Ratio
	τ_a	τ_s	τ	τ	τ_a	
New	3.8	2.4	6.2	0.002	0.022	400:1
Old	1.9	1.6	3.6	0.027	0.15	1400:1

Source: Ref. 23.

The transmission was calculated as $I/I^0 = e^{-\tau_a}$ or $e^{-\tau_s}$ (in view of the strongly elongated phase function, an assumed value of τ_a gives a transmission value that is closer to the real one).

As seen from Table 4.8, an assumption of the worst nuclear war scenario causes a decrease of insolation at the surface to 15% at noon, not to 1% as obtained by Crutzen and Birks. It follows, in particular, that Crutzen and Birks have largely overestimated the effect of intensification of erithemal UV radiation. Barton and Paltridge (23) obtained estimates from which it follows that 50 days after the beginning of a war in early June the dose of erithemal UV radiation near 55°NB doubles (it does not increase by a factor of 9 to 10 as suggested by the findings of Crutzen and Birks).

Assuming a uniform distribution of smoke aerosol over the Northern Hemisphere, Crutzen and Birks drew their conclusion about the cessation of rains caused by the increasing content of condensation nuclei. However, since the distribution of smoke must be horizontally inhomogeneous, a differential warming of the atmosphere occurs in various regions, which intensifies large-scale atmospheric circulation. A similar situation takes place in the case of monsoon formation, when the latent heat released to the atmosphere over land maintains the land-to-sea atmospheric circulation. A large thermal inertia of the oceanic mixed layer excludes the possibility of rapid cooling. Therefore, the moisture, as always, will get from the ocean to the continents, providing rainfall which must be redistributed by large-scale atmospheric circulation.

In reply to Barton and Paltridge's critical comments on the overestimated assessments of the effects of postnuclear forest fires on the radiative regime and climate, Crutzen (24) notes that although Barton and Paltridge are in principle correct, they have obtained a number of overestimated values—in particular, an estimation of the rate of coagulation. In fact, a doubling of an average radius of particles due to coagulation can only take place in about a month (it should be borne in mind that the efficiency of coagulation at each collision is not equal to unity). Apparently, the lifetime of aerosols varies between 5 and 30 days depending on regional climate conditions and altitude (averaging about 10 days).

The absorption coefficient for new elemental carbon (EC) is about 10 m^2/g, with an assumed lifetime of particles of 5 days. With a longer lifetime the absorption coefficient will decrease to 5 m^2/g in a month, and in this case the time-averaged value will be about 7.5 m^2/g. Since, however, the aerosol content must exceed by several times that obtained earlier, the solar radiation transmission will remain nearly the same, as follows from initial estimates. Even a 15% decrease of the atmospheric transmission obtained by Barton and Paltridge (with the sun in zenith) is of great importance. With an assumed global mean value of the solar zenith angle to be 60°, this is equivalent to a 2% transmission.

When calculating the absorption cross sections for EC, Barton and Paltridge have not taken into account the possibility of growing cross sections resulting from the formation of the surface film of water, oil, or other liquids. As for the initial scenario, which forms the basis for the assessment of the effect on ozone and the intensification of erythemal UV radiation, it must be considered obsolete, and therefore new estimates by Barton and Paltridge require attention.

Crutzen (24) analyzed the contributions by smoke aerosols from urban and forest fires. Although in this case the mass of combustibles is difficult to estimate reliably, a value of 40 kg/m^2 can be assumed as a conservative estimate for the urban regions. It is assumed that total aerosols from the urban fires will be 3×10^{14} g, 27% of the mass falling on EC. Initially, the urban aerosol layer will cover 6×10^{13} m^2 (the entire band 30–60°N, except for the Pacific area). Then the average mass concentration will be 7×10^{-4} g/m^2 and number density 2×10^4 cm^{-3}. The average radius 0.15 μm is assumed to correspond to the maximum concentration, which is equivalent to an indirect incorporation of the effect of particles' coagulation.

The aerosol rainout is an unclear aspect of the problem. Thus far, two characteristic times of rainout have been considered: 10 and 30 days. With the radiative warming of the smoke cloud top reaching 30° per hour due to solar radiation absorption, the possibility may arise of developing intensive convection in the above-cloud atmosphere and hence of a change in conditions for the aerosol rainout from the stratosphere.

Even with the characteristic time of rainout 10 days, calculations show that during the first week after the beginning of a nuclear war the transmission will drop down to the 1% level, and during the second week it will drop to less than 10%. At a very long characteristic time, transmission at the 1% level will remain during 1 month (during this period, about 70% of the Northern Hemisphere area must be covered by smoke clouds).

With a supposed horizontal homogeneity of the global smoke cover, the resulting stable temperature inversion (e.g., the alteration of the normal tropospheric temperature lapse rate) must suppress convection, cloud formation, and aerosol rainout, which determines the probability of a long lifetime for aerosols. Although results of analysis of the possible impact of a global nuclear war on the atmosphere and climate remain largely uncertain because of the lack of knowledge about many important processes, there is no doubt that conditions will be created that are extremely unfavorable for life on Earth.

Results of numerical modeling of the possible climatic impact of postnuclear urban and forest fires largely depend on the following: (i) on data related to the amount of smoke, the height of its location, its rainout intensity (rains occur at an early stage in the formation of a rising smoke column); and (ii) on the size distribution and optical properties of aerosols determined by coagulation and sedimentation.

In this connection Penner et al. (25) estimated the coagulation of particles in the process of the evolution of smoke as well as the amount of condensed water, based on a generalized hydrodynamic model of the propagation of smoke produced in combustion engines (atmospheric conditions are prescribed for spring and fall midlatitudes).

One of the major objectives of numerical modeling was the analysis of the interaction between smoke and water vapor, aimed at assessing the role of the processes of condensation nucleation, the "sticking" of smoke particles to water droplets, and the processes of washing out. Initial calculation results showed that the altitude of a smoke cloud depends largely on the amount of water condensed within the cloud, especially during large-scale fires.

The process of coagulation in the rising smoke veil can cause considerable changes in its particle size distribution: small particles are rapidly removed by large particles. However, during the first hour of the veil's rise and its horizontal diffusion, only small changes in the optical characteristics of smoke take place, provided that an initial mass concentration of smoke does not exceed 5×10^{-8} g/cm^3. Later, the coagulation causes a strong decrease in the cross section of extinction, which can reach not less than 40% in the most dense part of the veil during the first week. This points to the importance of considering coagulation when assessing the climatic impact of nuclear explosions.

Of great value is the schematic interactive modeling of the climatic effect of the postnuclear smoke aerosol undertaken by MacCracken and Walton (26). The modeling was based on the combined use of the OSU 3-D model and GRANTOUR 3-D aerosol diffusion model, taking into account aerosol rainout. No account was taken of the transformation of the aerosol size distribution (and its optical properties), processes of dry deposition, or homogeneous and heterogeneous coagulation. It was assumed that particles are wind-driven and removed by rains (within a $5° \times 5°$ cell). A schematic parameterization at the atmospheric cleansing is a major factor determining the conditional character of the model. The smoke particle concentration (varying in the process of diffusion) is used in calculations of the solar radiation starting with prescribed aerosol ejections with a constant smoke particle mixing ratio in the layer from 0 to 11 km in four regions (the western and eastern United States, Europe, West Asia), with subsequent account of diffusion and rainout during 30 days.

Calculations of the climatic impact of smoke aerosols were made at a prescribed constant sun elevation corresponding to July conditions. The model GRANTOUR is based on dividing the global atmosphere into 10^4 particles of equal volume (this means that spatial averaging is not less than over several hundred kilometers) driven by winds. The smoke particle size distribution is prescribed by dividing them into two categories with diameters less and more than 1 μm; large-sized aerosol rainout is four times faster. The cross section of extinction for small (large) particles is 6.7 (2.6) m²/g.

Numerical modeling was made for two cases taking interactivity into account, and two cases without it (in the latter two cases a fixed homogeneous distribution of smoke aerosol in the Northern Hemisphere atmosphere at an aerosol optical thickness of 2.4 is prescribed as the initial one). The cumulative mass of smoke ejections is taken to be 150 Tg and 15 Tg. To the first of these versions corresponds a contribution by urban fires, according to the TTAPS scenario, with a total yield of 5000 megatons.

Calculations showed that in one day the spatial distribution of smoke is very inhomogeneous: in some regions of Europe and Asia maximum aerosol optical thicknesses reach 50. Patchy spatial structure of AOT remains until the tenth day, although this model of diffusion may underestimate this inhomogeneity because of the schematic cleansing of the atmosphere and because of large spatial averaging.

By day 20 smoke becomes homogeneously distributed over the Northern Hemisphere (except for the low latitudes) and enters the Southern Hemisphere equatorial and subtropical latitudes (with the use of control experiment data in calculations of smoke transport there is no smoke input to the Southern Hemisphere even by day 30.

Calculations of SAT for the control experiment for the period 1 to 10 days for a smoke mass of 150 Tg revealed a temperature drop over the continents by 10 to 15°C, which increases to 20 to 30°C by days 21 to the 30. Data on SAT for the "interactive" case reflect the effect of spatial inhomogeneity of smoke: there is a sharp drop in SAT in the centers of smoke accumulation (e.g., the SAT in Asia dropped by 25°C by days 11 to 20), but without that sharp decrease in other regions. An average drop of SAT over the Northern Hemisphere continents after 1 month does not exceed 15°C. With the mass of ejection 15 Tg there is a drop in SAT by several tenths of a degree only in the regions of smoke accumulation (the hemispheric mean AOT in the control case is 0.24).

The patchy effect of the spatially inhomogeneous smoke aerosol must cause a strong variability in weather conditions. For example, calculations of SAT temperal variations in the central United States revealed a rapid succession of coolings and warmings with an amplitude of about 10°C, which only starts to attenuate by the end of the 30-day period. Calculations made for the western U.S. coastline revealed a smoothed course of attenuation of the SAT drop due to the buffering effect of the ocean. No doubt, an appearance of a pulsating regime of the SAT variability should entail more harmful ecological consequences than a persistent temperature decrease with subsequent recovery.

These results (relevant only to some aspects of the problem) indicate the numerous and serious uncertainties in the assessment of possible climatic impacts of nuclear war. These uncertainties underscore the need for an extensive program of future research.

In connection with the development of the "nuclear winter" concept, of the highest priority are the following problems relevant to aerosols: (i) interactive aerosol transport in the atmosphere; (ii) aerosol lifetime in the atmosphere; (iii) temporal variations of aerosol properties determined by their aging and gas-to-particle conversion: and (iv) smoke and cloud aerosol interaction.

There are many uncertainties in current ideas about the development of forest and urban fires, especially from the viewpoint of estimating the rate of smoke particles rainout from the atmosphere due to various processes. Interesting ideas about this problem were reported by Cotton (27) discussing the results of a workshop on the problem of aerosol washout by cloud particles and rains. Apparently, the removal will be very fast at the state of the development of convection resulting from explosions and large-scale urban fires (the amount of remaining smoke particles ranged between 5 and 10%).

Based on the use of a 3-D model of a convective storm, Cotton (28) showed that in conditions of Denver, Colorado, for example, a convective storm must occur in springtime, which will surpass, by power, the natural process of formation of cumulonimbus clouds by a factor of about 2. As a result, a large amount of soot particles gets into the upper troposphere and stratosphere. Estimates of reliability of this conclusion are, however, hindered by uncertain information about aerosol removal by clouds and mesoscale vortices. An interesting analysis of the origin of the urban fires was undertaken by Kang et al. (29).

In Carrier's opinion (30), there is a set of uncertain estimates: combustible fuel (the coefficient of uncertainty is not less than 2); the share of burned fuel (not less than 2); the share of particles remaining as a submicron fraction (not less than 2); and the share of burned fuel transformed into smoke (not less than 3). It follows that the total coefficient of uncertainty is, at the least, 36.

Penner et al. (31) considered a model of the fire smoke veil formation, taking into account varying strength of fires and environmental conditions. Calculations showed that in the cases considered, the stratosphere gets little smoke in the absence of an extremely powerful fire under conditions of a strongly unstable atmosphere.

Ditchburn (32) made serious critical remarks on the estimates obtained by TTAPS for the amount (and size distribution) of smoke and dust as well as on results of calculations of the amount of solar radiation reaching the Earth's surface. Ditchburn believes that a powerful convective storm is impossible, and therefore less than 1% of the smoke can reach the stratosphere. Hence the estimates by TTAPS of the aerosol absorption have been overestimated by a factor of about 5. In this case even at a cumulative yield of 5000 megatons the "nuclear winter" will not occur, but it may take place at a yield of 10^4 megatons.

The climate models used to assess the effects of nuclear explosions on the atmosphere must be seriously improved. No doubt, one of the key problems here is a substantiation of applicability, under conditions of an abnormally disturbed atmosphere, of the physical processes parameterization schemes developed for normal atmosphere conditions.

Undoubtedly, enthusiasm about the "nuclear winter" concept has diverted attention from consideration of the role of variations in the gas composition of the atmosphere from the point of view of both the impact on the atmospheric greenhouse effect and on solar radiation absorption by the atmosphere. Analysis of variations in the global biogeochemical cycles are of paramount importance.

In conclusion we emphasize again, however, that despite numerous and serious uncertainties in the estimates of possible climatic impacts of a nuclear war, there is no doubt that catastrophic changes in the atmospheric composition would result. Compared to these changes, the effects of a doubled CO_2 concentration seem insignificant. Strong temporal and spatial variabilities of climate, which exclude the possibility of normal industrial acitvities and life on the planet, would be a major feature of the global ecological catastrophe that would follow nuclear war.

REFERENCES

1. NOAA, 1985. *Interagency Research Report for Assessing Climatic Effects of Nuclear War,* report to the Office of Science, Technology and Policy prepared by the National Climate Program Office. Washington, D.C.: NOAA, 60 pp., appendices.

2. AMS, 1985. *Extended Summaries of the 3rd Conference on Climate Variations and Symposium on Contemporary Climate: 1850–2100, Jan. 8–11, 1985, Los Angeles, Calif.* Boston: American Meteorological Society, 188 pp.

3. P. M. Kelly, P. D. Jones, T. M. L. Wigley, C. M. Goodess, R. S. Bradley, and H. F. Diaz, 1985. The extended Northern Hemisphere surface air temperature record: 1851–1984, *Ext. Abstr. 3rd Conf. On Climate Variations and Symp. on Contemporary Climate: 1850–2100, Jan. 8–11, Los Angeles, Calif.* Boston: American Meteorological Society, pp. 35–36.

4. R. S. Chen, 1985. Surface air temperatures over land and ocean, 1949–1972, *Ext. Abstr. 3rd Conf. on Climate Variations and Symp. on Contemporary Climate: 1850–2100, Jan. 8–11, Los Angeles, Calif.* Boston: American Meteorological Society, pp. 39–40.

5. A. Robock, 1985. Detection of volcanic, CO_2, and ENSO signals in surface air temperature, *Ext. Abstr. 3rd Conf. on Climate Variations and Symp. on Contemporary Climate: 1850–2100, Jan. 8–11, Los Angeles, Calif.* Boston: American Meteorological Society, pp. 78–80.

6. M. C. MacCracken, 1985. The challenge of understanding contemporary climate, *Extr. Abstr. 3rd Conf. on Climate Variations and Symp. on Contemporary Climate: 1850–2100, Jan. 8–11, Los Angeles, Calif.* Boston: American Meteorological Society, p. 10.

7. H. W. Ellsaesser, 1985. Do the recorded data of the past century indicate a CO_2 warming? *Ext. Abstr. 3rd Conf. on Climate Variations and Symp. on Contemporary Climate: 1850–2100, Jan. 8–11, Los Angeles, Calif.* Boston: American Meteorological Society, pp. 87–88.

8. M. I. Hoffert, S. Gaffin, Z. Y. Wang, C. T. Hsieh, and T. Volk, 1985. Interannaul oscillations in past and future temperature records: models and data analysis, *Ext. Abstr. 3rd Conf. on Climate Variations and Symp. on Contemporary Climate: 1850–2100, Jan. 8–11, Los Angeles, Calif.,* Boston: American Meteorological Society, pp. 118–119.

9. W. L. Gates and G. L. Potter, 1985. The response of a coupled atmospheric GCM and mixed layer ocean model to doubled CO_2, *Ext. Abstr. 3rd Conf. on Climate Variations and Symp. on Contemporary Climate: 1850–2100, Jan. 8–11, Los Angeles, Calif.* Boston: American Meterological Society, p. 132.

10. G. A. Meehl and W. M. Washington, 1985. Tropical response to a doubling CO_2 with an atmospheric GCM coupled to a simple mixed layer ocean model, *Ext. Abstr. 3rd Conf. on Climate Variations and Symp. on Contemporary Climate: 1850–2100, Jan. 8–11, Los Angeles, Calif.* Boston: American Meteorological Society, pp. 130–131.

11. M. E. Schlesinger, Y.-J. Jan, and W. L. Gates, 1985. The role of the ocean in the CO_2-induced climate change: a study with the OSU coupled atmosphere-ocean general circulation model, *Ext. Abstr. 3rd Conf. on Climate Variations and Symp. on Contemporary Climate: 1850–2100, Jan. 8–11, Los Angeles, Calif.* Boston: American Meteorological Society, pp. 133–134.

12. T. L. Bell and A. Abdullah, 1985. Detecting global climatic change predicted by climate models, *Ext. Abstr. 3rd Conf. on Climate Variations and Symp. on Contemporary Climate: 1850–2100, Jan. 8–11, Los Angeles, Calif.* Boston: American Meteorological Society, pp. 89–90.

13. W. J. Gutowsky, Jr., G. Molnar, and W.-C. Wang, 1985. Feedback effects of sensible and latent heat fluxes during climate change, *Ext. Abstr. 3rd Conf. on Climate Variations and Symp. on Contemporary Climate: 1850–2100, Jan. 8–11, Los Angeles, Calif.* Boston: American Meteorological Society, pp. 107–108.

14. R. C. J. Somerville, 1985. Climate stabilization by cloud optical thickness feedbacks, *Ext. Abstr. 3rd Conf. on Clim. Variat. and Symp. on Contemp. Climate: 1850–2100, Jan. 8–11, Los Angeles, Calif.* Boston: American Meteorological Society, pp. 101–102.

15. R. G. Watts, 1985. Climatic transients caused by variations in the thermohaline circulation, *Ext. Abstr. 3rd Conf. on Climate Variations and Symp. on Contemporary Climate: 1850–2100, Jan. 8–11, Los Angeles, Calif.* Boston: American Meteorological Society, pp. 83–84.

16. M. L. Black and E. J. Pitcher, 1985. Effects of changes in ground wetness on the climate of a general circulation model, *Ext. Abstr. 3rd Conf. on Climate Variations and Symp. on Contemporary Climate: 1850–2100, Jan. 8–11, 1985, Los Angeles, Calif.* Boston: American Meteorological Society, pp. 91–92.

17. V. Ramaswamy, 1985. Climatic implications of cloud-aerosol radiative interaction due to increasing pollution, *Ext. Abstr. 3rd Conf. on Climate Variations and Symp. on Contemporary Climate: 1850–2100, Jan. 8–11, Los Angeles, Calif.* Boston: American Meteorological Society, pp. 103–104.

18. D. J. Wuebbles, A. J. Owens, and C. H. Hales, 1985. Trace gas influences on climate from 1850 to 1980, *Ext. Abstr. 3rd Conf. on Climate Variations and Symp. on Contemporary Climate: 1850–2100, Jan. 8–11, Los Angeles, Calif.* Boston: American Meteorological Society, pp. 81–82.

19. C. Covey, S. H. Schneider, and S. L. Thompson, 1984. Global atmospheric effects of massive smoke injections from a nuclear war: Results from general circulation model simulations, *Nature* **308**, 21–25.

20. R. D. Cess, G. L. Potter, and L. W. Gates, 1985. The climatic impact of a nuclear exchange: Sensitivity studies using a general circulation model, *Ext. Abstr. 3rd Conf. on Climate Variations and Symp. on Contemporary Climate: 1850–2100, Jan. 8–11, Los Angeles, Calif.* Boston: American Meteorological Society, pp. 4–5.

21. A. Robock, 1984. Snow and ice feedbacks prolong effects of nuclear winter, *Nature* **310**, 667–670.

22. S. G. Warren and W. J. Wiscombe, 1985. Dirty snow after nuclear war, *Nature,* **313**, 467–469.

23. I. J. Barton and G. W. Paltridge, 1984. Twilight at noon overstated, *Ambio* **13** (1), 49–51.

24. P. J. Crutzen, 1984. Darkness after a nuclear war, *Ambio* **13**(1), 52–54.

25. J. E. Penner, L. C. Haselman, Jr., and L. L. Edwards, 1985. The dynamics and microphysics of large-scale fires, *Ext. Abstr. 3rd Conf. on Climate Variations and Symp. on Contemporary Climate: 1850–2100, Jan. 8–11, 1985, Los Angeles, Calif.* Boston: American Meteorological Society, pp. 2–3.

26. M. C. MacCracken and J. J. Walton, 1985. The effects of interactive transport and scavenging of smoke on the calculated temperature change from large amounts of smoke, *Ext. Abstr. 3rd Conf. on Climate Variations and Symp. on Contemporary Climate: 1850–2100, Jan. 8–11, Los Angeles, Calif.* Boston: American Meteorological Society, pp. 6–7.

27. W. R. Cotton, 1985. *Workshop on Precipitation Scavenging and the Nuclear Winter.* Report to SCOPE Unit, SCOPE-ENUWAR H1.02.85, 9 pp.

28. W. R. Cotton, 1985. A simulation of the cumulonimbus response to a large firestorm—implication to a nuclear winter, *Ext. Abstr. 3rd Conf. on Climate Variations and Symp. on Contemporary Climate: 1850–2100, Jan. 8–11, Los Angeles, Calif.* Boston: American Meteorological Society, p.1.

29. S.-W. Kang, Y. A. Reitter, and A. N. Takata, 1985. *Analysis of Large Urban Fires,* SCOPE-ENUWAR H1.03.85, 11 pp.

30. G. F. Carrier, 1985. Nuclear winter—the state of science, *Issues Sci. Technol.,* Winter, 114–117.

31. J. E. Penner, L. C. Haselman, and L. L. Edwards, 1985. Buoyant plume calculations, *UCRL Preprint 90915,* Lawrence Livermore National Laboratories, Livermore, Calif., 9 pp.

32. R. W. Ditchburn, 1985. *Nuclear Winter,* SCOPE-ENUWAR H1.04.85, 14 pp.

ADDITIONAL REFERENCES

1. B. Bolin, B. R. Döös, J. Jäger, and R. A. Warrick (Eds.), 1986. *The Greenhouse Effect, Climatic Change, and Ecosystems.* SCOPE 29, New York, Chichester: John Wiley and Sons.
2. T. P. Barnett, 1986. Detection of changes in the global troposphere temperature field induced by greenhouse gases, *J. Geophys. Res.* **D91**(6), 6659–6667.
3. G. T. Bates and G. A. Meehl, 1986. The effect of CO_2 concentration on the frequency of blocking in a general circulation model coupled to a simple mixed layer ocean model, *Mon. Wea. Rev. 114(4)*, 687–701.
4. T. L. Bell, 1986. Theory of optimal weighting of data to detect climatic change, *J. Atmos. Sci.* **43**(16), 1694–1710.
5. A. Berger, 1986. Nuclear winter, or nuclear fall? *EOS* **67**(32), 617–621.
6. K. Bernhardt, P. Hupfer, and E. A. Lauter, 1986. Säkulare Änderungen in der atmosphärischen Umwelt der Menschen, *Sitzb. der Akad. der Wisc. der DDR. Mathem.-Phys.-Techn.* **4**(49). [page numbers].
7. J. P. Blanchet, J. Heintzenberg, and P. Winkler, 1986. Radiative heating during an intense pollution episode in Hamburg, FRG, *Contrib. Atmos. Phys.* **59**(3), 559–574.
8. R. D. Bojkov, 1986. The 1983 and 1985 anomalies in the ozone distribution: possible result of weak volcanic and of strong circulation effects, *MAP Handbook* **20**, [page numbers].
9. R. D. Cess, G. L. Potter, S. J. Chan, and W. L. Gates, 1985. The climatic effects of large injections of atmospheric smoke and dust: a study of climate feedback mechanisms with one- and three-dimensional climate models, *J. Geophys. Res.* **D90**(7), 12937–12950.

10. R. D. Cess, 1985. Nuclear war: illustrative effects of atmospheric smoke and dust upon solar radiation, *Climatic Change* 7(2), 237–252.

11. W. R. Cotton, 1985. Atmospheric convection and nuclear winter, *American Scientist* 73(3), 275–280.

12. C. Covey, S. L. Thompson, and S. H. Schneider, 1985. Nuclear winter: a diagnosis of atmospheric general circulation model simulations, *J. Geophys. Res.* D90(3), 5615–5628.

13. P. J. Crutzen and J. Hahn, 1985. Atmosphärische Auswirkungen eines Atomkrieges, *Phys. unserer Zeit* 16(1), 3–15.

14. P. F. Demchenko and A. G. Ginzburg, 1986. The effect of radiation on the vertical development of a turbid atmospheric layer, *Meteorology and Hydrology* [Vol.](8), 51–57.

15. R. E. Dickinson and R. J. Cicerone, 1986. Future global warming from atmospheric trace gases, *Nature* 319(6049), 109–115.

16. H. Flohn, 1986. Singular events and catastrophes now and in climatic history, *Naturwissenschaften* [Vol.](73), 136–149.

17. S. R. Gaffin, M. I. Hoffert, and T. Volk, 1986. Nonlinear coupling between surface temperature and ocean upwelling as an agent in historical climate variations, *J. Geophys. Res.* C91(3), 3944–3950.

18. D. A. Gillette and E. O. Box, 1986. Modeling seasonal changes of atmospheric carbon dioxide and carbon 13, *J. Geophys. Res.* D91(4), 5287–5304.

19. A. S. Ginzburg, D. S. Golitzyn, and A. A. Vasiliev, 1985. Global consequences of a nuclear war: a review of recent Soviet studies, *SIPRI Yearbook 1985,* Chapter 4. London: Taylor & Francis, 107–125.

20. G. S. Golitsyn and A. S. Ginzburg, 1986. Atmospheric consequences of a nuclear catastrophe: a serach for natural analogs, in [editor's name], *Cybernetics, Cryosphere and Global Problems.* Moscow: Nauka, 78–92.

21. A. Golombek and G. G. Prinn, 1986. A global three-dimensional model of the circulation and chemistry of $CFCL_3$, CF_2Cl_2, CH_3CCl_3, CCl_4, and N_2O, *J. Geophys. Res.* 91(3), 3985–4002.

22. V. G. Gorshkov and S. G. Sherman, 1986. Atmospheric CO_2 and destructivity of the land biota: seasonal variations, *Il Nuovo Cimento* 9C, Ser. 1(4), 902–917.

23. V. G. Gorshkov, 1986. Atmospheric disturbance of the carbon cycle: impact upon the biosphere, *Il Nuovo Cimento* 9C, Ser. 1(5), 937–952.

24. J. L. Gras, E. K. Bigg, C. G. Michael, M. A. Adriaansen, and R. Swinton, 1986. Stratospheric aerosol at 34°S following the 1982 El Chichon and Galunggung eruptions, *Tellus* 38B(1), 67–73.

25. M. C. G. Hall, 1986. Feedback in the climate: accuracy of first-order estimates, *J. Atmos/ Sci.* 43(4), 397–398.

26. L. D. D. Harvey, 1986. Effects of ocean mixing on the transient climate response to a CO_2 increase: analysis of recent model results, *J. Geophys. Res.* D91(2), 2079–2718.

27. A. Henderson-Sellers, 1986. Increasing cloud in a warming world, *Climatic Change* **9**(3), 267–310.

28. J. R. Holton, 1986. A dynamically based transport parameterization for one-dimensional photochemical models of the stratosphere, *J. Geophys. Res. D91*(2), 2681–2686.

29. Yu. A. Izrael, 1985. On the choice of basic factors for calculations of the geophysical and ecological consequences of a possible nuclear war, *Doklady USSR Acad. Sci.* **281**(4), 821–825.

30. P. D. Jones, S. C. B. Raper, R. S. Bradley, H. F. Diaz, P. M. Kelly, and T. M. L. Wigley, 1986. Northern hemisphere surface air temperature variations: 1851–1984, *J. Clim. and Appl. Meteorol.* **25**(2), 161–179.

31. B. A. Kagan, V. A. Riabchenko and A. S. Safrai, 1986. Modeling of a non-stationary response of the ocean-atmosphere system to increasing CO_2 concentration in the atmosphere, *Izv. USSR Acad. Sci.* **22**(11), 1131–1141.

32. I. L. Karol, 1986. On possible anthropogenic changes on the gas composition and temperature of the atmosphere by the year 2000, *Meteorology and Hydrology* [Vol.](4), 115–123.

33. P. M. Kelly and J. H. W. Karas, 1986. No place to hide: nuclear winter and the third world, *Earthscan Press Briefing Document 43.*

34. S. S. Khmelevtsov (Ed.), 1986. *Volcanoes, Stratospheric Aerosol and the Climate of the Earth.* Leningrad: Gidrometeoizdat.

35. E. P. Velikhov (Ed.), 1986. *Climatic and Biological Consequences of a Nuclear War.* Moscow: Nauka.

36. J. B. Knox, 1985. *Climatic Consequences of Nuclear War: New Findings 1985,* UCRL93768. Livermore, CA: Lawrence Livermore Nat. Lab.

37. K. Ya. Kondratyev, V. A. Ivanov, D. V. Pozdnyakov, and M. A. Prokofyev, 1985. Natural and anthropogenic aerosols: a comparative anaylsis, *Pontif. Acad. Sci. Ser. Var.* [Vol.](56), 281–303.

38. K. Ya. Kondratyev, 1985. Volcanoes and climate, *Progr. in Sci. and Technol., Meteorol. and Climatol.* Series No. 14. Moscow: VINITI.

39. K. Ya. Kondratyev, 1986. Natural and anthropogenic climate changes, *Progr. in Sci. and Technol., Meteorol. and Climatol.* Series No. 16. Moscow: VINITI.

40. E. B. Kraus, 1986. The smoke plume from the bombed city of Dresden: a personal recollection, *Climatic Change* **8**(3), 225–230.

41. G. Kukla, J. Gavin and T. R. Karl, 1986. Northern Hemisphere peak warmth in the early 1980's: fact of fiction? *Proc. First WMO Workshop on the Diagnosis and Prediction of Monthly and Seasonal Atmospheric Variations over the Globe,* WMO Tech. Doc. N87(1), Geneva: WMO.

42. M. Lai, S. K. Dube, P. C. Sinha, and A. K. Jain, 1986. Potential climatic consequences of increasing anthropogenic constituents in the atmosphere, *Atmos. Environ.* **20**(4), 639–642.

43. T. S. Ledley and S. L. Thompson, 1986. Potential effect of nuclear war smokefall on sea ice, *Climatic Change* **8**(2), 155–172.

44. R. C. Malone, L. H. Auer, G. A. Glatzmaier, M. C. Wood, and O. B. Toon, 1985. Influence of heating and precipitation scavenging on the simulated lifetime of post-nuclear war smoke, *Science* **230**(4723), 317–319.

45. R. C. Malone, L. H. Auer, G. A. Glatzmaier, M. C. Wood, and O. B. Toon, 1986. Nuclear winter: three-dimensional simulations including interactive transport, scavenging and solar heating of smoke, *J. Geophys. Res.* **D91**(1), 1039–1054.

46. P. C. Manins, 1985. Cloud heights and stratospheric injections resulting from a thermonuclear war, *Atmos. Environ.* **19**(8), 1245–1256.

47. G. IO. Marchuk, K. Ya. Kondratyev, V. V. Kozoderov, and V. I. Khvorostyanov, 1986. *Clouds and Climate.* Leningrad: Gidrometeoizdat.

48. G. A. Meehl and W. M. Washington, 1986. Tropical response to increased CO_2 in a GCM with a simple mixed layer ocean: similarities to an observed Pacific warm event, *Mon. Wea. Rev.* **114**(4), 667–674.

49. Royal Society of Canada, 1985. *Nuclear Winter and Associated Effects: A Canadian Appraisal of the Environmental Impact of Nuclear War.* Ottawa: Royal Society of Canada.

50. B. L. Otto-Bliesner and D. D. Houghton, 1986. Sensitivity of the seasonal climate of a general circulation model to ocean surface conditions and solar forcing, *J. Geophys. Res.* **D91**(6), 6682–6694.

51. E. M. Patterson, C. K. McMahon, and D. E. Ward, 1986. Absorption properties and graphitic carbon emission factors of forest fire aerosols, *Geophys Res. Lett.* **13**(2), 129–132.

52. J. Peterson, 1986. Scientific studies of the unthinkable: the physical and biological effects of nuclear war, *Ambio* **15**(2), 60–69.

53. A. B. Pittock, T. P. Ackerman, P. J. Crutzen, M. C. MacCracken, C. S. Shapiro, and R. P. Turco, 1986. *Environmental Consequences of Nuclear War: Volume I, Physical and Atmospheric Effects.* SCOPE 28. Chichester: John Wiley and Sons.

54. World Meteorological Organization, 1986. *Report of the International Conference on the Assessment of the Role of Carbon Dioxide and of Other Greenhouse Gases in Climate Variations and Associated Impacts,* 9–15 October 1985, Villach, Austria. WMO Tech. Note N661. Villach, Austria: WMO.

55. V. Ramaswamy and J. T. Kiehl, 1985. Sensitivities of the radiative forcing due to large loadings of smoke and dust aerosols, *J. Geophys. Res.* **D90**(3), 5597–5614.

56. C. N. R. Rao and T. Takashima, 1986. Solar radiation anomalies caused by the El Chichon volcanic cloud: measurements and model comparisons, *Quart. J. Roy. Meteorolog. Soc.* **112**(474), 1111–1126.

57. M. E. Schlesinger, 1986. Equilibrium and transient climatic warming induced by increased atmsopheric CO_2, *Climate Dynamics* **1**(1), 35–52.

58. J. Servant, 1986. The burden of sulphate layer of the stratosphere during volcanic "quiescent" periods, *Tellus* **38B**(1), 74–79.

59. J. Shukla, 1986. Physical basis for monthly and seasonal prediction, in WMO *Proc. First WMO Workshop on the Diagnosis and Prediction of Monthly and*

Seasonal Atmospheric Variations over the Globe. WMO Tech. Note N87. Geneva: WMO.

60. A. Slingo and P. Goldsmith, 1985. Nuclear winter: calculations of the shortwave radiative effects of soot aerosols, *Dynamical Climatology* Tech. Note 24.

61. R. D. Small and B. W. Bush, 1985. Smoke production from multiple nuclear explosions in non-urban areas, *Science* **229**(4712), 465–469.

62. I. N. Sokolik, T. A. Tarasova, and E. M. Feigelson, 1986. Optical characteristics of a smoke-loaded atmosphere and the radiative heating, *Meteorology and Hydrology* [Vol](11), 53–61.

63. R. S. J. Sparks, J. G. Moore, and C. J. Rice, 1986. The initial giant umbrella cloud of the May 18th, 1980, explosive eruption of Mount St. Helens, *J. Volcano and Geotherm. Res.* **28**(3-4), 257–274.

64. J. R. Stearns, M. S. Zahniser, C. E. Kolb, and B. P. Sandford, 1986. Airborne infrared observations and analyses of a large forest fire, *Appl. Opt.* **25**(15), 2554–2562.

65. G. L. Stenchikov, 1986. Numerical modeling of nuclear winter with the aerosol distribution taken into account, *Doklady USSR Acad. Sci.* **287**(3), 498–602.

66. J. R. Trabalka and D. E. Reichle (Eds.), 1986. *The Changing Carbon Cycle: A Global Analysis.* Berlin: Springer-Verlag.

67. S. L. Thompson and S. H. Schneider, 1986. The nuclear winter debate: comment and correspondence, *Foreign Affairs* **65**, 171–178.

68. M. C. MacCracken and F. M. Luther (Eds.), 1985. *Projecting the Climatic Effects of Increasing Carbon Dioxide.* DOE/ER-0237. Washington, D.C.: U.S. Dept. of Energy.

69. S. L. Thompson and S. H. Schneider, 1986. Nuclear winter reappraised, *Foreign Affairs,* **64**, 981–1005.

70. M. Ya. Verbitsky and D. V. Chalikov, 1986. *Modeling the Glaciers–Ocean–Atmosphere System.* Leningrad: Gidrometeoizdat.

71. K. Ya. Vinnikov, 1986. *Climate Sensitivity.* Leningrad: Gidrometeoizdat.

72. K. Ya. Vinnikov, P. Ya. Groisman, K. M. Luchina, and B. A. Golubev, 1987. Changes in the mean air temperature of the Northern Hemisphere between 1871 and 1985, *Meteorology and Hydrology* No. 1, 45–55.

73. R. K. R. Vupputuri, 1986. The effect of ozone photochemistry on atmospheric and surface temperature changes due to large atmospheric injections of smoke and NO_x by a large-scale nuclear war, *Atmos. Environ.* **20**(4), 665–680.

74. W.-C. Wang, D. J. Wuebbles, W. M. Washington, R. G. Isaacs, and G. Molnar, 1986. Trace gases and other potential perturbations to global climate, *Revs. Geophys.* **24**(1), 110–140.

75. W. M. Washington and G. A. Meehl, 1986. General ciculation model CO_2 sensitivity experiments: snow–ice albedo parameterizations and globally averaged surface air temperature, *Climatic Change* **8**(3), 231–242.

76. R. T. Watson, M. A. Geller, R. S. Stolarski, and R. F. Hampson, 1986. *Present State of Knowledge of the Upper Atmosphere: An Assessment Report—Processes That Control Ozone and Other Climatically Important Trace Gases.* NASA Ref. Pub. 1162. Washington, D.C.: NASA.

77. W. H. White, C. Seigneur, D. W. Heinoid, L. W. Richard, W. E. Wilson, and P. T. Roberts, 1986. Radiative transfer budgets for scattering and absorbing plumes: measurements and model predictions, *Atmos. Environ.* **20**(11), 2243–2258.

78. World Meteorological Organization, 1986. *Workshop on Comparison of Simulation by Numerical Models of the Sensitivity of the Atmospheric Circulation to Sea–Surface Temperature Anomalies,* 9-12 December 1985, NCAR. WMO/TD-138, WCP-121. Geneva: WMO.

Index